The Design and Tuning of Competition Engines

The Design and Tuning of Competition Engines

BY

Philip H. Smith

SIXTH EDITION REVISED BY

David N. Wenner

ROBERT BENTLEY, INC.
872 Massachusetts Avenue
Cambridge, Mass. 02139

First Edition 1954
Second Revised Edition 1957
Third Revised Edition 1963
Fourth Revised Edition 1967
Fifth Revised Edition 1971
Sixth Revised Edition 1977
10 9 8 7 6 5 4 3 2 1

Library of Congress Catalog Card No. 77-78834
ISBN 0-8376-140-1

Manufactured in the United States of America

Acknowledgments

Basic to the requirements of a comprehensive book on any subject is suitable illustrative matter. Philip H. Smith, deceased, the original author of this book, and T.C. Millington, who edited and revised the Fifth Edition, were fortunate to have cooperation from many sources. Some of their original illustrations remain in the present edition—notably drawings that were provided by IPC Transport Press Ltd., British publishers of *Autocar, Automobile Engineer,* and *Motor Trader.* In addition, one or more illustrations were provided by the following:

The Alexander Engineering Co. Ltd., Haddenham, England
Robert Bosch G.m.b.H., Stuttgart, Germany
V.W. Derrington Ltd., Kingston on Thames, England
Ken Lowe Accessories Ltd., Slough, England
London Art Tech (*Motor Sport*), London, England
Joseph Lucas Ltd., Birmingham, England
Dr.-Ing. h.c.F. Porsche, Stuttgart, Germany
Renold Chains Ltd., Manchester, England
Rubery Owen and Co. Ltd., London, England
Shorrock Superchargers Ltd., Wednesbury, England
Wade Engineering Ltd., Brighton, England
Soc. p. Az. Ed. Weber, Bologna, Italy
Mr. Douglas Armstrong, Reading, England
Mr. Michael Porter, Brooklyn, N.Y., U.S.A.

In the preparation of this Sixth Revised Edition, it has been our pleasure to use a number of illustrations made available through the courtesy of American Motors, Chrysler Corporation, the Ford Motor Company, and General Motors—America's "Big Four". Special thanks are also due the Champion Spark Plug Company of Toledo, Ohio, whose excellent *Service Corner* publication has been the source for several of our illustrations. The vast support given to racing by these American companies is well known to all.

Our thanks are also extended to Volkswagen of America, Inc., and British Leyland Motors, Inc., with whom it has been the present author's privilege to work in the preparation of official service manuals. Both of these representatives of European manufacturers maintain very active racing programs in the U.S.A. Our appreciation is likewise extended to Fiat, both for making available the basic art for our dust jacket, and for bringing high-performance overhead camshaft engines within reach of the average motorist.

It is with particular pleasure that we express our appreciation to Datsun, Honda, Mazda, and Toyota—four auto manufacturers based in Japan who have in recent years brought new excitement to racing and to highway motoring in America. Only one or two illustrations have been provided by each of these carmakers, but their contributions to the modern automobile have been monumental. Toyota has lent great financial support to Grand Prix racing in America—in addition to building fast and durable cars. Honda and Mazda have made outstanding contributions to racing and are especially noteworthy for having developed radically new engines—showing that they represent the forefront of automobile engineering today. Datsun, of course, wins races (which is what this book is really about), and has been given an entire chapter as an indication of its important place in American competition.

Finally, we are indebted to the Moser Engine Corporation of Monterey Park, California for providing us with photos of their beautifully designed four-cam racing engine. Undoubtedly, this is the last word in the development of America's favorite competition powerplant, the small-block Chevrolet V8.

Preface

Revised Sixth Edition

This sixth edition of *The Design and Tuning of Competition Engines* remains a carefully-proportioned amalgam of the timely and the timeless. It includes both the laws of thermodynamics, as immutable today as in 1954 when Philip H. Smith included them in the first edition of the book, and a wealth of information on present-day engines and the tuning methods that form the cornerstone of racing in America.

When the first edition was published, the sidevalve engine still occupied an important place in motor competition. In America the immortal Ford flathead V8 was a major staple in all forms of racing, and in England the 1098-cm^3 Ford Ten was propelling unknowns such as Colin Chapman and Eric Broadley to their first tastes of racing glory. The overhead camshaft engine—which today is used in well over half the makes of cars sold in the United States—was, in 1954, a piece of pure exotica reserved for Grand Prix cars and a few expensive Gran Turismo machines.

Not only have many of the "glory" engines of 1954, 1957, 1963, 1967, and 1971 passed into obsolescence and obscurity, most of the racing classes that these engines participated in have also fallen by the way—superseded by new racing classes that are more in tune with the times. Who could have predicted in 1963 that the underpowered Volkswagen would find a history-making place in Formula Vee competition—not to mention in drag racing, where outputs of 200 bhp are not unheard of? For that matter, who could have predicted in 1971, when the fifth edition was published, that there would be a racing class for stock Volkswagens–with watercooled engines mounted at the front?

But though the engines used in racing have changed, the fundamental principles have not. Because of this, some of the obsolete engines are mentioned here; occasionally they were the last applications of principles that may someday be revived. The 1935-1940 V16 Cadillac, for example, is discussed not because it is a significant competition engine today (it is not; its sole claim to fame was as the powerplant for the Southern California Timing Association's fastest "lakester" back in the late 1940s), but because it is representative of all 45° V16 engines—a cylinder arrangement that could find a place on the racing scene in the not-distant future.

This book is not a "how-to-do-it" book. It is better described as a "how-to-know-what-you're-doing" book. That is, there are many engine tuning methods and many design features that the racing enthusiast recognizes as effective in competition. Consequently, such a person, in preparing an engine for racing, may decide that forged aluminum connecting rods are necessary–because he has seen forged aluminum connecting rods in other competition engines. This book will help him to make his decision more wisely by telling him when forged aluminum rods are necessary and, for that matter, when they are unnecessary or undesirable.

The Design and Tuning of Competition Engines is particularly helpful for aspiring speed tuners, who may be well informed in matters of practical mechanics, but lack comprehensive engineering training. Many such people have gone on to great accomplishments in racing, and it is our hope that this book will be an important step for many others in their quest for a sound theoretical foundation. Auto racing aside, we need look no further than the Wright brothers for an historical precedent that shows the manner in which the self-taught can tread the path to success.

Contents

Introduction / The Competition Engine in America

The Competition Engine Defined

What exactly is a competition engine? It is not, as the general public fancies, the same thing as a high-performance or hot rod engine. A competition engine is first of all one that is competitive *in a particular kind of racing or in a particular racing class.*

For example, the humble VW 1200 engine, offering at best about 45 bhp per liter, would scarcely be competitive in most kinds of racing. Yet, by virtue of its being the only engine permitted by the rules of Formula Vee, it *is* a competition engine. Consequently it is appropriate in a book of this kind to talk about the VW 1200. On the other hand, it would be inappropriate to waste space on the Rover 2000 or Mercedes Benz 250C engines; they are unquestionably powerful and sophisticated, but they are never used in American racing.

Yesterday's racing engine cannot be classified as a competition engine today either. In the previous edition of this book, it seemed indispensable to discuss the Hillman Imp and the 260 cubic inch Ford V8, which were being campaigned, respectively, in the SCCA's D Sports Racing and B Production classes. Alas, how the mighty are fallen. The appearance of either engine is rare in road racing these days.

Development Potential

And so it can be seen that time is the natural enemy of all competition engines—not only in critical fractions of seconds that it must annihilate during a race but also in irretrievable hours of competitive life that tick away while the car lies idle in the garage. What, then, can account for the brief life spans of some engines and the incredible longevity of others? Why are some engines, which enjoyed no great advantage over their competition during the past decade, still competitive today against even more sophisticated rivals?

These questions can best be answered by surveying a list of those few unbeatable engines of fifteen years ago that are still competitive in the late 1970s. None, it will be seen, has survived unchanged. Their long life in competition is entirely the result of a peculiar amenability to long-term development. This development potential is a competition engine's sole and ultimate defense against the remorseless progress of time.

To be successful, a competition engine that lacks development potential must reach its full potential immediately so that a respectable number of racing wins can be attained before the design becomes obsolete. In this context it should be remembered that debugging a design is not the same thing as development. In fact, time wasted in correcting basic design and manufacturing faults invariably means less time available for effective development. Examples of engines that could have succeeded—had they been running right before they became obsolete—are too abundant to catalog here.

It can be argued with considerable success that a "sensible" design is better than a "sophisticated" design, both in terms of immediate success and development potential. Nowhere is this more apparent than in Formula I. The first two seasons of the present formula were dominated by the SOHC Repco-Brabham V8, which was a development of an Oldsmobile production engine; it was a simple design that "worked" while more sophisticated engines were being "debugged." From that point on Formula I has virtually belonged to the Cosworth-Ford, a relatively uncomplicated V8 engine designed for extensive parts interchangeability and based on a four-cylinder Formula II engine that was itself a development of a production Ford Cortina engine. Meanwhile, an incredible number of sophisticated, expensive and complex sixteen-cylinder and twelve-cylinder designs have come and gone, leaving behind them a trail of shattered hopes and broken bank accounts.

Know-how

The designer's know-how has as much to do with an engine's immediate success as development potential has to do with its continued success. Keith Duckworth, the designer of the Cosworth-Ford Grand Prix engine, had at his disposal considerable know-how gained through development work with four-cylinder Fords—which suggests that success is more readily obtained by a designer who is starting out with something more than a blank piece of paper and a headful of theoretical knowledge. It is obvious from studying the history of Formula I that, when it comes to engine design, evolution (development and the know-how derived from it) has always had a better record than revolution (innovation based on abstract theory).

An existing tradition of know-how is also indispensable when the designer or tuner must begin with a production engine, either for economic reasons or because it is mandatory by the rules of a racing class. For example, it is conceivable that the Mercedes Benz 450 V8 could be developed into a "world beater" drag racing powerplant. But no one has tried it. Consequently any tuner who sets out to develop this engine for drag racing will have to start from scratch. On the other hand, many tuners have worked for many years developing drag engines based on the big-block Chevrolet V8—creating a tremendous tradition of know-how that can be drawn upon by any person who decides to prepare such an engine for competition.

Which brings us to the matter of cost. Chevrolet engines are available in grand profusion from any junkyard; Mercedes engines are expensive from any source. Moreover, a fantastic array of speed equipment is readily available for the big "Chev", whereas expensive tooling and pattern making would be necessary before the first piece of speed equipment could be obtained for the big "Merc".

Economic Considerations

The extensive competition success of Chevrolet, Ford, and Volkswagen engines should not be credited solely to the merits of their designs. In large part, they owe their winning ways to their low cost and ready availability—which has made possible rapid and extensive development by countless speed tuners. The proponents of competing powerplants find it very difficult to catch up without an influx of sponsorship money from their engine's manufacturer. But, as nearly every major automaker has found, winning at all costs usually means a very costly engine. So the sponsorship often dries up once the manufacturer has reaped the publicity benefits, and the engine fades back into obscurity.

Summing up, the successful competition engine is likely to have a majority of the following attributes: (1) it is suited to the rules of its class and is among the best available engines for its time, (2) it has the capability for immediate success and the development potential for continued success, (3) its design is based on the know-how acquired through development work on similar engines, and (4) it is a comparatively economical engine to build so that modification costs do not preclude effective development.

Blueprinting—What It Is

Blueprinting is the basis of most competition engine preparation. Specifically it is the remanufacturing of an engine to bring it into precise conformity with the designer's intentions and to advance various dimensions to the limits of the factory's tolerance range so that the powerplant is suited for high-rpm operation.

In rare instances factories have made available actual blueprints for their engines. Most blueprinting, however, is done without blueprints. Instead a complete and accurate list of engineering specifications is obtained so that each engine component can be remachined to its ideal dimensions—correcting the errors and variations that are common among mass-produced parts. Technically the blueprinted engine conforms to factory specifications, making it eligible for production class racing. But it is incomparably faster than the production line version of the same powerplant.

Dyno Tuning

Blueprinting always begins with the complete disassembly of the engine—even if it is a brand-new engine fresh from the factory crate. Therefore, blueprinting should not be confused with dyno tuning, which can be done without disassembling the engine. In certain racing categories, such as the "pure stock" drag racing classes, no preparation other than dyno tuning is permitted by the rules of the class. The actual modifications are limited to minor adjustments and the replacement of small parts.

A typical dyno tune, which is performed with the car on a chassis dynamometer, usually includes selecting different carburetor jets or needles and, if possible, different carburetor venturis. Different springs will be installed in the ignition distributor's centrifugal advance mechanism, and the tension of the springs will be adjusted with the distributor installed on a distributor testing machine. The acceleration pump, the valve clearances, and the ignition timing may all be adjusted to nonstandard specifications. Different spark plugs, ignition coils, low-restriction air filters, and metallic-conductor high tension cables are usually substituted for the factory equivalents, and the exhaust system may be modified if the rules permit.

Blueprinting Fundamentals

The first step in blueprinting is the careful cleaning and inspection of every engine component. Actual machining usually begins with a rework of the main bearing and camshaft bearing bores in the block, which is called *align boring.* In production engines, these bearing bores are sometimes out of line with one another, tapered, out-of-round, or not parallel to the crankshaft centerline. Any such production errors will cause power-robbing internal friction and possible bearing failure when the engine is used in racing.

Next, the deck height is corrected (the deck is the block surface that the cylinder head bolts to). The deck is milled to make it exactly parallel to the crankshaft centerline—and at the precise distance from the crankshaft centerline that is indicated in the original blueprints. The cylinders are then bored to make them perpendicular to the crankshaft centerline and to the deck. In addition, the cylinders are usually bored to the maximum diameter permitted by the engine manufacturer or by the racing class rules. If factory pistons are not specified by the rules, racing pistons are usually substituted.

The clearances between parts are increased to the widest dimension permitted by factory tolerances, again to reduce friction. (During factory assembly, clearances are usually held to near the opposite end of the tolerance range in the interest of long service life during highway usage.) The bearing surfaces will be polished to a mirror finish, the cylinders honed to an absolutely uniform diameter throughout, and all reciprocating parts and rotating parts will be balanced—the latter with electronic equipment.

During assembly, the valve timing is carefully checked with a degree wheel installed on the crankshaft (Fig. 0-1)—whether a stock camshaft is used or whether the rules permit a reground or custom-made camshaft. Yet it is the cylinder heads that are

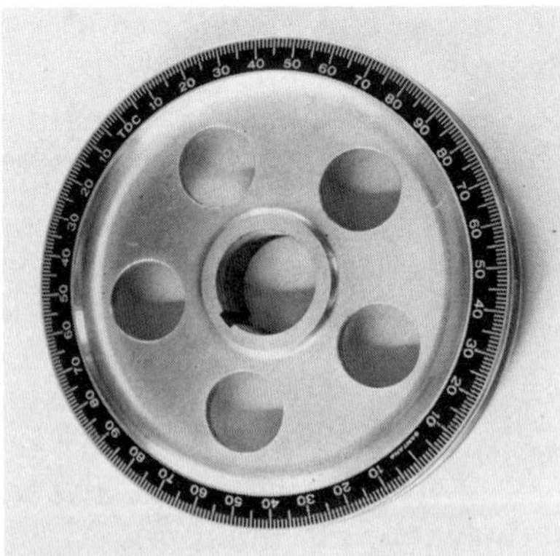

Fig. 0-1. Degree wheel for permanent installation on VW crankshaft. Precision valve and ignition timing is possible with this device.

most extensively checked and reworked. Each head is given a competition valve grind, the combustion chamber volume is carefully measured, and then the head is milled to obtain the precise compression ratio given in the factory specifications. If the rules permit it, the intake and exhaust ports may be machined slightly to align them perfectly with the manifolds.

Supertuning—What It Is

Supertuning is the modification of an engine to specifications other than those given by the manufacturer. The purpose, of course, is to obtain greater performance than is possible by blueprinting alone. Nevertheless all production components used in a supertuned competition engine are carefully blueprinted before any additional reworking begins.

Fitting larger valves, polishing and reshaping the cylinder head ports (Fig. 0-2) and combustion chambers, and installing different manifolds are all included in supertuning. In addition, the moving parts of the engine are nearly always replaced by components designed especially for racing. Only a few pieces, such as the crankshaft, the cylinder head castings, and the engine block, are derived from production components. In full-race modifications only the cylinder block itself is likely to be based on a stock part.

Fig. 0-2. Modified cylinder head for Chevrolet V8. Because ports have good shape and fairly ample size in most American V8 engines, modification consists mainly of contouring and smoothing surfaces, not enlarging ports greatly. Combustion chambers are similarly smoothed and polished and high-performance valve gear is fitted.

Supertuning Applications

Supertuning in its various degrees sires the majority of competition engines used in professional racing. In fact, the rules are often written to encourage wide use of "stock block" engines, for example, by permitting them greater piston displacement than is allowed for pure racing designs. The consideration is mainly economic. Though a supertuned engine is certainly not cheap, it is nevertheless considerably cheaper than an engine designed and built solely for racing.

Whether the preparation of a competition engine is confined to blueprinting or includes the highest degree of supertuning, the services of an expert machinist are indispensable. For this reason, one must never confuse supertuning with the mere bolting on of high compression heads, "tuned" exhaust headers, dual-carburetor manifolds, or any of the other widely sold hop-up parts that have made backyard hot rodding the greatest educational hobby in the United States. It is the constant failing of the hobbyist/street racer to install speed equipment on an engine that has not had the benefit of balancing or a good precision valve grind. This slipshod practice, if it is carried over into serious competition, produces the engines that look impressive in the paddock and disintegrate on the track or that use the same components as the winning car but nevertheless fail to finish anywhere near the front.

The theory and practice described in this book are intended as a guide for machinist/tuners who do supertuning work, for drivers who wish to plan the construction of engines that they intend to have built for their cars, and for those who want to design racing engines. It should be pointed out, therefore, that blueprinting and supertuning, which are commonly thought of as the province of the stock block racing engine builder, also have a vital place in the preparation of pure racing powerplants.

It is no secret that most of the racing engines sold on the open market are produced on a semi-mass production basis. Furthermore, their builders recognize that no competent team manager is going to use an engine in a race that has not been disassembled, inspected, and carefully reassembled to the standdards of his or her own skilled employees. Therefore, it is not uncommon for the finest and most expensive racing powerplants to arrive from their makers in a less than perfect state. In addition to correcting minor imperfections, installing different components that are personally favored (or are demanded by a sponsor), and then reassembling the engine with absolute precision, the tuner/engine builder may be called upon to carry out the kind of supertuning work that is necessary to fit manifolds, ignition components, or similar parts and systems that are often not included on the basic racing engine as delivered by its maker.

It is in the supertuning of stock block and production engines that the tuner/machinist has the greatest opportunity for applying his craft. Here, a thorough grasp of theory can lead to accurately guided experiments with combustion chamber shapes, compression ratios, supercharging, exhaust system design, and every other facet of the engine builder's art that could eventually lead to a highly superior competition engine. Engine builders who produce winners never have to worry about a lack of business—or an abundance of bill collectors—coming through the doors of their shops!

The Racing Engine

A true racing engine is developed from a prototype *designed solely for racing*. It is not a modified production car powerplant. Aside from rules limitations, which may limit the piston displacement, restrict the fuel used, specify the maximum supercharger boost, or proscribe supercharging altogether, the racing engine designer is free to draft an engine that has but one purpose—winning races.

Nevertheless, racing engine prototypes do require development. Fig. 0-3 shows a racing engine prototype that is markedly different from the engine that eventually entered successful competition (Fig. 0-4). The designer must at all costs create an engine that is good enough on paper to avoid the long gestation that is disastrous for a racing engine once it is "in the metal".

Fig. 0-3. First development of Ford V8 Indy engine, with double-choke carburetors centrally mounted.

Art or Science?

There is a maxim with a long tradition in racing that "what looks right generally is right". This is essentially true. But, as can be seen from the spotty records of some designers and the large number of racing engines that never win a race, a great deal depends on who is doing the looking. Thus, the dividing line between art and science is exceedingly thin, and Keith Duckworth, who admits to using intuition in designing combustion chambers, is not far removed from Leonardo da Vinci, who by "intuition" put just the right smile on Mona Lisa.

Those who have tried to duplicate the feats of Duckworth or da Vinci and have failed may suspect that these men were

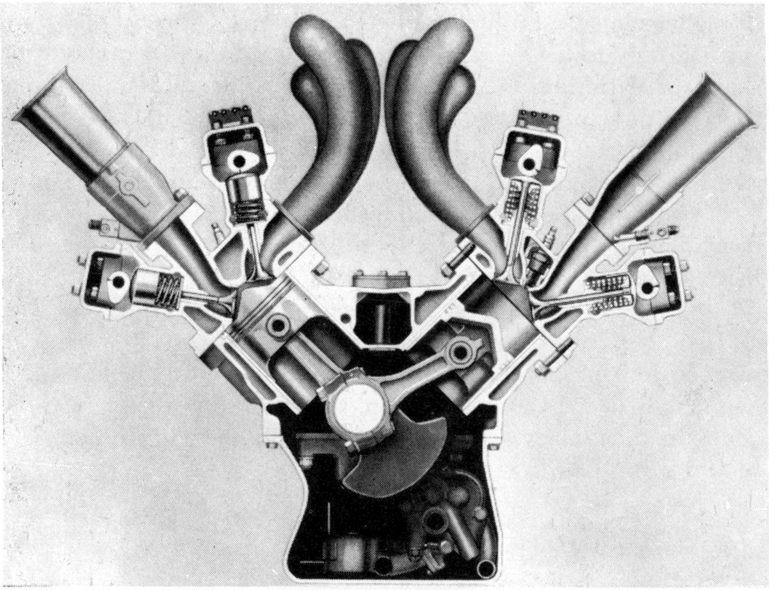

Fig. 0-4. End view of redesigned Ford V8 Indy engine, with fuel injection ducts between the cams and central exhaust system.

merely lucky. In public, however, they are more likely to say that there is a secret mathematics behind Duckworth's combustion chamber designs or Leonardo's compositions. We inferior mortals always find it consoling to believe that secret mathematics are responsible for great designs, for this would mean that any person could be a master designer once the mathematics have been learned. In truth, masterpieces—whether paintings or racing engines—owe far more to the genius that guides the artist/designer's eye than to any quality that can be imparted by education or a mastery of mathematics.

The Designer

The racing engine designer needs sufficient mathematics to understand books such as this one and adequate engineering knowledge to provide a familiarity with materials, stresses, and design fundamentals. It is also necessary to be an expert draughtsman and be prepared to work long hours at the drawing board. However, it is a racing apprenticeship of hot rodding, speed tuning, dyno tuning, blueprinting, and supertuning that is the *Gradus ad Parnassum* for the designer of racing engines.

As suggested earlier, the successful racing engine design is nearly always based on the know-how gained through successful development of an earlier engine. Seeking a theoretical "ultimate"—that is, doing things the "best" or "right" way *based on theory without practical precedent* and *regardless of cost and complexity*—is not, in many unfortunate cases, the path to competition success. The many elaborate but unsuccessful BRM engines, both of the now-distant past and the almost present, could in themselves provide all the examples that are necessary to prove the truth of this allegation.

The United States

A racing engine's developmental roots are important, and in America these roots have always extended literally into the dirt—whether the actual clay of the dirt tracks or the sand between the bricks at Indianapolis. There, in the last races before World War I, the Peugeot Grand Prix cars achieved their greatest wins in U.S. competition. From copies of these triumphant Peugeot engines, through derivatives such as the Miller, the lineage of American racing engines can be traced through an uninterrupted strain of DOHC, barrel-crankcase scions that persists today in the indomitable "Offy".

In other parts of the world, racing engine ancestry has been more diverse. Consequently Formula I and Formula II road racing—not to mention the various sports racing car classes that permit the use of pure racing engines—provide many more interesting models for aspiring designers than are commonly available on this side of the Atlantic. In Japan a fantastically successful racing technology has grown up around the motorcycle engine, and Honda has participated just enough in international auto racing to whet the interest of the rest of the world. Needless to say, their motorcycle-based tradition of competition engine design is far removed from the racing engine tradition of the United States, which at present is virtually moribund.

With one exception, no American auto manufacturer has initiated the design of a racing engine since the Auburn, Cord, Duesenberg Company died in the 1930s. The exception, of course, is Ford, whose Indy V8 (now known in developed form as the "Foyt") was launched long ago in 1964. With the death of eighty-two-year-old Leo Goossen near the end of 1974, there remained no person fully employed in the United States as a racing engine designer, and the rather dismal showing of American-based Formula I teams during the mid-1970s suggests that racing designers of genuine genius either are unavailable in North America or are present but lack the support of people with vision who will finance and support their efforts.

The Plan of This Book

This book has been divided into three parts. Part 1 is devoted to theory. It explores the scientific basis of engine operation—in particular, the design problems that are raised by high-speed operation.

Part 2 describes practice—the practical applications of the theory described in part 1. It is not, however, a how-to-do-it guide. It is practical in that it investigates the various design features and modifications that have proven effective for achieving high outputs from competition engines.

Part 3 covers the principal engines used in American racing today, including descriptions of how they are tuned for particular racing classes. Finally, there is an appendix to the book that gives various definitions, constants, and formulas that are useful to the racing engine designer and to the competition engine tuner who will find them invaluable for converting "book learning" into race track wins.

PART I

Theory

1 / Heat Engine Operation

Theoretical Considerations

The extreme conditions encountered in rocketry and space exploration have thoroughly upset some of our most cherished conclusions about the behavior of liquids and gases at very high temperatures and pressures. But despite the apparent progress of today's competition engines toward the outer limits attainable by reciprocating designs, the old basic laws governing pressures and temperatures remain unchanged insofar as they are applicable to this book.

This chapter begins with an examination of some of the traditional fundamentals of thermodynamics—a science that is essential to those whose interest in high-speed gasoline engines goes deeper than a mere knowledge of its obvious, or visual to the eye, operations. Our examination must be brief for reasons of space. But it is hoped that the reader's interest will be aroused, and those who wish to go further will find that a modern textbook on thermodynamics offers an invaluable opportunity for continued investigation.

Although a great deal of blueprinting and supertuning work can be done without any knowledge of theory simply by imitating the work of others, an understanding of thermodynamics is the key that unlocks doors to those uncharted regions of speed tuning where added power may be discovered. Theory itself remains constant; but designs are not constant and can be changed in order to apply known theory to a greater advantage in competition.

To avoid complexity in the first part of this book, a number of definitions, constants, and formulas have been relegated to the appendix. However, a familiarity with the following standard symbols and terms for certain quantities is indispensable for understanding this chapter:

P=pressure
T=temperature
V=volume
H=enthalpy
S=entropy
U=internal energy of gas
C=degrees Celsius (Centigrade)
F=degrees Fahrenheit
R=universal gas constant
Y=ratio of specific heats
c=specific heat
r=compression ratio
ase=air standard efficiency

Thermodynamics

Thermodynamics is the science of the relationship between heat and mechanical work. Its basis is in two laws, both of which have been firmly accepted for a long time because phenomena have invariably been found to occur in a manner consistent with them; but being empirical they can be expressed in many ways. The first law concerns the *conservation of energy;* although energy, as heat or work, can be interchanged between a system and its surroundings, the total energy of the whole remains unaltered even though its form may change. (The term *system* indicates a specific layout or machine—such as an engine—in which changes are taking place.) The following definitions are typical of the first law:

1. Heat and work are mutually convertible, the one into the other.
2. In all transformations, the energy resulting from the heat units supplied must be balanced by the external work done, plus the gain in internal energy resulting from the rise in temperature.

3. The total energy of an isolated system remains constant whatever changes may occur within (Clausius).

The second law deals with *heat transfer* and might seem to stress the obvious; but in applying heat engine theory one can easily lose sight of the inevitability of certain losses if this law is not kept firmly in mind. The following definitions are typical of the second law:

1. It is impossible for an automatic machine, unaided by external power, to convey heat from a colder to a hotter body.
2. Irrespective of the design of an engine, only a fraction of the heat supplied can be converted into work; the remainder is rejected as heat at some lower temperature (Carnot).
3. Heat will not pass from a lower to a higher temperature reservoir without work being done on the system.

Scientists whose work was significant in the establishment of the two thermodynamics laws were Robert Boyle and H. J. Charles—familiar as originators of the gas laws that bear their names. Engine operation depends on a gas behaving in accordance with these laws, which relate to the effect of temperature, pressure, and volume changes and their interrelation.

Gas Laws

The first gas law, formulated by Boyle in 1662, is the earliest recorded observation on the behavior of gases. It states that, assuming the gas temperature is kept constant, the volume will vary inversely as its pressure. In other words, if we compress the gas into half its original volume or space at unvarying temperature and at a certain original pressure, its pressure will become double the original. Thus, if P = pressure, and V = volume,

$$PV = \text{a constant.}$$

The second gas law, formulated in 1787, is attributed to the French scientist Charles. (A similar independent statement was made by Joseph Louis Gay-Lussac in 1802.) It states that at constant pressure, the volume of a gas rises uniformly with rise in temperature from the absolute zero of −273°C (−460°F). It assumes that the volume at this temperature will be zero so that the volume of a quantity of gas will vary 1/273 of its volume at 0°C for every 1°C change in temperature (using the Fahrenheit scale, 1/492 for every 1°F), providing the pressure is kept constant. This is based on the assumption that at the absolute zero temperature of −273°C (−460°F) the volume will become zero. In practice, of course, gases that are cooled beyond a certain point liquefy and finally solidify.

What Charles's law tells us is that the gas volume increases by equal increments for equal increases in temperature. Because the volume is proportional to the absolute temperature, with V = volume and T = temperature,

$$\frac{V}{T} = \text{a constant.}$$

It will be apparent that if the gas is in a closed cylinder and unable to expand, thus keeping the volume constant and unaltered, any application of heat will alter its pressure in proportion to the rise in temperature. Thus

$$\frac{P}{T} = \text{a constant.}$$

Therefore, since PV and P/T both equal a constant, we can combine the two and obtain

$$\frac{PV}{T} = \text{a constant.}$$

The constant is denoted as R and is known as the universal gas constant: the expression

$$PV = RT$$

is called the ideal or perfect gas equation when dealing with (theoretical) gases that follow both gas laws with exact precision. Real-world "working gases" do not always follow these gas laws precisely. However, gases such as oxygen, hydrogen, nitrogen, and helium (represented in Fig. 1-1), which liquefy at very low temperatures under atmospheric pressure, come close to ideal behavior when at room temperature and atmospheric pressure—though they deviate considerably from the ideal at higher pressures. This relationship between *P, T,* and *V* prevails throughout the operating cycle and, of course, works both ways (that is, whether the pressure or the volume, or both, are increasing or decreasing).

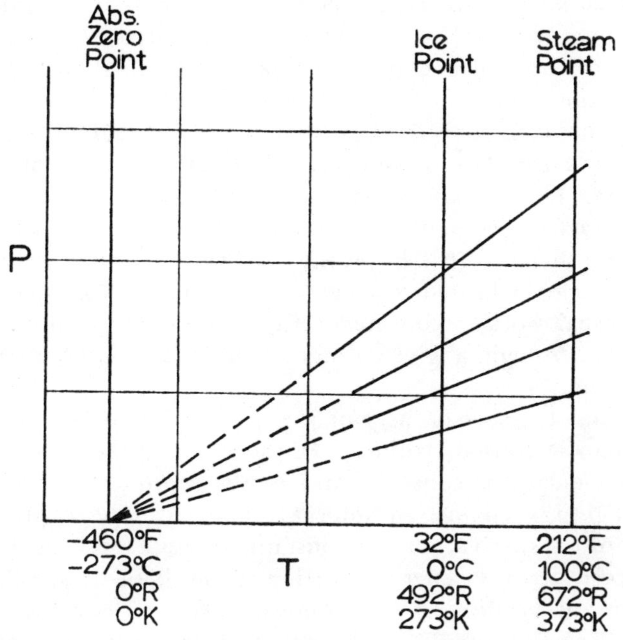

Fig. 1-1. The fall in pressure of several "working" gases closely follows the perfect gas equation, shown by the interrupted-line extension to absolute zero *T*.

Internal Energy

Energy must be expended for work to be done. Therefore, it is evident that gases must possess some such energy that can be used. The energy contained in a cylinder of gas under pressure will comprise (1) potential energy (gravitational), (2) kinetic energy, which could be shown, for instance, by thrust-causing movement of a piston, and (3) molecular energy that results from the movements, vibrations, and attractions of the gas molecules themselves. Taken together, these three forms of energy are known as the *internal energy* of the gas.

If a gas is neither doing work nor having work done to it—but is at the same time receiving or losing heat—*its internal energy will be increased or reduced by exactly that equivalent amount of heat.* (By the first law of thermodynamics, the *change* of internal energy in any process is equal to the difference of the heat gained and the external work done.) As far as engine theory is concerned, only the *changes* in internal energy are of any significance; the *total* internal energy is of no importance. These changes can result only from an alteration in the characteristics of the system since the internal energy is a function thereof. If the pressure remains constant when the working gas is heated (the gas being allowed to expand and do external work), more heat will be required to raise its temperature through a given range than if it is prevented from expanding.

Thus, the specific heat of a gas is greater at constant pressure than at constant volume. In the first case stated above, the internal energy is constant, and in the second case it changes. Although the change in internal energy can be expressed in terms of its specific heat at constant *volume,* it is also possible to express it as an energy function in terms of its specific heat at constant *pressure.* This is known as the *enthalpy* of the system. (Terms formerly used, such as *total heat* and *heat constant,* are now obsolete.) Enthalpy is denoted by H and internal energy by U; thus enthalpy can be expressed

$$H = PV + U.$$

The Specific Heat of Air

In a gasoline engine, air constitutes the bulk of the working gas (or mixture). The behavior of the air that is being subjected to the changing conditions of the operating cycle of an engine is, therefore, a useful guide to engine performance.

At constant pressure, the specific heat of dry air is generally accepted as .2374 Btu per pound weight. If a quantity of air is allowed to expand as its temperature rises, in accordance with Charles's law, it will do work by exerting pressure against the surrounding atmosphere. For example, if the quantity of air is 1 pound in weight, its volume at normal temperature and pressure will be 12.387 cubic feet. If its temperature is raised by 1°F, it will expand by 1/492 of its volume against atmospheric pressure. That is equal to 14.7 × 144 pounds per square foot, assuming atmospheric pressure at 14.7 psi. Multiply this by the volume of 12.387 cubic feet, and then by 1/492. The result is work done against the atmosphere in foot pounds.

$$\text{Thus:} \quad \frac{14.7 \times 144 \times 12.387}{492} = 53.29 \text{ ft. lb.}$$

To convert this to Btu we divide it by 778.

$$\text{Thus:} \quad \frac{53.29}{778} = .0685 \text{ Btu.}$$

If the volume of air is not allowed to expand against atmospheric pressure (that is, the volume is maintained unaltered), the pressure of the air will rise, but there will be no external work performed. The specific heat required (.2374 Btu per pound at constant pressure) will therefore (at constant volume) be less by the amount calculated above: .2374 – .0685 = .1689 Btu.

There are two values of specific heat for air:

$$c^p = .2374 \text{ Btu per lb.}$$
$$c^v = .1689 \text{ Btu per lb.}$$

The relationship between the two values is very important in engine calculations. By simple calculation, it will be seen that, for air,

$$\gamma = \frac{.2374}{.1689} = 1.406.$$

Expansion and Compression

Expansion and compression can be classified as *isothermal* and *adiabatic.*

Isothermal expansion or compression assumes that Boyle's law is exactly conformed to; thus, no change in temperature takes place. This would mean that, during expansion under constant temperature, the internal energy of the air, which is proportional to the absolute temperature, would also remain unchanged. Therefore, the work necessary for expansion would have to be supplied from a source of external heat. Further, during isothermal compression, the heat generated in the air would have to be allowed to escape as quickly as it was generated, and the heat flowing away during compression would be equal to the heat supplied from outside during expansion.

Adiabatic expansion or compression assumes that no heat flows either to or from the air during the operation. Thus, the air would gain or lose internal energy as the temperature is raised by compression and lowered by expansion, and the amount of internal energy is proportional to the quantity of external work put into or done by the air. See Fig. 1-2.

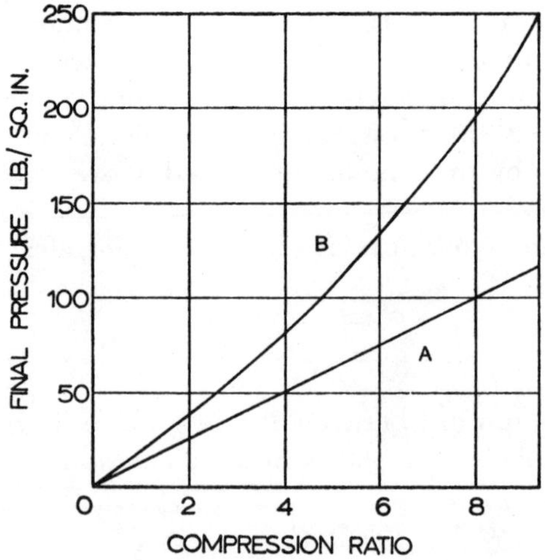

Fig. 1-2. Rise in pressure with isothermal (**A**) and adiabatic (**B**) compression.

Returning for a moment to the isothermal condition, suppose that the original volume V is compressed to a smaller volume V_1. The original pressure P will then rise to a pressure P_1:

$$PV = P_1V_1 = \text{a constant.}$$

If, however, the operation is carried out adiabatically, the equation becomes

$$PV^n = P_1V_1^n.$$

With isothermal conditions, $n = 1$, because the internal energy of the air is unchanged. Under adiabatic operation, the air gains or loses internal energy. Thus n is equal to γ, or to 1.406.

In actual practice, the value of n lies somewhere between 1 and 1.406; it is dependent on the characteristics of such factors as the fuel added to form a combustible mixture and the conditions of heat flow in the engine. If, for instance, the mixture is admitted to the cylinder of an engine that is already warm through running, heat will flow from the cylinder to the mixture. Thus, in the early stages of induction—including the first part of the compression stroke—this heat will continue to flow, warming the mixture.

Therefore, n will temporarily exceed the normal value of γ. As compression continues and the mixture's temperature first equals and then exceeds the cylinder's temperature, the heat flow will first stop and then reverse direction. At the later stages of compression, heat will be lost from the mixture to the cylinder walls, causing n to fall below γ by an amount dependent on the temperature difference between the gas and the metal. The extent of the mixture's heat loss will obviously be influenced by the area of metal exposed to the gas, the ratio between volume and internal surfaces, the density and amount of movement of the gas, and so on.

Experimental Determination

An interesting experimental method was developed by Clément and Désormes for determining the ratio of specific heats. Their apparatus consists of an insulated vessel (shown in Fig. 1-3). The vessel is equipped with a three-way valve and a U-tube manometer that contains a fluid for indicating the pressure differential between the air in the vessel and the atmosphere. Initially the vessel is charged with air that is under slightly more than atmospheric pressure, and the valve is closed.

If the valve is opened to the atmosphere and again closed, the pressure in the vessel will become atmospheric because of the adiabatic expansion of the gas. This gives the condition

$$PV = P_1V_1.$$

If the vessel is allowed to stand for a time with the valve closed, the temperature will again rise to its original value, and the pressure will also rise because V remains constant. Hence, we now obtain the condition

$$PV^n = P_1V_1^n.$$

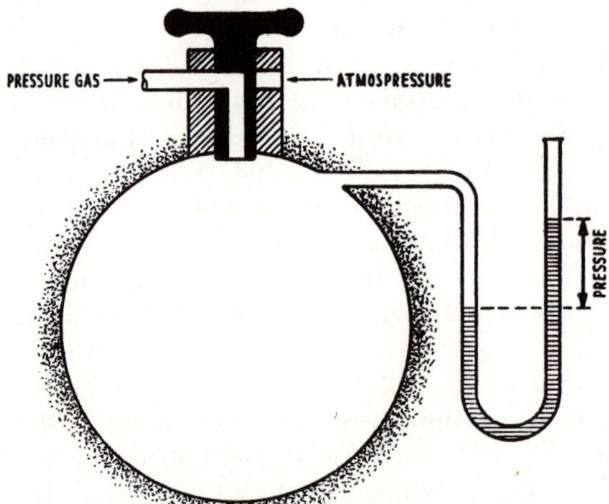

Fig. 1-3. Clément and Désormes apparatus for measuring ratio of specific heats.

Entropy

Entropy, denoted by S, is another of those functions that cannot be precisely defined. A dictionary definition of entropy might be "a measure of a system's (thermal) energy that is unavailable for conversion into mechanical work during a natural process." It is some property of a gas's state that determines the relationship between the rise in gas temperature and the increase of heat units.

One pound of water at the freezing point at atmospheric pressure is generally considered to have an entropy of zero. But in practical terms it is only *changes* in entropy that matter. Published tables are available that show the specific entropy for certain gases over a range of pressures and temperatures. The significance of entropy is that any *increase in entropy* implies *a decrease in the heat that is available for doing work.*

The diagram given in Fig. 1-4 shows a "reversible" cycle, that is, one in which heat can be converted into work or work into heat. This cycle is indicated by line *ABCDA.* If, as between process *ABC* and process *CDA,* the heat absorbed is represented by the area *ABCFEA* and the heat rejected by area *CFEADC,* the work done during the cycle is represented by the area *ABCD.*

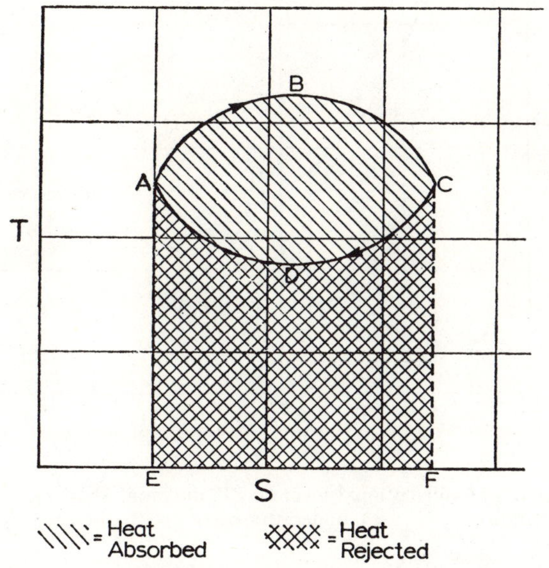

Fig. 1-4. Temperature-entropy diagram showing absorption and rejection of heat in reversible cycle.

PV *and* T *Diagrams*

The usual *PV* diagram (Fig. 1-5) shows the external work done in foot pounds when the volume *V* of a gas increases from V_1 to V_2. If, however, we use *T* as the vertical axis instead of *P*, and *S* as the horizontal axis instead of *V*, as shown in Fig. 1-6, the area under the line T_2 will indicate the number of heat units supplied to the gas. Suppose that this area is 1,000 Chu (a unit of heat) and that *T* is constant at 500°C (absolute). With isothermal expansion, the curve will be flat, and $S_1 - S_2$ will equal 2 units of entropy in length and $2 \times 500 = 1000$ Chu. In this case, one unit of entropy would be the value of the increase in entropy caused by the reception of a number of heat units equal in amount to the absolute temperature at which the heat is received.

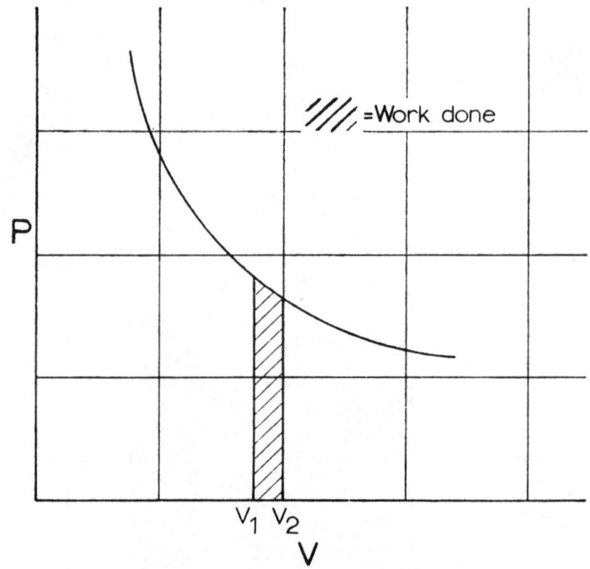

Fig. 1-5. The conventional form of *PV* diagram, showing work done (indicated by area under the curve) with increase in *V*.

We have shown that in isothermal expansion or compression the change of entropy is equal to the heat added or withdrawn, divided by the absolute temperature. In adiabatic expansion or compression, entropy is the one factor that does not change—though the pressure, the temperature, the enthalpy, and the internal energy are all changing. Thus, the temperature-entropy diagram for the isothermal condition will be a straight line alongside the entropy axis. For adiabatic conditions, the entropy is constant; the line will now lie vertically alongside the temperature axis. Therefore, a "closed circuit" made up of both conditions would be shown on a graph as a number of straight lines at right angles to one another.

In summing up the relationships between *PV* and *TS* diagrams as applied to a cycle using an ideal gas, we can state as an example that the former could show that work in foot pounds is equal to the mean force in pounds, multiplied by the space range in feet (or the mean pressure in psi, multiplied by the volume range in cubic feet). The *TS* diagram shows that the heat units taken in are equal to the average absolute temperature, multiplied by the entropy.

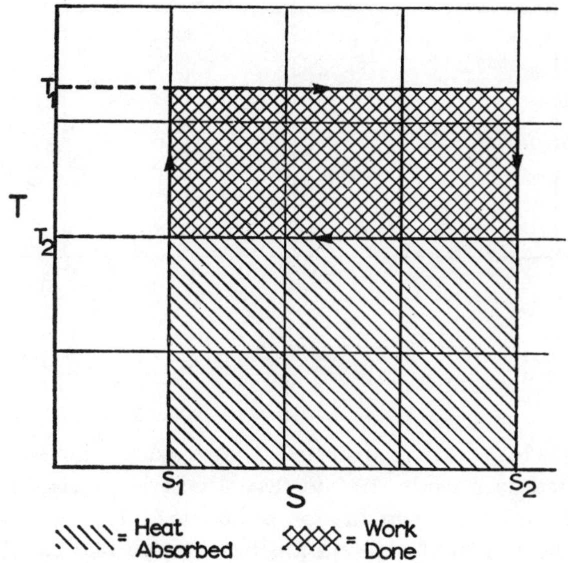

Fig. 1-6. With *TS* in place of *PV*, area under line T2 represents heat units supplied.

Thermal Efficiency

The thermal efficiency of a cycle is the difference between the heat units supplied and the work done. The *PV* diagram in Fig. 1-7 shows a cycle that approaches practicability; it is reversible and comprises two isothermal and two adiabatic processes. Along the line *ab,* a quantity of gas takes in heat from the outside and expands at constant pressure from *a* to *b*. It next expands adiabatically, with falling temperature, from *b* to *c*. Compression from *c* to *d* at constant pressure is followed by adiabatic compression from *d* to *a*.

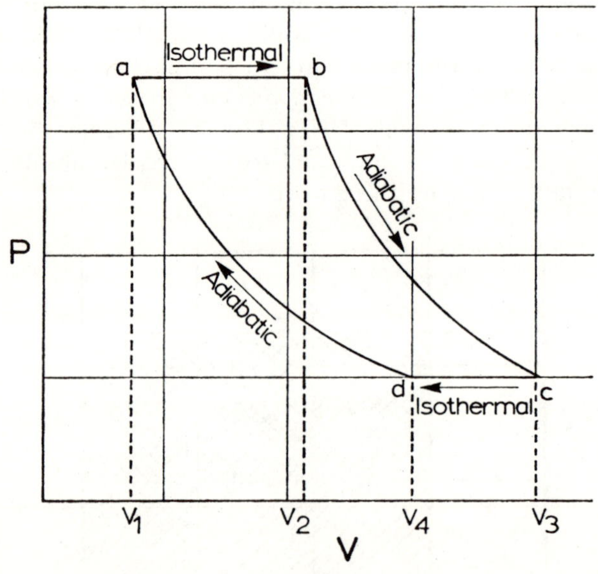

Fig. 1-7. Reversible ideal cycle based on Carnot, with two isothermal and two adiabatic processes.

From *a* to *b* heat is absorbed from outside, and from *c* to *d* it is rejected to outside. In the two adiabatic processes, *bc* and *da,* heat is neither taken in nor passed out. The work done is thus represented by the differences between *ab* and *cd* (the heat absorbed minus the heat rejected). The heat rejected, however, is not necessarily wasted but could be reused—though the percentage of rejected heat actually converted into work could be considered unimportant even theoretically were it not for the fact that the temperature at which rejection takes place (at *cd*) might be lower than that at which heat is supplied (at *ab*). As the second law of thermodynamics confirms, the heat will not flow naturally between *cd* and *ab*.

In 1824 Sadi Carnot gave conditions for an "ideal" heat engine cycle, which, though obviously unattainable, gives a concise exposition of the effect of various phenomena. Carnot visualized a quantity of an ideal gas in a cylinder with a weightless piston. The cylinder was capable of associating with two thermal sources, each of unlimited capacity and at different temperatures.

If Carnot's ideal engine is applied to Fig. 1-7, the cylinder comes into contact with the "hot" thermal source at *a,* causing the gas to expand isothermally from $V1$ to $V2$, at constant temperature. The cylinder is removed from contact with the "hot" course at *b,* and, with thermal insulation being assumed, there is adiabatic expansion from *b* to *c,* with the temperature falling. At *c* the cylinder comes into contact with the "cold" reservoir, and from *c* to *d* isothermal compression takes place from $V3$ to $V4$; because the temperature must remain constant, heat is lost to the "cold" source. The cylinder is removed from contact with the "cold" source at *d,* and, again assuming thermal insulation of the cylinder, adiabatic compression takes place from *d* to *a,* with a rise in temperature.

It is evident that the efficiency of the Carnot engine depends on the temperature at which heat is supplied and rejected by the hot and cold sources respectively. Complete thermal efficiency could obviously be obtained only if the cold reservoir was of infinite capacity at the absolute zero temperature, which is, of course, impossible.

The Otto Cycle

The four-stroke spark-ignition engine operates on a cycle usually credited to Otto (1876). His ideal cycle—in which *air* is assumed to be the working gas—is interesting to compare with the Carnot cycle. The Otto cycle is sometimes known as the *constant volume* cycle. The four strokes that make up the Otto cycle are familiar in their practical application to automobile engines. Let us consider these strokes in relation to the ideal cycle given in Fig. 1-8.

The horizontal line *ab* represents the induction stroke at constant (atmospheric) pressure. From *b* to *c,* adiabatic compression takes place, with no change in entropy. With the explosion intiated at *c,* there is a further pressure rise at constant volume to *d,* followed by adiabatic expansion with constant entropy to *e.* The exhaust valve opens at *e,* the pressure dropping at constant volume to *b,* and this being followed by the exhaust stroke at constant pressure from *b* to *a.*

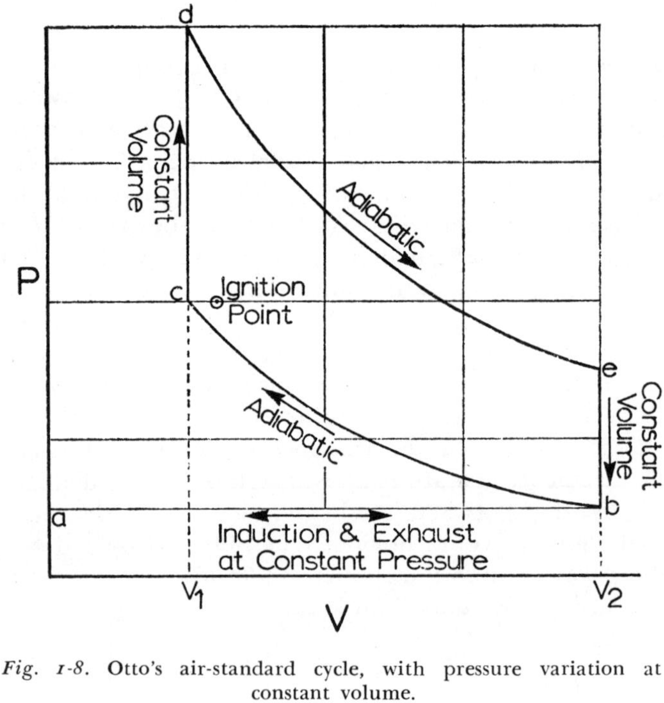

Fig. 1-8. Otto's air-standard cycle, with pressure variation at constant volume.

Fig. 1-8 shows that the conventional induction and exhaust strokes are not of significance as such—a consideration that has made possible the application of the Otto cycle to the "stroke-less" Wankel rotary engine. The important processes are the compression from *b* to *c,* the expansion from *c* to *e,* and the rejection of the remaining heat from *e* to *b*. During *cd,* heat is added; therefore, the work must be the equivalent of the difference between the heat that is added at *cd* and the heat that is rejected at *eb*.

Based on the absolute temperatures and pressures and the respective volume of the air at points *bcd* and *e,* the thermal efficiency of the cycle is equal to

$$1 - \left(\frac{1}{r}\right)^{\gamma - 1}.$$

Since, with air, γ is equal to 1.406 (usually shortened to 1.4) the formula becomes

$$1 - \left(\frac{1}{r}\right)^{0.4}$$

where r = the compression ratio. In this case, the efficiency is known as *air standard efficiency,* or ase.

The ase shows what percentage of the external heat added to the air is converted into work. (In actual practice this heat is produced by combining the air with fuel to form a combustible mixture.) It will be evident from a study of the above formula that the compression ratio governs the ase and that the higher the value of *r,* the greater the ase. The value of being able to work out the ase for any particular engine lies in the fact that it forms a yardstick against which the results obtained in practice can be assessed.

Air Standard Efficiency, percent at different compression ratios

r	*ase*	*r*	*ase*
6.0	51.16	7.5	55.34
6.2	51.8	8.0	56.47
6.4	52.38	9.0	57.4
6.6	53.0	10.0	60.2
6.8	53.55	11.0	61.7
7.0	53.98	12.0	62.9

To the designer, ase represents "top end" efficiency, or, more exactly, the efficiency of the engine's cylinders and cylinder head (s) in converting the heat energy of the fuel into work on the pistons. If he chooses a compression ratio of 10:1, the perfect or ideal engine would convert over 60 percent of the fuel energy into power. The nearer that the actual thermal efficiency of the engine being designed can be made to approach this figure, the more power will be produced per unit of fuel mixture delivered to the inlet valves—and the more the designer can congratulate himself on his cylinder and head layout. But power is not the only point to think about. Another important consideration is that any heat in the mixture that is liberated *but not converted into work* must somehow be got rid of.

Indicated and Brake Horsepower

Most readers are familiar with the definitions of *indicated horsepower* (ihp) and *brake horsepower* (bhp). (Although metrication is making *kilowatts* the power rating system of the future, it seems advisable for the purposes of this book to keep the traditional units of measure; they undoubtedly will be grasped more easily.) Indicated horsepower represents the horsepower produced on the pistons. It is calculated from the cylinder expansion pressure, which is measured by a device called an *indicator*—hence the name. This pressure is stated in pounds per square inch as *mean effective pressure* (mep). In obtaining mep, the pressures throughout the cycle are used, and in each case the pressure is the average, or mean, obtaining on that particular stroke. An example will explain matters.

Assuming that atmospheric pressure is 15 psi:

Induction pressure, say 12 psi (absolute)
Compression pressure, say 40 psi (absolute)
Expansion pressure, say 185 psi (absolute)
Exhaust pressure, say 18 psi (absolute)

The expansion pressure is the only positive stroke, the other three being employed in pumping work, or negatively.

Thus the effective pressure is 185 − (12 + 40 + 18), or 115 psi. If, having obtained the pressure on the pistons, we combine with this the distance and the time factors required, we can obtain a result in foot pounds per minute, and thus horsepower.

If p = expansion pressure (imep) in psi
l = length of stroke in feet (or fraction thereof)
a = area of one piston in square inches
n = number of working (expansion) strokes per minute of the engine,

the multiplication of the above four quantities will give the result in foot pounds per minute. If this is divided by 33,000, the result will give ihp. Therefore,

$$\text{ihp} = \frac{p \times l \times a \times n}{33{,}000}$$

Brake horsepower is the actual power available at the engine's crankshaft or flywheel, and it is called *brake* horsepower because, in the early days, it was measured by applying a braking force to a pully or a drum on a shaft being driven at a 1:1 ratio by the engine. Modern "brakes" are known as dynamometers, and the power is absorbed electrically or hydraulically.

Horsepower Ratings

A number of horsepower rating systems have been in use during the past quarter-century. Until recently, passenger car engines in the United States were rated by "SAE horsepower." Since 1972, ratings have been given in "SAE net horsepower." Both of these measurements are derived from a dynamometer test (Fig. 1-9) of a sample, or prototype, engine. The old system was grossly misleading because the test was made without any auxiliaries (such as the alternator, the air cleaner, an air pump, the fan, and so forth) and without the kind of exhaust system that is actually used on cars. The "net" rating is somewhat more informative, being measured with all of the auxiliaries in place.

For the purpose of this book, horsepower should be considered as the scientific quantity described by James Watt (1736–1829). That is, one horsepower is equal to 33,000 ft. lbf/min. (or 550 ft. lbf/sec.). Though the term *horsepower* is no longer accepted by the scientific community, the international General Conference of Weights and Measures has honored it indirectly by making the watt (or kilowatt—watts × 1000) the SI (*Systeme International,* or "modernized metric system") unit of power.

To convert horsepower (550 ft. lbf/sec.) to kilowatts, multiply by .746. Metric horsepower can be converted to kilowatts by multiplying by .736. Thus, the 82 DIN (metric) horsepower engine used in the 1977 VW Scirocco could also be rated as a 78 SAE net horsepower engine (DIN hp × .953 = SAE hp) or as a 60 kilowatt engine (DIN hp × .736 = kilowatts).

Fig. 1-9. Racing engine mounted on dynamometer stand. Notice driveshaft with universal joints that transmits engine output from flywheel to dynamo.

Mechanical Efficiency

Since ihp represents the power at the engine's pistons and bhp represents the power at the engine's crankshaft, it follows that the difference in the two values is caused by the mechanical losses consequent in the conversion of the expansion pressure into rotating motion. The percentage of ihp available as bhp is known as the *mechanical efficiency* and is obtained thus:

$$\text{Mechanical efficiency } (\%) = \frac{\text{bhp} \times 1000}{\text{ihp}}.$$

It will be appreciated that the losses mentioned earlier include not only piston and bearing friction but also the power required to drive the valve gear, the ignition apparatus, the oil pump, and other auxiliaries. Of no small importance in very high-rpm engines is the wind resistance of the moving parts inside the confines of the crankcase. Designers have always recognized, abstractly at least, that a valveless rotary engine would have superior mechanical efficiency. The Wankel rotary engine has recently made this a reality. The Wankel also has an aerodynamically "smooth" internal shape that suggests another direction for future development.

Factors Governing Expansion Pressure

In general, the expansion pressure depends on the quantity of mixture inducted into the cylinders, the efficiency of its conversion into work on the pistons, and the percentage of the mixture that is actually so converted. These three factors are known, respectively, as *volumetric efficiency, combustion efficiency,* and *thermal efficiency.*

It is usual to combine the above three factors into a single efficiency figure, *thermal efficiency,* and on engine test runs, the mechanical efficiency is also usually combined with them; the resulting overall figure is known as *brake thermal efficiency.* This useful and important figure shows the merits of the engine concerning its ability to convert fuel heat energy into crankshaft bhp.

The indicated thermal efficiency is the practical counterpart of ase, and useful information can be obtained by comparing the two. In practice, the figure obtained on test is considerably lower than the ase for an equivalent compression ratio. There are several reasons for this:

1. The ratio of the specific heat of air (γ) is less than 1.4 because of the necessity for mixing fuel with the air.
2. Heat is lost during the compression stroke, thus reducing the value of T immediately before expansion.
3. The heat is not added instantaneously at tdc (top dead center), but the mixture begins to burn before tdc and continues to burn for some part of the expansion stroke. Because the pressure is dropping constantly during this stroke, the fuel burned at the lower pressures will not be used to its fullest advantage.
4. To discharge the burned gases completely after expansion, it is necessary to open the exhaust valve a long way before bdc (bottom dead center) of the stroke while there is still appreciable pressure in the cylinder. Thus, the effective expansion ratio is in practice always less than the compression ratio because the effective stroke ends when the exhaust valve opens.

If this list is compared with the assumptions made in calculating the ase, they certainly seem to represent formidable sources of energy loss. Nevertheless, it is easily possible on a well-designed engine to obtain a thermal efficiency, in terms of foot pounds on the pistons, of 38 percent, so that the result obtained is only 28 percent less than the theoretical ideal.

Calculating Thermal Efficiency

A simple example will make clear the requirements for calculating thermal efficiency in practice. The engine in this example develops 30 bhp steadily on test for a period of 1 hour, during which time it consumes 17 pounds of gasoline that is known to have a calorific value of 18,000 Btu per pound. By reducing both the "input" and the "output" to a common quantity (Btu), we get:

Fuel consumed = 17 lb. Heat consumed = $17 \times 18{,}000 =$ 306,000 Btu.

Bhp developed = 30. Time = 60 min. = $30 \times 60 \times 33{,}000$ ft./lb. per hour.

As 1 Btu = 778 ft./lb., heat available as work at crankshaft =

$$\frac{30 \times 60 \times 33{,}000}{778} = 76{,}350 \text{ Btu.}$$

Thus, brake thermal efficiency $= \dfrac{76{,}350}{306{,}000} \times 100 = 25\%$.

It is usual to include volumetric efficiency in the overall thermal efficiency, but the former can also be measured separately. This might well be desirable when valve gear modifications or port shape modifications are concerned. The connection of the air intake to an accurate measuring device, such as a type of gasholder, would enable the volume inhaled to be measured over any required number of induction strokes, and comparison with the theoretical swept volume of the engine then gives the percentage volumetric efficiency.

The volumetric efficiency will not remain constant in a variable-speed engine since the valve timing must necessarily be a compromise, as will the port areas. A volumetric efficiency of around 80 percent is, however, obtainable on a good engine with overhead valves over most of the useful speed range. Much higher figures are obtained on racing engines, especially those that make use of supercharging or the resonance effects available from tuned manifold systems.

If required, the mechanical efficiency can be measured on the dynamometer by running the engine at the same speed at which maximum bhp is developed and then cutting off the ignition to each of the cylinders one at a time. The reduction in bhp under these conditions represents the ihp of the cylinder cut out, and by adding all the "debits" together, the total ihp—and thus the mechanical efficiency—can be obtained. The figure for this is again not constant; it is less at higher rpm because of greater friction. It is usually above 80 percent, however, and particular attention to free running of components may be rewarded by a figure approaching 88 percent.

A much more accurate method on the same principle is to cut out all of the spark plugs in such a way that a misfire occurs at regular intervals in each cylinder. In this way, all of the cylinders continue to function, and thus the temperature over the whole of the engine is kept constant. The reduction in power under various frequencies of misfiring can be closely assessed, with the advantage that all the engine systems are functioning virtually as in normal running. This is far from being the case when one cylinder at a time is completely out of action.

Combustion Efficiency

This topic is raised mainly as a matter for speculation—not as something to be explored theoretically. Insofar as competition engines are concerned, combustion efficiency seems to be the one aspect of thermal efficiency that is given the least consideration. By far the most emphasis is given to improving volumetric efficiency. Unfortunately this preoccupation often blinds the tuner or designer to something equally important: how well the combustion chamber "burns".

Poor combustion efficiency should be the first thing looked into when an engine that has high volumetric efficiency, from a theoretical point of view, fails to achieve the desired (or projected) output. In actual practice, however, it is usually the last factor to be investigated. More often than not, development work gets bogged down in the futile redesigning of camshafts, spark advance curves, and induction or exhaust systems. In such circumstances, the wise tuner will set aside, at least temporarily, considerations of volumetric efficiency and look more closely at the combustion chamber design.

In addition to shortcomings in combustion efficiency, deficient thermal efficiency is also sometimes masked by combining volumetric efficiency, combustion efficiency, and thermal efficiency into a single thermal efficiency figure. For example, in recent years the Wankel engine has come under criticism for its poor fuel economy. Have the engineers working with this design been so preoccupied with its excellent mechanical efficiency that they have neglected to observe something equally fundamental to the engine's efficiency? Should not more attention be given to the heat loss that is consequent to having the burning mixture transported along a wide expanse of (comparatively cool) chamber wall? It would seem that a closer look at the second law of thermodynamics is advisable in this regard.

2 / The Production of Power

Bmep and Torque

Torque, the turning effort on the crankshaft, is independent of engine speed. Its value depends on the pressure on the pistons, the piston area, and the length of the piston stroke. Since the cylinder bore and piston stroke dimensions are fixed, any increase in mep (mean effective pressure) means an increase in torque, and vice versa. Thus, the mep curve and the torque curve follow the same line on a performance graph. That torque can be plotted as a curve is not necessarily contradictory to the first sentence of this paragraph.

As already stated, imep is obtained directly from an indicator measurement of the cylinder expansion pressure. On the test bed, it is usual to calculate the pressure from the bhp and refer to it as *brake mep* or *bmep*. The figure obtained is imaginary, because it allows for mechanical losses in the engine; it does, nevertheless, enable useful comparisons to be made (as between designs) in almost all relevant aspects.

If the bhp is known, the bmep can be calculated as follows:

If $l =$ the length of the stroke in feet (or fraction thereof)

$a =$ the area of one piston in square inches

$n =$ the number of power strokes per minute,

$$\text{bmep (in pounds per square inch)} = \frac{\text{bhp} \times 33{,}000}{l \times a \times n}.$$

The makeup of the above formula will be apparent if it is compared with the final calculation made in chapter 1.

To obtain the torque from the bmep, we can use a constant since, for any engine, the only two variables affecting torque are the swept volume (piston displacement) and the bmep—and all other mathematical calculations can be resolved into the constant:

$$\text{Torque (in lbf./ft.)} = \frac{\text{bmep} \times \text{swept volume in cm}^3}{2{,}473}.$$

For the makeup of this formula, consult the appendix.

Power Production

If the torque of the engine remains constant, the bhp will increase in direct proportion to the speed, and the only limitation on power will be the mechanical construction of the engine in its ability to withstand high rpm. As a matter of fact, the mechanical construction in general *does* determine the rpm limit, but this limit is, nevertheless, accompanied by a falling off in torque resulting from the decreased bmep and increased frictional losses.

The bmep falls off as the rpm increases because of the inability of the engine to inhale a full charge in the diminishing time that is available. Thus, while the torque curve is reasonably flat, the bhp curve will climb approximately *pro rata* with the rpm. As the bmep and torque fall off, the bhp curve will begin to droop, and then it will flatten out until a point is reached when increasing the rpm does not produce any more power. The mechanical limitations at top revolutions mainly concern the stresses arising from the inertia of the reciprocating parts, particularly the pistons and the valve gear. As one can see by referring to the ideal engine cycle, a high piston speed aids the adiabatic process, making for high thermal efficiency. Still, the mechanical limit has always to be considered.

The bmep obtained from an engine is greatly influenced by the compression ratio, which is obvious from a consideration of the fundamental principles governing the air standard efficiency. There are, however, factors that limit the compression ratio that can be used. First, a very high compression ratio may necessitate a combustion chamber shape that is not conducive to high thermal efficiency. Second, there is the danger of detonation.

Compression Ratio and Detonation

Normally the fuel/air charge in the combustion chamber burns with a smoothly advancing flame front. Detonation occurs when the fuel/air charge, or a portion of it, ignites instantaneously throughout, causing an explosion rather than progressive combustion.

For example, an overheated exhaust valve may ignite the charge at a point remote from the spark plug. The flame front that advances from this uncontrolled point of ignition may trap an unburned portion of the charge between itself and an opposing flame front advancing from the spark plug. The unburned portion of the charge, thus "squeezed", rapidly increases in temperature until it ignites spontaneously and nearly instantaneously, causing detonation. Similarly, if the compression ratio is very high, or the fuel octane low, the entire charge may detonate because of the temperature rise caused by compression pressure alone.

It is important to differentiate between preignition and detonation. If the fuel/air mixture in the combustion chamber is ignited by a "hot spot", such as an overheated spark plug electrode, before a spark actually occurs, there is untimed ignition, or preignition. The flame front may advance smoothly through the charge but the engine "pings" because ignition has taken place too early. If preignition takes place early in the compression stroke, or if it takes place at several points—or at a point remote from the spark plug—there may be sufficient pressure and temperature rise to produce detonation.

The shock waves set up by preignition and detonation give rise to a characteristic metallic sound. This sound can vary in magnitude from a mild "ping" caused by preignition at low speeds and wide throttle openings, to such violent manifestations caused by detonation that the engine pulls up dead—most likely with holes in its pistons.

The choice of fuel has a great influence on detonation. Inferior (low octane) fuels demand comparatively low compression ratios if detonation is to be avoided. High octane gasoline and the special alcohol-based fuels used in some kinds of racing allow very high compression ratios. In fact, the compression ratios of "alki burning" racing engines are often limited more by considerations of valve-to-piston clearances than by fuel octane.

To appreciate what happens during detonation, it must be understood that, if the temperature of an explosive mixture is raised sufficiently, the mixture will eventually explode spontaneously (the temperature concerned is known as the *spontaneous ignition temperature*). Before exploding, however, there is an interval called the *ignition time-lag,* which in the case of a mixture of gasoline and air is about 0.0033 seconds. If the piston, approaching tdc of the compression stroke, takes longer than this period of time before the spark plug operates, there will be a premature explosion, the magnitude of which depends on the extent of the symptoms. Obviously, the longer the ignition time lag, the less liability there is of spontaneous ignition.

Even if the mixture is correctly fired at the "timed" point, and with an apparently smooth flame spread, there may still be a loss of power through detonation. The burned mixture behind the flame front will be at high temperature and pressure, causing compression of the unburned gas ahead of the flame; this can raise the temperature of the latter above the spontaneous ignition point. It is thus necessary for the flame front to move right through the unburned charge before the end of the ignition time-lag period; otherwise the violent counterpressures set up will lead to overheating and excessive mechanical stressing.

With any particular fuel, the brake thermal efficiency will decrease if the compression ratio is raised above the optimum figure for that fuel. Thus, high octane fuel must be used to reap the ase advantages of high compression. Octane improvements extend the ignition time-lag period of a fuel, thereby decreasing the fuel's liability to spontaneous ignition. The octane of the fuel permitted by the racing class rules is therefore of great importance since any limitation on compression ratio is tantamount to a limitation on thermal efficiency.

Fuel technology is a formidable subject in itself, and hardly within the scope of this book. However, because of recent reductions in the octane of pump gasoline, the subject is assuming ever greater importance for designers and tuners of competition engines. At present, there are a number of octane-improving additives on the market that can be mixed with pump gasoline to make today's fuel as suitable for racing as were the high-octane "super premium" gasolines of the 1960s.

In improving the octane rating, a compound is added to the fuel that diminishes the tendency to detonation. Usually, this compound is tetraethyllead. Large gasoline companies test their fuels on a variable-compression engine in the laboratory, the compression being increased to the highest point possible without detonation. The fuel is graded by an octane number based on the percentage of tetraethyllead that is necessary to achieve detonation-free operation at a specific compression ratio.

It is important to note that the calorific value of high octane fuel may be inferior to that of more ordinary gasolines. The latter have a calorific value of about 18,000 Btu per pound, whereas the same weight of high octane alcohol-based racing fuel may well produce only about half this figure, so that a very much greater quantity of fuel is required to produce the same amount of heat. This, of course, explains the abnormal increase in size of carburetor jets, venturis, fuel lines, and so forth required when such fuels are used. The much higher thermal efficiency obtainable naturally results in a net power gain, and the high latent heat of evaporation assists in lowering internal temperatures—a good point so long as the fuel is not used deliberately to this end. Nevertheless, engines that race with high-boost supercharging or turbocharging frequently pass a great deal of unburned fuel through their combustion chambers, mainly to keep the pistons from melting.

Pressure Increase with Compression Ratio

On competition engines, the object of increasing the compression ratio is obviously to increase the torque, though increased fuel economy is an important consideration for passenger cars. Particularly when large power output increases are contemplated for a supertuned production engine, it is important to know ahead of time what the consequences will be in terms of pressure increases. If, for example, a 50 percent increase in bhp is aimed for and the maximum rpm remains the same, this means a 50 percent increase in bmep or an increase, say, of from 100 psi to 150 psi in compression pressures.

The graph in Fig. 2-1 shows cylinder pressures at various compression ratios. The pressure indicated is that at the end of the compression stroke and assumes adequate cylinder filling (good volumetric efficiency and wide-open throttle). At the moment of ignition, the pressure will rise to a maximum of three or four times the compression pressure. Fig. 2-2 shows the approximate bhp figures obtained on engines of similar construction using, first, 72-octane pump gasoline and, second, an alcohol-base fuel that allows a 12:1 compression ratio.

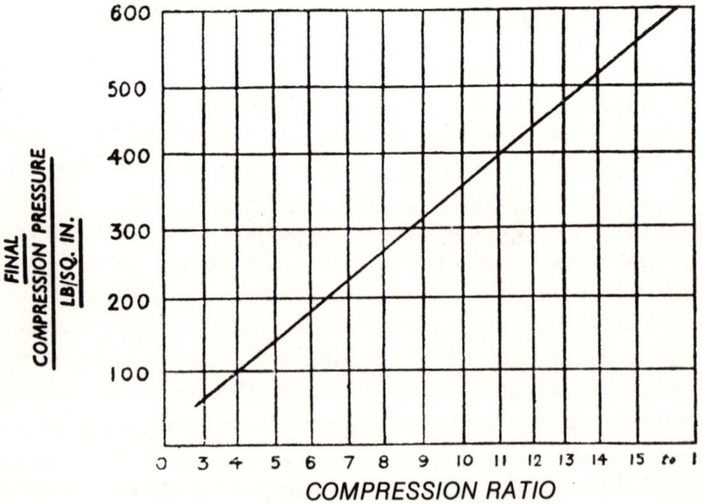

Fig. 2-1. Increase of pressure prior to ignition with increase in compression ratio.

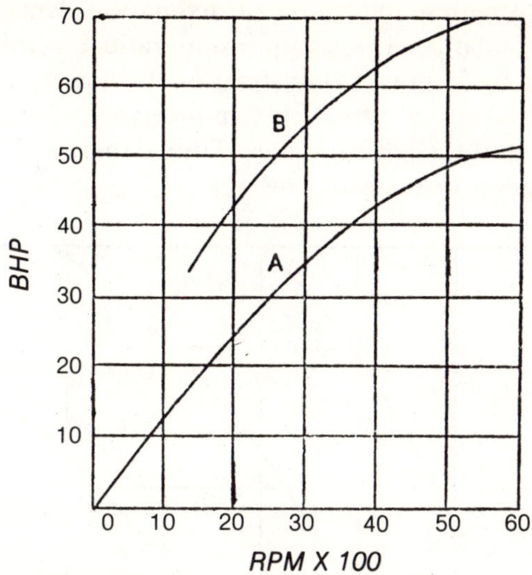

Fig. 2-2. Power output with compression ratios graded to suit fuel. (A) medium octane, 7:1; (B) high octane, 12:1.

The final compression pressure is governed largely by the volumetric efficiency. Assuming induction at atmospheric pressure, and $\gamma = 1.3$,

$$\text{final compression pressure} = 14.7 \times r^{1.3}.$$

In practice, of course, induction will be at a pressure somewhat lower than atmospheric pressure so that any pressure calculated as above will be safe. Obviously supercharging, which produces induction at above-atmospheric pressures, is another thing altogether.

When an engine has been blueprinted or supertuned to produce even a moderate increase in torque, it is necessary that detonation is not allowed. What would be innocuous "pinging" on a lower compression ratio can, if judged by the same degree of audibility, have serious results if permitted at higher compression ratios. Unfortunately, with a wide-open, unmuffled racing exhaust system, even the loudest of "pings" is usually inaudible—making it doubly important to run the engine conservatively at first and then to check the spark plugs frequently for even the slightest indication of detonation damage. Usually, given a particular fuel, a compression ratio on which the engine is perfectly happy is only fractionally lower than the one at which all kinds of troubles can intervene. As can be seen from Fig. 2-3, the slightly higher (and dangerous) ratio gives only a negligible increase in power.

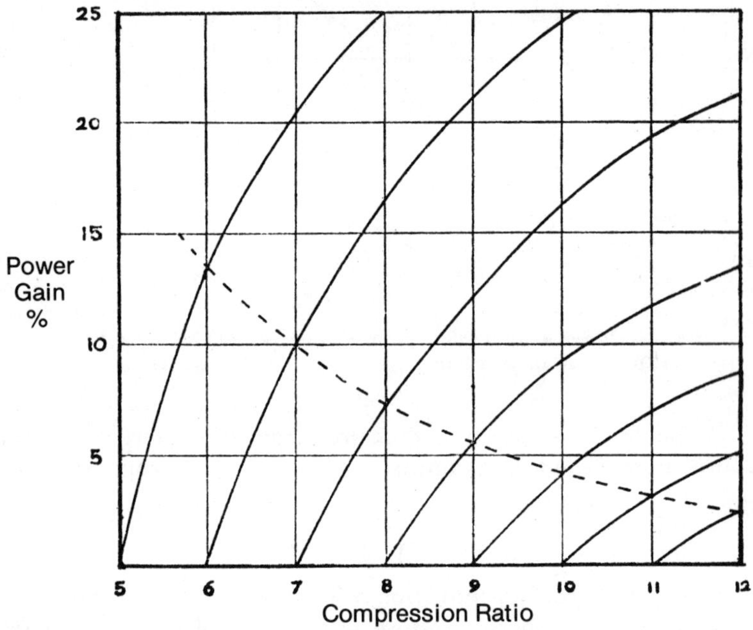

Fig. 2-3. Progressive decrease in power gained as a result of raising compression ratio.

Heat Loss

In comparing the average thermal efficiency figure with the ase for an equivalent compression ratio, it is evident that even in the very best designs a large proportion of the fuel energy goes to waste. We have already indicated some of the reasons for a possible 30 percent discrepancy between actual and ase figures. One important point is the inevitable early opening of the exhaust valve, which accounts for most of the loss. For example, if we assume an indicated thermal efficiency of 35 percent, it means that there is a loss of 65 percent; of this loss nearly 40 percent will go to exhaust. The remaining 25 percent is carried away in the cooling medium.

The heat loss to the cylinders, and thus to the cooling water (or air, or oil), takes place roughly as 5 percent during initial combustion, 6 percent during expansion, and the remaining 14 percent during exhaust. Of this last-named loss, none can possibly be recovered, so we are left with 11 percent to be accounted for during the combustion and expansion process.

Obviously, the combustion/expansion stroke, even with no heat loss at all, could never be 100 percent efficient. The efficiency of the processes depends on the compression ratio and is governed by basic thermodynamic laws. It can safely be said that the ignition process accounts for a loss of 5 out of the 11 percent mentioned and that the process in itself must lose half of this, the jacket loss therefore accounting for 2.5 percent.

As for expansion, it might be thought that a serious heat loss could take place as the cylinder wall is uncovered by the descending piston because of the large area of metal exposed. It should be borne in mind, however, that the pressure is dropping all the time. Thus, the heat loss toward the end of expansion is not so serious since its retention would have added little to the work done before the exhaust valve opened. In fact, it is doubtful if more than one-fifth of the 6 percent lost during expansion could be saved even with no jacket loss whatever. The total saving, 3.7 percent, would be very small, even if all the heat could be trapped. This represents a gain in thermal efficiency from 35 to 38.7 percent. Actually the gain would be less because of the higher operating temperature of the mixture, and its consequent increase in specific heat would cause a decrease in thermal efficiency. As a result, the net gain would probably only be about 2 percent. We are, of course, considering only piston engines here.

The significance of the preceding analysis lies in the fact that high thermal efficiencies are sometimes claimed for engines on the basis of the use of a compact form of combustion chamber; the idea is that such a shape presents the minimum area for loss of heat to the cooling jackets. However, our analysis suggests that this conclusion is not so, and the probability is that the high thermal efficiency figures obtained by these engines largely result from the good "breathing" characteristics that usually go with a compact combustion chamber shape.

Water Temperature

Engines run better when the cooling water is hot. The reason frequently given is that the increased temperature leads to higher pressures and less heat loss. Actually the improvement is mostly caused by the improved carburetion and mixture distribution afforded by the warm induction arrangements and by the decreased piston friction and improved ring sealing that is a result of the thinning of the oil on the bores. There are very little data drawn from experience with air-cooled engines to contradict this conclusion.

As the weight of mixture inhaled into the cylinders is proportionate to its absolute temperature, it is obvious that with cool cylinders a greater weight of charge will be drawn in than when the temperature is higher. This will more than compensate for any increase in heat loss to the water jacket or cooling fins, and as far as the ihp is concerned, there will be a definite gain. However, the extra piston oil drag with cold oil cancels out this gain so that the bhp figure will be lower when the engine is cold; the decreased mechanical efficiency more than compensates for the increased volumetric efficiency.

In showing that the gain in thermal efficiency would be extremely small with all combustion and expansion heat losses completely cancelled (if this were possible), we have considered an engine that is already thermally efficient—a figure of 35 percent having been chosen. Actually there are many sports car and passenger car engines that do not operate at anything near this figure, though by dint of extravagant fuel consumption they can be tuned to perform reasonably well for production class competition. But performance obtained in this way is objectionable, not merely because it wastes fuel, but for another far more important reason.

For the engine to be efficient, it must burn its fuel to the utmost possible advantage from the point of view of power production. If as big a percentage as possible of the heat energy is converted to work on the pistons, good design will take care of the disposal of the remainder, which in any case is unavoidably wasted. If the percentage converted to work is reduced, a greater proportion has to be carried away to waste, and that extra can spell trouble in the course of long periods of full-load operation.

Valve Size and Gas Velocity

The dimensions of valves and ports must necessarily be something of a compromise. For the desirable amount of turbulence—which is necessary to prevent stagnant areas of mixture, in particular adjacent to the cylinder walls—high gas speed through the intake valve is required. On the other hand, an excessive gas speed means reduced volumetric efficiency and increased pumping work.

In general, reasonably high gas velocity is the most important consideration. If this sets up a degree of turbulence that adequately scours the walls, combustion and expansion heat losses will be minimized, combustion will be complete, and mep will be satisfactorily high. In the kind of engine with which we are concerned, with overhead valves, the volumetric efficiency should not suffer in view of the very free entry afforded for the gas. Therefore, though it is desirable to reduce pumping work to the minimum, the increasing of port areas and valve sizes should not be done haphazardly—particularly if by doing so the shape of the combustion chamber is altered since turbulence obviously depends largely on the compactness of the latter.

The exhaust valve should cause little concern in regard to power loss. The relatively early opening, when there is still a high pressure (at least 70 psi above atmospheric) in the cylinder, ensures a rapid exit of the spent gas so that the energy required to pump out the residue is negligible. Most of the gas is expelled when the piston is around bdc, and therefore the back pressure on the exhaust upstroke will not exceed an average of more than 2 or 3 psi. It may even be considerably less over a good speed range as a result of the buildup of energy in the escaping gas columns (unless the muffler and exhaust system are outrageously inefficient).

Nevertheless, it is sometimes necessary for larger exhaust valves to be fitted in supertuned American V8 engines that are used in drag racing—particularly if they are being run at high supercharger boosts. The use of semiexplosive nitromethane fuels causes these engines to develop enormous cylinder pressures, which ensure that the velocity of the exiting gases is satisfactorily high despite the large valve and port diameters.

The fact that large-diameter intake valves made of exhaust valve steels have been specified by various carmakers for certain American V8s provides the speed tuner with a source of huge valves that can be installed in the exhaust sides of the combustion chambers. Though some auto manufacturers maintain a competition department, which can inform tuners about the existence of big valves suitable for exhaust service (in addition to disseminating other valuable engineering data), it is nevertheless important for the machinist/tuner who builds competition engines to be aware of the materials specifications for all the parts that might be available to meet his or her design needs.

Although an intake gas velocity above about 160 feet per second is liable to lead to an increase in pumping losses and a decrease in the weight of charge drawn in, the velocity through the exhaust valve may be up to 50 percent above this figure before any measurable back pressure occurs. Even doubling the back pressure only decreases the mep by a like amount. A negative induction pressure of 2 psi, on the other hand, represents a reduction from 14.7 psi to 12.7 psi on unsupercharged engines, or about 14 percent. This is a much more serious matter because it reduces the mep by approximately the same percentage (14 percent).

3 / *Problems of High-speed Operation*

Valve Timing

Various assumptions are made when calculating the theoretical thermal efficiency of an engine having a certain compression ratio. Among these are that each of the four phases of the cycle occupies 180° of crankshaft rotation and that the induction and the exhaust strokes take place at atmospheric pressure. In practice, of course, even on wide-open throttle and at maximum torque, with the engine turning over at perhaps half its peak rpm (when there is plenty of time for the mixture to fill the cylinders), the induction pressure will be barely equal to atmospheric. The exhaust stroke begins when there is still a high pressure in the cylinder and a proportion of expanding gases representing useful power is forcibly ejected to waste.

In practice the valve timing is arranged so that the inlet valve, instead of closing at bdc, remains open for a considerable portion of the following compression stroke; the exhaust valve also remains open after tdc of the exhaust stroke—this is the "overlap" period between the two valves. The object of such valve timing is to take advantage of the inertia of the gas columns as an aid to better filling of the cylinders. If the intake valve opens at tdc, the piston will be well down on the induction stroke before the mixture in the intake port and pipe begins to move; thus the valve is usually timed to open earlier so that it is fully open when the piston actually begins to descend. See Fig. 3-1.

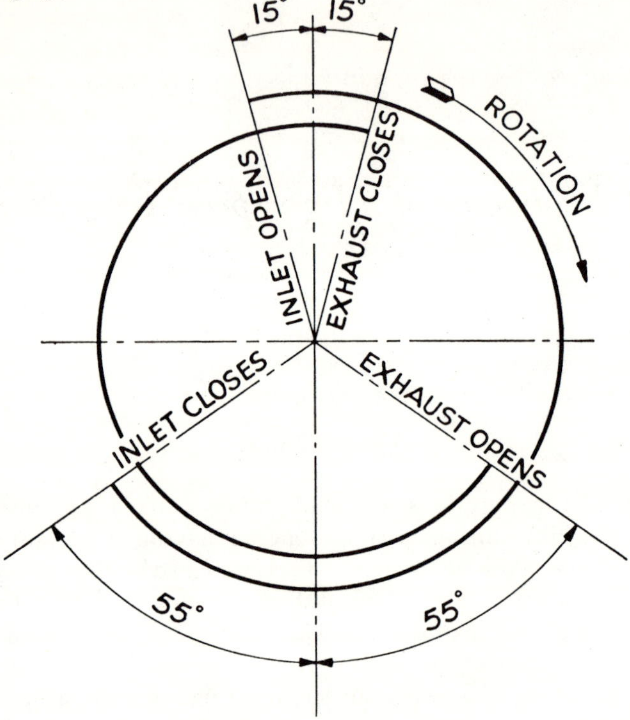

Fig. 3-1. Typical valve timing for showroom stock sports car. This diagram represents crankshaft rotation during exhaust and intake strokes only.

The piston stops momentarily at bdc, but the rapidly moving gas column does not stop. It is not only attempting to catch up on the slightly lower pressure inside the cylinder (compared with atmospheric), but now has considerable momentum that continues to propel it into the cylinder and will do so even when the piston begins to rise on the compression stroke. Not until there is danger of a reversal of the gas flow back through the intake valve is the valve closed, and this may be 50° or more after bdc.

Obviously the valves need to open quickly and close quickly at the timing points to obtain the maximum possible gas flow for the greatest possible duration. The large American V8 passenger car engines, which in blueprinted and supertuned form see wide use in competition, have (or can be given) excellent breathing characteristics. But, being pushrod OHV designs, the valve operating mechanism has rather substantial mass, which can create problems for the tuner.

The inertia developed by heavy valve gear at high rpm limits the speed with which the valves can be opened or closed. Therefore, light alloy rocker arms (Fig. 3-2), tubular aluminum pushrods, roller tappets, and similar hot rod components are usually a necessity. Naturally the tuner who works with overhead camshaft designs is not plagued by such considerations.

Fig. 3-2. Aluminum rocker arm for use on American V8 engine. These components for supertuned drag racing engines replace stock pressed steel units.

The "valve timing" (or, more accurately, the port timing) of the Wankel rotary engine is another thing altogether. In designs that use peripheral intake ports, such as the NSU Spider and RO 80 engines, the intake port never closes; in passing the port, the rotor's apex seal merely shuts off the intake gas flow to one chamber while simultaneously opening it to another chamber. The same is true of the exhaust port. Consequently the gas columns never come to a full halt from which they must then be restarted.

On Wankel rotary engines with side intake ports, as on Mazdas, it is possible to "tune" the intake port timing by altering the shape of the port. Whereas peripheral ports tend toward large nominal overlap periods, side intake ports can be designed to reduce overlap considerably, though they do this at the expense of opening late—at tdc or even afterward. If the port is sharp edged so that it is uncovered very rapidly, almost instantaneous opening can be obtained. Not only does this overcome the seeming loss that might result from late opening, it has the advantage of creating a pressure drop in the combustion chamber that subsequently accelerates induction. It is unfortunate that, at present, the rules of most racing classes dictate against Wankel engines; the competition development potential seems very promising.

Maximum Charging

When the exhaust valve timing is examined we may find that the valve opens perhaps 50° or so before bdc of the expansion stroke. This apparent wastage of expansion pressure, and a useful portion of the stroke, is unavoidable; if the valve opening is delayed there simply is not time to expel the burned charge. The task of doing so would then devolve on the piston on its upward exhaust stroke. This would be unacceptable; the negative work involved mechanically in pushing out the exhaust would entail a loss far greater than any gain made by continuing the expansion stroke for a longer period.

At tdc of the exhaust stroke, the intake valve has just opened. The high-speed exit of spent gases through the exhaust valve has, however, been continuing for some time so that there is negligible pressure existing in the cylinder. The designer may expect the energy of the exhaust gas column to assist in clearing the waste products from the combustion chamber; in fact, there is no other way to obtain this scavenge since it is unswept mechanically by the piston. Thus, the exhaust valve is kept open for a period after tdc when the piston is accelerating on the intake stroke. If the intake valve is well open by this time, there is little tendency for the exhaust flow to reverse direction on a wide throttle opening unless exhaust valve closing is delayed too long. If the intake manifold is under low pressure (that is, any throttle opening other than wide open), it will be evident that flow reversal can, and in fact does, take place at certain combinations of engine speed, load, and throttle opening.

The valve timing has thus to be very much of a compromise, and the more flexible the engine must be, the less liberties can be taken with extended opening periods and overlap (when both valves open together at tdc of the exhaust stroke, as shown in Fig. 3-3). It is quite usual on engine tests to find exhaust residuals left in the combustion chambers, in quantities that vary from cylinder to cylinder, at certain loads—even when the valve timing is such that it promotes good scavenging. On the other hand, traces of fresh mixture can many times be detected in the exhaust outlet, showing that the scavenge is effectively preventing the fresh charge from going its proper way into the cylinder. These faults generally show up only at reduced throttle openings, and this may account for the fact that fuel economy on light loads is, in many cases, not so good as might be expected.

On small throttle openings, the whole induction system will be operating at a pressure very much below atmospheric. This is true to a degree even on supercharged engines. Under these conditions, assuming normal valve timing, three of the intake valves in a four-cylinder engine will either open or be open during any half-revolution of the crankshaft. For example, if the intake valves are timed to open at 10° before tdc and to close 60° after bdc, then, with the piston of No. 1 cylinder starting on its intake stroke, its intake valve will be fully open. However, the intake valve of No. 2 cylinder, whose piston is starting from bdc of its compression stroke, will still be open and will remain open for a further 60°. Finally, the intake valve of No. 3 cylinder, which is exhausting, will open just as this piston arrives at tdc. Going back again to No. 1, this intake valve will still be fully open as the piston arrives at bdc. Fig. 3-4 will make this relationship clear.

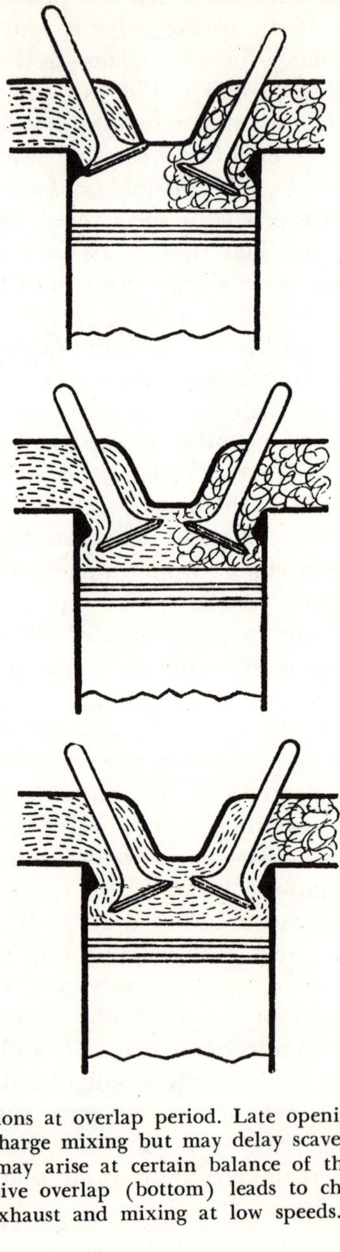

Fig. 3-3. Conditions at overlap period. Late opening intake valve (top) prevents charge mixing but may delay scavenge. Ideal situation (center) may arise at certain balance of throttle openings and rpm. Excessive overlap (bottom) leads to charge loss down exhaust and mixing at low speeds.

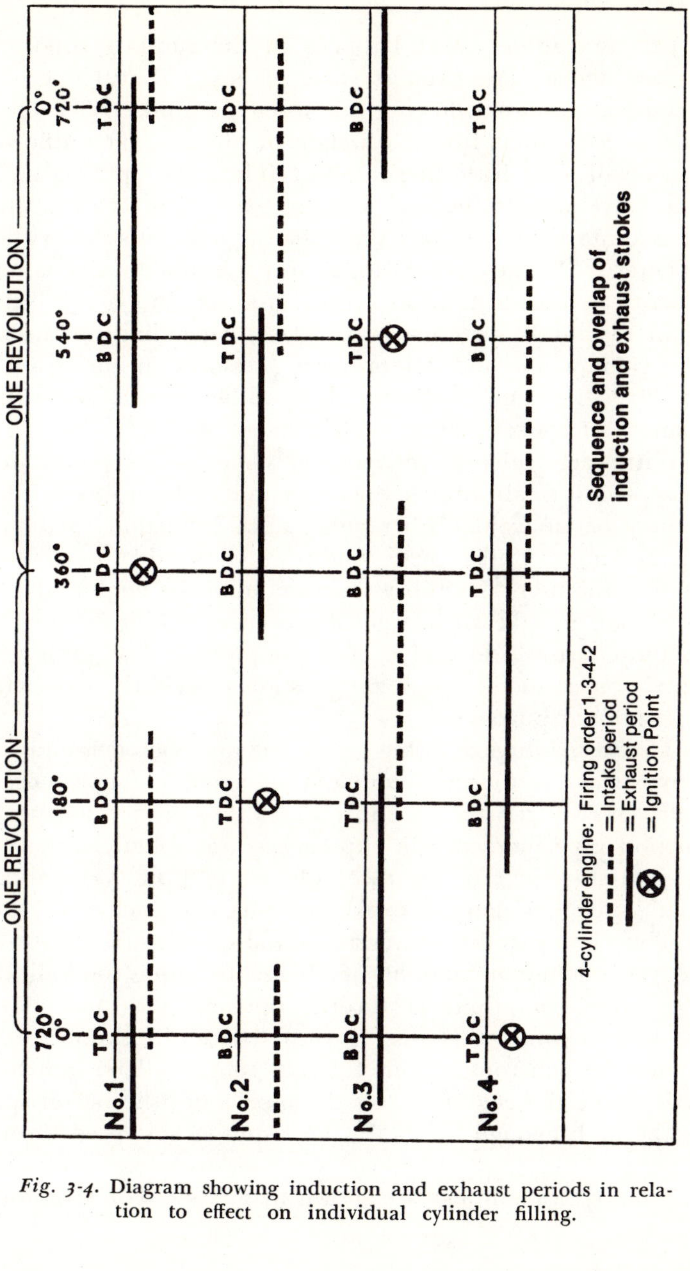

Fig. 3-4. Diagram showing induction and exhaust periods in relation to effect on individual cylinder filling.

Overlap Snags

Let us consider what happens in the running conditions described above. The exhaust valve of No. 1 is still open and the residual gases are flowing out when its intake valve opens. At low speeds, and with little weight of mixture, the outflowing exhaust will have little inertia effect. Thus, the opening of the intake valve against the low pressure existing in the induction pipe is quite likely to draw the exhaust back into the cylinder momentarily. In the case of No. 2, only the intake valve is open so there can be no dilution of the mixture. But a reversal of flow in the intake port may take place, depending on the balance of induction and compression pressures in the cylinder. Very much the same conditions apply to No. 3 as to No. 1, that is, danger of contamination with exhaust residuals.

With early inlet valve opening, some loss of power must be accepted at small throttle openings and medium speed. Idling also may be unavoidably irregular since induction vacuum is at its maximum under these conditions. It would therefore appear that the more flexibility we want from the engine, the less able we are to take advantage of gas energy. This is true. But it is quite feasible to arrive at a compromise that gives good torque over a useful speed range and is perfectly satisfactory for all normal purposes.

The possibility of using "pipe energy" becomes most attractive when considering competition engines. These engines are driven regularly on wide throttle so that the pressure in the induction pipe more nearly approaches atmospheric, and there is a greater weight of mixture inhaled to form a high-speed "bung" traveling down the exhaust pipe. In such conditions, the valve timing and the overlap period can be arranged to induce the maximum amount of charge to enter the cylinders under given conditions of throttle opening and rpm. But it must be recognized that, for any real effect, the throttle opening must be generous, the useful rpm range will be limited to speeds where the engine is "on the pipe", and the piston speed must be high enough to obtain the requisite gas speed through the ports.

Some typical timings are shown in Fig. 3-5. For a thorough

discussion of "tuned" pipes, and the influence that valve timing has on them, the reader is directed to *Scientific Design of Exhaust & Intake Systems* by Philip H. Smith and John C. Morrison, available from the publisher of this book and from selected booksellers.

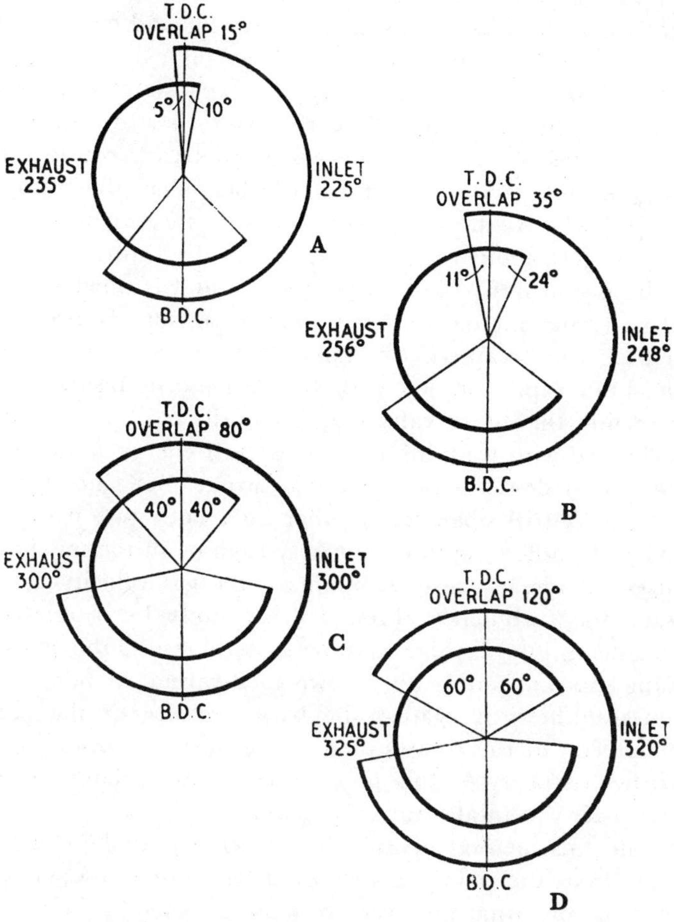

Fig. 3-5. Comparison of valve timings. (A) Passenger car engine, approximately 45 bhp/liter @ 4500 rpm; (B) stock sports car engine, approximately 60 bhp/liter @ 5500 rpm; (C) racing sports car engine, approximately 90 bhp/liter @ 7000 rpm; (D) racing engine, approximately 150 bhp/liter @ 8000 rpm.

High-speed Operation

Taking a look at the specs for a typical "sports" engine designed to take advantage of gas flow, we may find that its intake valve opens 40° before tdc and closes at 80° after bdc—a generous opening period of 300°. The exhaust timing will be similar to that of the intake, closing 40° after tdc and opening 80° before bdc, resulting in an overlap of 80° at exhaust tdc. Because the exhaust valve opens early, the expansion phase being of only 100° duration, the cylinder pressure is still high when the valve opens. As a result, the gas escapes with considerable energy; the bulk of it gets away before bdc when the piston is traveling slowly. As the piston accelerates on the exhaust stroke, it probably just about catches up with the remaining gas until, when the piston has passed its peak speed and is slowing down owing to crank angularity, the cylinder pressure is becoming subatmospheric—because of the momentum of the exhaust gases in the outlet pipe, coupled with the decrease of piston speed. At this point, the intake valve opens, and the combustion chamber is charged with fresh mixture by exhaust energy alone. After tdc the piston descends on the intake stroke. But since the exhaust valve is still open for another 40°, both piston descent and exhaust outflow combine not only to draw in the maximum weight of charge but also to maintain the gas velocity that is necessary for turbulence through what must be a relatively large-diameter intake valve and port. As a result, the mass of incoming mixture builds up its own momentum; at bdc it still continues to flow in against the back pressure of the rising piston for 80° of the compression stroke until the piston speed approaches its peak. At this point, when there is danger of gas flow reversal, the intake valve is closed.

While this method of inducing a weighty intake charge is excellent from the point of view of sheer torque at high rpm, it might be somewhat extravagant because there is a grave possibility of losing a good deal of charge down the exhaust pipe—particularly during the early opening intake valve phase when the piston is still ascending. It is, in fact, remarkable how thermally efficient an engine designed along these lines can be made, the figures standing more than favorable comparison with the best passenger car engine designs. If the direction of flow of the incoming mixture is taken care of, little loss of charge need take place. Though the valve opening periods and overlap seem large, the time intervals at high speed are short; as the charge weight increases, so does its inertia.

In considering the effects obtained from taking advantage of the rapid movement of the intake and exhaust gas columns, we have so far visualized what might be described as "bungs" of gas acting somewhat after the manner of auxiliary pistons. This concept is satisfactory, but in practice, the actual behavior of the gas is not quite so simple.

Pulsating Pressure

The gas flow, far from proceeding smoothly, actually takes the form of a series of pulsations, or pressure waves, that alternate as positive and negative pressures above and below the mean. This is in accordance with physical laws that have been known for many years and that have been applied in a practical way for centuries—in pipe organs and musical wind instruments, for example.

The pulsations in induction and exhaust systems can both help and hinder power output. By careful and scientific design, it is possible to harness them to obtain higher torque over the fairly wide speed range needed by sports car engines. If a narrower rpm band is acceptable, as in many kinds of racing, the benefit to power can be made even greater, though this extra power will fade away below a certain minimum engine speed, which will, of course, be quite high.

For an outstanding example of what can be done by scientifically harnessing the pulsating pressure, we need look no further than the typical single-cylinder, four-stroke racing motorcycle engine. The pipework on these engines is simplicity itself. It consists of a short intake stub from the carburetor to the intake port and an exhaust pipe of suitable length and varying diameter as called for. It has already been explained how, with a high-speed exhaust gas exit and a lengthy overlap period, the combustion chamber can be scavenged of residuals and charged with fresh mixture while the piston is still approaching tdc on the exhaust stroke. However, there is more to the impressive power obtained from a modern racing "single" than good scavenging; the source of its high output amounts quite literally to a supercharge.

Charge loss down the exhaust pipe is likely to take place if the exhaust valve closing is delayed unduly on the intake stroke. In the case of the sports engine previously described, the exhaust valve closed 40° after tdc. Now, imagine that by utilizing very high gas velocity and a correct design of open pipe for the specified conditions of engine speed and throttle opening, we can continue to draw a fresh intake charge down the exhaust pipe even when the piston is approaching its maximum speed on the intake stroke. We are obtaining a cylinder rapidly filling with mixture, plus a further volume of fresh mixture in the exhaust pipe—between the exhaust pressure wave "bung" and the exhaust valve.

At first glance, we might conclude that good mixture is about to be sent to waste. But suppose we can induce the exhaust pressure wave to reverse direction at the right moment? The extra mixture in the exhaust pipe will then be pushed back into the cylinder through the still-open exhaust valve, and thus it will be added to the charge that is still flowing in through the intake valve (the velocity of which will prevent its reversal). As soon as there is danger of exhaust residuals following the mixture back into the combustion chamber, the exhaust valve is closed, and we are left with what amounts to a supercharged cylinder—free of mechanical loss and without benefit of a blower.

High Power; Narrow Speed Range

One of the first outstandingly successful engines to apply the above theory was a 499-cm^3 single-cylinder British racing engine of the late 1930s, which achieved 50 bhp at 6700 rpm—an unheard-of figure for its day. To further shock its contemporaries, it had an intake valve duration of 320°, an exhaust valve duration of 325°, and an overlap period of 125°. The seemingly fantastic nature of these figures is best illustrated by considering that, in the 720° that constitute the engine cycle, there is a period of only 200° during which neither valve is open. And, in this 200° period, both compression and expansion phases take place. Subsequently engines of a similar kind were developed to peak at between 7500 and 8000 rpm with approximately *pro rata* increases in power.

If not only gas velocity but pressure waves are to be taken advantage of in the search for power, there are many often-conflicting factors involved. These soon become a siren choir, luring the designer ever on toward an engine with an absolutely minimum speed range; for to apply these various principles ideally, one would need a constant-speed engine with a fixed throttle opening. (The more cynical drivers may aver that some engines behave very much as if this ideal had been realized.)

In practice the designer must permit at least some degree of flexibility, and this search for flexibility has in recent years made turbocharging an increasingly attractive alternative to the "totally cost-free" supercharging obtained solely from "tuned" pipes. But in many kinds of racing, turbocharging is not an available option because of rules considerations. The obvious course—particularly in preparing a car for production class competition—is to concentrate on developing a good "tuned" exhaust system.

The exhaust pipe's effect is influenced not only by its diameter and length but also by the shape of the outlet. The frequency of the pressure pulsations can be altered by the use of a diffuser exit, and experimenting along these lines may prove advantageous. As in the musical instrument industry, the technique of pipe-end design is mainly a matter of trial and error, and it is possible that shapes other than a plain trumpet or megaphone may be evolved to greater effect. Typical of such is the divergent-convergent shape (or reverse cone), which is claimed to increase the range of rpms over which the supercharge takes place. This shape has been greatly exploited and developed on the two-stroke engines used in the SCCA's D Sports Racing class. Any suspicion that the reverse cone pipe end was inspired by the bell of an English horn is quickly dispelled when one hears a pack of these DSR engines in "full song"!

The main object of the inlet pipe (again considering the

single-cylinder motorcycle powerplant) is to obtain sufficient gas velocity to allow the pressure to build up in the cylinder without causing a flow reversal in the intake port. Normally the closer the carburetor is to the intake valve, the better—and in cases of doubt, this is a cast-iron rule. It is, however, possible to obtain an increase of power by extending the intake length between the valve and the carburetor choke, providing that the exhaust pipe dimensions are correct. The object is to increase the intake velocity so that the exhaust valve closing can be further delayed. Admittedly this extra length of inlet ducting will "store" considerable fuel/air mixture. But it can be assumed that the additional pipe length will not induce too much precipitation of fuel on the pipe walls at the gas speeds and the temperatures involved.

The use of a long intake pipe between the atmosphere and the carburetor choke is of doubtful utility unless the engine is already tuned to take full advantage of the gas vibration in other ways. For engines that work over the normal highway driving torque range, long intakes of this kind often affect the carburetion adversely, leading to erratic pickup and inferior performance from cold. In some instances, tuners have resorted to long intake pipes or "velocity stacks" to prevent mixture expulsion on unfavorably designed engines that have excessive exhaust back pressure or particularly late-closing intake valves.

The considerable increases in maximum bhp that were first obtained from single-cylinder engines, as a result of harnessing acoustical phenomena, required a long period of time and much painstaking experimentation. It is not difficult to lay down a theory about what happens using the formulas given in *Scientific Design of Exhaust & Intake Systems.* In fact, a workable design, with suitable dimensions for the intake and exhaust tracts and piping, can be laid out on paper largely from theoretical considerations. The maximum benefit, however, can be obtained only by experiment—preferably with the engine on a dynamometer.

When one considers all the possible variations, not only in pipe length and diameter, but in valve and port areas, piston speeds, valve timing, valve lift, carburetor size, and so on, it will be appreciated that long hours in the dyno room and much hard labor in the welding bay and the machine shop have gone into producing the results that have been seen in all kinds of racing where tuned systems are permitted. Magic results should not be expected after removing the muffler from the family bus, substituting a short pipe with a chromed megaphone, and screwing in a larger carburetor jet.

To apply the single-cylinder technique to the multicylinder engines with which we are most concerned, it is necessary to treat each cylinder individually—though mainly insofar as the intake side is concerned. This means a separate carburetor (or venturi) and a separate intake port for each cylinder. With this arrangement, each intake tract is direct to the valve, and variations in length for experimental purposes are readily accommodated. It will be evident that some advantage may be gained by increasing the intake charge momentum to allow the closing of the intake valve to be delayed somewhat longer than normal after the start of the compression stroke; however, the fullest use of these "ram tuning" principles cannot be obtained unless the exhaust side is also modified.

Unfortunately the rules of many racing classes do not permit the use of a separate carburetor venturi or a separate injection pipe for each cylinder. It is usual in sports car classes to have one carburetor serving two cylinders of a "four" or three cylinders of a "six". In NASCAR racing, a single four-barrel carburetor must be used to its best advantage in feeding a big V8. The success that tuners have had with these engines is often the result of arranging the intake manifolding so that equally spaced aspiration periods are obtained.

For example, with four cylinders inline, one venturi feeds a manifold pipe that serves cylinders 1 and 4; the second venturi feeds a manifold pipe that serves cylinders 2 and 3. By crossing over from one bank to the other of a V8, the four venturis of the typical NASCAR carburetor can be made to supply equally spaced aspiration periods for an eight-cylinder engine. Some awkward lengths and bends can be involved, which increase gas friction and encourage fuel deposition (the inertia of the fuel droplets in the mixture frequently causes them to be "thrown out" against the manifold walls as the mixture moves through sharp bends at high speeds). If these problems are severe on an engine—for example, because the stock low hood profile must be retained—it is sometimes necessary to feed adjacent cylinders from a single venturi even though the aspiration periods are then uneven.

In the case of a six-cylinder inline engine, equal aspiration periods can be obtained with two carburetors by feeding adjacent cylinders (1-2-3 and 4-5-6). This is also an excellent layout from the standpoint of short and compact piping. Sixes with three carburetors (1-2, 3-4, 5-6) will have uneven aspiration periods on the outer pairs but may benefit from the shorter intake manifold pipes. The lure of short, or at least equal length, manifold pipes is responsible for the "tarantula-type" intake manifolds that are used on V8s in drag racing.

Exhaust Systems

For engines that have an individual carburetor venturi or injection pipe for each cylinder, it would seem logical to use a separate exhaust pipe for each cylinder—thus enabling the pipes to be "tuned" using the principles derived from single-cylinder practice. This is not always the best technique, though it has been used successfully at times. In most applications, it is possible to improve performance by combining the exhaust pipes at some point, which allows the exhaust pulsations of one cylinder to help those of the others.

As far as normal layouts are concerned, the main requirement is to prevent interference between cylinders that have overlapping exhaust strokes and to provide sufficient length in each exhaust branch to give a useful extractor effect before the pipes merge. There may be some difficulty in accommodating a system that has branch pipes of suitable length feeding into a single tailpipe, and division of the system is frequently resorted to in order to simplify matters. Dividing the system may, in fact, even offer advantages if the engine is required to operate frequently in the mid-range of rpm, instead of only at maximum revs.

A common example is four-cylinder engines such as those used in the Triumph Spitfire Mark II and the Ford Cortina GT (Fig. 3-6)—two imported cars that were sold in the United

States during the middle and early 1960s when emission controls were unheard of and performance was king. These engines are fitted with tuned headers that have one Y-branch from cylinders 1 and 4 and another from cylinders 2 and 3. The two branches then join and lead into a common tailpipe. This is the so-called tri-Y system and gives equal time intervals in exhaust expulsions with little or no possibility of backflow into the cylinders. These systems have the further advantage of giving good mid-range power—which makes them especially suitable for sporting machines that must double as everyday transportation. A similar system for an inline six-cylinder engine would consist of two trifurcating branches for cylinders 1-2-3 and 4-5-6. However, on a "six", the practice is sometimes to keep the branches separate, with two individual tailpipes.

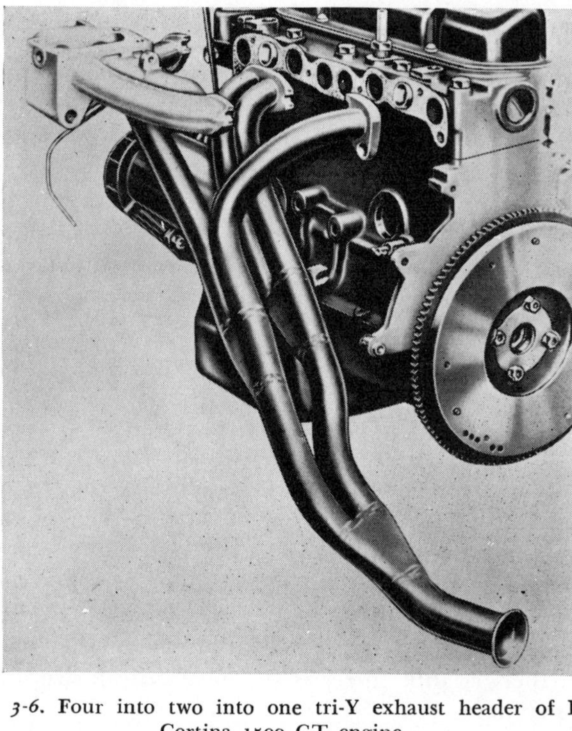

Fig. 3-6. Four into two into one tri-Y exhaust header of Ford Cortina 1500 GT engine.

V8 engines present a particular problem because of the firing orders that are necessary with the normal kind of crankshaft, which has four crankpins spaced at 90° intervals of rotation. To obtain equally spaced exhaust expulsions with two 4-into-1 headers, it is necessary to design a complicated "bundle of snakes" system that has pipes crossing over from one bank of cylinders to join those of the other. For this reason, many V8 racing engines use a "flat" 180° crankshaft that is similar to the crankshaft of a four-cylinder engine. A certain amount of secondary imbalance vibration must be accepted, but with a "flat" crank it is possible to fire cylinders on alternate banks so that each bank can be treated as an inline "four" with four evenly spaced exhaust expulsions.

Because the exhaust gases exit from the cylinders under considerable pressure, an exhaust system that restricts gas movement has a much less disadvantageous effect than would be the case were the same condition to prevail in the induction system. This is the reason why many engines operate quite well with exhaust manifolding that was designed more to facilitate foundry work (or to cram an oversize engine into a small space) than for any power-producing potential. That most production engines can gain several bhp by the substitution of tuned headers is less a credit to the design of the special exhaust system than a strong condemnation of the standard design.

Mechanical Stresses

The main source of operating stresses in the conventional engine is the reciprocating action, which sets up bearing loads quite unlike those encountered in purely rotating motion. (This is a consideration that has made the Wankel rotary engine attractive.) In stepping up power output for competition, it is usual to increase the upper rev limit for the engine and also, through the use of a higher numerical final drive gear ratio, to increase the engine revolutions per mile. These modifications unavoidably increase the mechanical stresses on the engine's moving parts.

In a conventional engine, the crankshaft main bearing

loads can be fairly well balanced so that wear is evenly distributed around the bearing shells and the journal. But the reciprocating loads are particularly severe at the connecting rod big ends. The major source of stress is the reversal of the piston movement at tdc and bdc; it introduces an alternating tensile and compressive load in the connecting rod, which is transmitted to both the piston pin and the crankpin as a shear stress.

These inertia loads caused by the reciprocating mass are independent of any other loads that may come from the actual pressures on the pistons during the working cycle. If we consider the case of an engine's being driven by an electric motor, with all the spark plugs removed, the reversal of forces at tdc and bdc will be virtually the only stresses with which the reciprocating parts have to contend. Nevertheless, if the electric motor were speeded up, the loads caused by reciprocating mass alone would be found to increase in an alarming way. In fact, the increase in stress is proportional to the square of the engine rpm.

For example, at 6000 rpm, the loading is four times that at 3000 rpm. It is not difficult to see why big end failure may occur if the revs are pushed even a small amount past the tachometer redline. Thus, an engine that travels happily for mile after mile at 4500 rpm should not necessarily be condemned if it suffers a bearing failure after a brisk hour at 5000. The extra 500 rpm has increased the big end bearing loading by nearly 25 percent.

If we are to increase the rev limit for an engine that is to be used in competition, some steps must be taken to minimize the increase in mechanical stresses. Of first importance is the balancing of all the engine's moving parts. Reciprocating components must be of equal weights for each cylinder to reduce tortional loadings on the crankshaft and to avoid the additional stress that can be caused by ordinary imbalance vibrations. The rotating parts should be balanced dynamically with the help of electronic equipment. For any major increase in operating rpm, lightweight reciprocating parts must be used in place of the standards parts. Forged aluminum connecting rods, for example, are almost universally used in supertuned drag racing engines. The lightening of valve gear components is also important.

Power Loading

The actual operational cycle of the engine has a pronounced effect on the inertia stresses. For instance, on the intake stroke, the connecting rod assembly will be under tension caused by inertia, friction (ring drag), and the "suction" effect of drawing in the mixture for the first part of its movement. For the remainder of the stroke, the assembly is being slowed down by the crankshaft, causing a compressive load as the reciprocating weight attempts to overrun the uniform crank speed. The piston, of course, is still subject to the suction effect, so that the degree of compressive load on the connecting rod depends on the throttle opening as well as on rpm.

On the compression stroke, the initial loading is obviously compressive. But during the latter part of the stroke, piston deceleration exerts a powerful inertia tensile loading on the rod. The extent of the changeover from compression to tension depends, of course, on compression pressure (which again means throttle opening) and engine speed (which determines the deceleration inertia load).

The firing or power stroke puts a compressive load on the connecting rod, which opposes the tensile inertia load. The extent to which the two forces act depends on the throttle opening (that is, on the strength of the expansive force) and the engine speed. When the exhaust valve opens, the pressure is removed from above the piston, which is now being slowed by the crankshaft. Finally, on the exhaust stroke, the load changes from compressive to tensile, as in the other strokes.

Complications of Bearing Loads

The foregoing is no more than a brief outline of what the big end bearings and the materials of an engine's reciprocating parts have to withstand. But we have covered the subject sufficiently to show that the combination of loading caused by the inertia of reciprocating components and the stresses imposed by the engine operating cycle is extremely complicated. At this point, it will be interesting to see how a third factor, driving conditions, further affects the loading.

Generally, bearing failure occurs at high rpm and wide throttle openings. But often the failure is caused by conditions that are independent of the engine's speed and the power being developed. Serious loads can be imposed on bearings while driving sedately at relatively low rpm. Because damage can thus take place under conditions that might not seem taxing for the engine, it is important to investigate the reasons why.

On an average passenger car engine, the maximum torque comes in about half the maximum speed of which the engine is capable. It is at this point of full torque that the expansive pressure on the power stroke is highest, so that a considerable load, the result of the power production, is put on the big ends. At these seemingly innocent revs, the inertia force is admittedly moderate. But consequently *there is less of this force to counteract the thrust of the power stroke.* Furthermore, this full-torque effect persists even at very low rpm because there is plenty of time for the cylinders to become fully charged.

All of these conditions add up to the fact that "lugging" the engine on wide throttle at low rpm is bad for the bearings and can lead to serious overloading if persisted in. Of course, engines with generous bearing areas are less prone to damage, and engines that have roller bearing big ends can be quite susceptible to damage; the available contact surfaces are limited, and the necessary supply of lubricating oil is reduced by the low revolutions. As the rpm builds up—even on the same throttle opening—the rising revolutions, far from increasing the load, actually decrease it because of the counteracting inertia forces.

At or above the maximum engine rpm, the inertia forces take command. The torque is necessarily falling off at this point (unless the engine is supercharged or has an exceptionally efficient induction system) because of valve restrictions and an impedance to mixture flow that is a consequence of the high speed of operation. Thus, the balancing load on the bearings is reduced, and they are subjected to their maximum stress.

High oil temperatures can also contribute to high-rpm bearing failure. Sustained high speeds obviously mean increased heat, and if the engine's lubricating oil reaches an unusually high temperature, its viscosity will decrease—just when viscosity should be ample to maintain the oil film that separates the highly loaded bearing surfaces. Modern multigrade oils are much less subject to high-temperature viscosity loss than were the engine oils of twenty years ago. Nevertheless, many competition engine tuners see fit to use a viscosity-improving agent, such as the ubiquitous STP®.

Any minor rupture of a bearing's oil film will at once increase the heat generated in that bearing so that a situation is created that very soon results in failure if the conditions are persisted in. This is the reason why short bursts of high speed are not harmful, even on a well-worn engine with low oil pressure, whereas indiscreet flogging for miles on end will wreck the bearings of the best-maintained powerplant.

Extra Heating

So far we have considered only the additional loading induced by higher rpm in combination with extra piston thrust. The next item to consider is the extra heat developed by supertuned or supercharged engines. Any measures taken to increase engine power by higher cylinder pressures must inevitably release more heat in the combustion chambers. A higher compression ratio does this—more so when multiple carburetors, improved manifolding, or a blower allow the engine to inhale a larger quantity of mixture.

The average production engine can take care of the extra heat without much difficulty because it is dissipated to the atmosphere by the cooling system and the engine oil. But in some cases increased cooling capacity may be necessary. Cylinders and cylinder heads with deeper cooling fins may be needed on air-cooled engines; on liquid-cooled engines, the water pump speed may have to be increased. An oil radiator can easily be added to any four-stroke engine, and the oil capacity can be increased by a deeper sump or by changing to a dry sump system with a large storage tank. In any case, when more heat is added to the loadings, the total requirement in extra construction strength and ruggedness may be quite appreciable.

It is possible, for example, for cylinder head gasket failure to become a problem if the head and block mating surface areas are not adequate for the increased heat conduction requirements or if there are too few cylinder head bolts or studs to withstand the extra pressure. Cylinder head castings may distort because of heat and pressure and actually lift between the widely spaced studs—though these troubles are encountered mainly in highly supertuned or supercharged powerplants. Providing that the necessary and logical modifications are made to the components that must carry the extra stresses, there is no reason to suppose that stepping up the power output of a basic design of engine need have any adverse effect. The necessity of improved sealing—and, above all, improved heat transfer from head to block—is a factor that dictates such practices as O-ringing of cylinders and the use of solid embossed steel or shim-type gaskets.

4 / *Mechanical Construction of High-power Engines*

Basic Materials in Construction

The first requirement for ensuring reliability under conditions of continuous high-speed operation is that the materials chosen be of the correct type and adequate for the duty of any particular task. The advances made in foundry and metallurgical techniques during the past two decades have ensured that failure of a component is comparatively rare under normal conditions. It is still possible for a crankshaft or a connecting rod to break under the abnormal stress produced by racing, but these failures arise from many causes other than faulty materials or dimensional errors.

There are two main classes of materials used in engine construction: *ferrous metals* and *nonferrous metals.* The former comprise those with an iron base, such as cast-iron, mild steel, and alloy steel. The latter can be casehardened through processes such as carburizing, cyaniding, nitriding, carbonitriding, or other special treatments if required for particular duties. A fairly recent newcomer to the ferrous range is nodular or high-strength ductile iron; various compounds—such as magnesium, cerium, calcium, lithium, sodium, or barium—are added just before pouring to achieve a high degree of toughness and strength.

This material is widely used as an alternative to steel for cast crankshafts; Meehanite is one trade name.

Present-day foundry techniques allow castings to be not only far more intricate than was once possible, but also extremely thin and lightweight. The heavy iron engine block castings of former times were not made thick merely for reasons of strength. The greater thickness was mainly necessary because reliable ways to locate the cores for the cylinders and water jackets had not been developed. Adequate "safety" thickness therefore had to be left for machining; otherwise a high percentage of wasted castings would result when the cylinder boring tool cut through into imprecisely located portions of the water jacket. Today's thin-wall iron castings have little weight penalty over aluminum castings when used for the engine blocks of production cars.

Nonferrous Metals

Aluminum alloy is the most widely used nonferrous metal in engine construction. Magnesium alloy has also been widely used, particularly in the crankcases of the Volkswagen and Porsche air-cooled engines. Aluminum and magnesium alloys combine adequate strength with lightness, with magnesium being particularly lightweight for its strength. On the whole, aluminum alloys are easier to cast and machine than iron. A third, and perhaps most important, property of aluminum is its high thermal conductivity—a factor that has always made it attractive for the construction of cylinder heads of engines that are otherwise constructed of ferrous metals. Chrysler, for example, has cast a few aluminum heads for the "Hemi" engine, and these are highly prized by drag racers. VW uses aluminum heads on the cast-iron blocks of its water-cooled models. Not to be overlooked are the "rare" aluminum Chevrolet blocks that have been used in Can-Am racing.

For large castings that carry little stress, ordinary diecast or sandcast aluminum is excellent. This material is widely used for oil tanks, timing covers, overhead camshaft covers, and oil pans. Insofar as oil containers are concerned, the virtues of aluminum from the standpoint of heat conductivity are probably overrated in comparison with, say, an oil pan of pressed steel. However, an aluminum sump does combine strength with lightness, and this is a requirement in sumps with extra oil capacity, which are frequently desirable for competition applications.

The thermal conductivity of aluminum really shows to advantage when this material is used for cylinder heads. It will be apparent that the temperature range of the cycle has an important bearing on thermal efficiency and that, although rapid heat dissipation at certain high-temperature phases in the cycle is essential to prevent overheating, the retention of heat at other phases is desirable. Thus, a material that will rapidly transfer the heat between the mixture and the cooling liquid or air, in whichever direction is required, will make for high thermal efficiency as well as reliability under sustained high loadings. A material of lower thermal conductivity characteristics, on the other hand, will tend to retain the heat within itself, leading to local superheated areas (hot spots) during conditions of high temperature operation.

For very high pressures, heads of aluminum bronze alloy are sometimes used because the mechanical strength is in this case equal to that of cast-iron. Because the typical makeup of the alloy is about 86 percent copper, 10.5 percent aluminum, and 3.5 percent iron, valve seat inserts, normally required with aluminum heads, may be dispensed with. Nevertheless the thermal conductivity is far superior to that of iron. Barronia, a copper-tin base alloy, is another successful material that has been used without valve seat inserts. Some aluminum alloys are noted for outstanding wear resistance. In the VW overhead camshaft engines, for example, the use of such an alloy has eliminated the need for camshaft bearing shells, the camshaft running directly in the cylinder head material.

Light alloy crankcases are encountered on almost all engines that have their cylinders (VW, Porsche) or their blocks (Offy) separate from the crankcase. Unique is the Chevrolet Vega engine, which has an aluminum block and an iron cylinder head—with no iron liners for the cylinders; previous designs, such as the one shown in Fig. 4-1, have had iron cylinder liners.

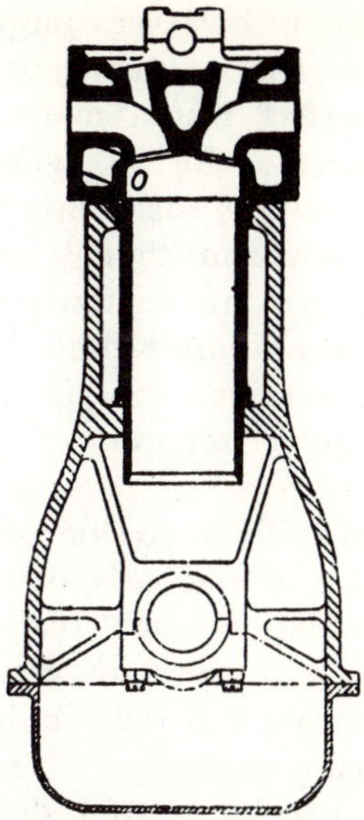

Fig. 4-1. Light alloy crankcase and cylinder block casting, in conjunction with wet cast-iron cylinder liners on 2-liter AC engine. The cylinder head is also of cast-iron.

When iron liners are used in light alloy blocks, which is the case in most racing engines, they are usually of the wet type (wet liners are familiar to anyone who has worked with Triumph sports car engines). A kind of construction used with dry liners is to cast the aluminum around the iron liners in the mold; the liners in this case have a specially finished exterior to form a mechanical interlock, such as threading or a roughened, sandpaper-texture pockmarking. Both wet and dry liners in aluminum blocks make for commendably light powerplants and, though the manufacturing operations are to some extent increased in complexity, the durability is praiseworthy in comparison to experiments that have been made using plated or chemically treated aluminum cylinder bores.

The Chevy "six" finally abandoned iron pistons in the 1950s and, that milestone fortunately behind us, aluminum alloy pistons are now universally used in all engines—for both highway driving and competition. Production car pistons are usually of diecast aluminum; however, forged aluminum is invariably used for the pistons of racing or supertuned engines.

Aluminum is a perfect piston material, as much for its heat-conducting properties as for its lightness. Even in forged aluminum, racing pistons are sufficiently robust in section to have considerable weight, and though some stronger "space age" material might reduce this section enough to save weight, no metal other than aluminum would be more suitable for ultrarapid heat conduction from the piston crown to the cylinder walls. This is of great importance; otherwise the pistons could easily melt.

Forged steel is generally used for connecting rods. However, the weight penalty is obviously of great concern in this application. Drilling holes through the connecting rod web has been tried—in most cases without very desirable results. A better, if more costly, solution is the tubular steel rods used for many years in Offy racing engines.

Connecting rods machined from duralumin and similar materials have been used, as have diecast aluminum rods in smaller engines. Today, however, forged aluminum connecting rods are being used increasingly. Rods of this kind are almost universally used in large-displacement drag racing V8s where, because of the large size of the rods, a weight reduction has invaluable benefits from the standpoints of reducing inertia loadings. The popularity of forged aluminum rods with speed tuners is fostered also by the wide use of nodular cast-iron connecting rods in American production car engines—not the best of materials for high-rpm purposes.

Fatigue Failure

Failure of highly stressed parts was once frequently caused by an actual fault in the metal. Now, with X-ray, Magnaflux, Zyglo, and other electronic inspection procedures—mandatory forms of regular inspection in some racing classes—these failures no longer occur. Consequently breakages in modern competition engines are almost invariably caused by metal fatigue. This is the "tiring" of the metal, along molecular or crystal lines, caused by abnormal stresses and resulting in the development of a crack. The detection of these fatigue cracks is, in racing, the main use of the Magnaflux and, for nonmagnetic metals, the Zyglo processes.

A fatigue crack starts at a creeping pace, then spreads with accelerating rapidity until breakage occurs. The process can take a second or years, depending on the overstrength margin. The greater the overstrength margin in the component, the less likely there is to be fatigue—and the greater the possibility that the "creeping crack" will be detected during a teardown between races instead of progressing to an immediate cataclysmic failure during the race itself. Where reciprocating parts are concerned, superfluous weight of metal is undesirable. Unfortunately these parts are the very ones in which fractures are most serious and frequent. Old age alone leads to changes (crystallization) in the metal structure that lessen its resistance to fatigue—a point to watch where vintage class racing car engines are concerned.

The painstaking elimination of places that are likely to encourage the start of fatigue cracks is an important part of good design and of supertuning machine shop work. Typical danger points are at the bottoms of screw threads, junctions of bolt heads with their shanks, and at any other sharp corners or points where there are abrupt changes in section. Accidental nicks, scratches, or file marks can be starting points for cracks. Conversely a highly polished surface is a distinct discouragement to breakage. It has been established by testing that an accidental scratch on a polished surface causes a 15 percent reduction in fatigue resistance, while the refinishing of a normal "production standard" smooth surface to a high polish will increase fatigue resistance by 2 percent.

The examination of a fracture can often provide useful information. The final breakage point is usually discernible by a rough spot at the break, the early part of the "creeping crack" being almost polished in appearance, with curved lines radiating back to the starting point. This semipolished surface is caused by the working together of the surfaces before the final parting, and the start of the trouble is usually traceable to the commencement of the curved lines. Investigation by an expert metallurgist can often give a clue to the direction of the force that caused the breakage and thus can help in determining whether an abnormal load in the normal direction or some unexpected additional stress was responsible.

In many respects, the design of the competition engine follows closely that of its more sober counterpart. In fact, many of the components used in stock engines can be employed with equal success in engines of somewhat greater power output, providing that they can cope—or be made to cope—with the extra stresses involved. A study of chapters 15 through 21, which deal with the specifications of typical competition engines, will help to indicate how various manufacturers deal with particular aspects of design and how tuners go about making them even more stress resistant.

Valve Operation

One of the more remarkable aspects of engine development is the way in which the spring-closed poppet valve has retained its supremacy and continues to meet most requirements of modern high-speed operation. No alternative system has yet appeared that has stood the test of time in the hands of the private motorist, though sleeve valves, rotary valves, slide valves, cuff valves, and other kinds of valves have had a run at some period —not without success. Mechanically closed valves, called *desmodromic valves,* have had some use in racing engines, and the Wankel engine manages a four-stroke cycle with no valves at all—thus profiting from one of the best features of two-stroke piston engine design.

The very simplicity of the poppet valve makes its operation one of those things that is rather taken for granted. When checking valve clearances, for instance, the car owner will note how the rocker arm or the cam lobe moves on its appointed way as he or she rotates the crankshaft by hand and can imagine how it will go on doing this under all operational conditions. In actual fact, high engine rpm, allied with valve timing that is intended to produce rapid opening and closing, brings up many problems for the competition engine designer that are not so much of a worry with more leisurely highway driving.

Valve Float

It is well known that *valve float,* sometimes called *valve crash* or *valve bounce,* in many cases limits the maximum rpm and the power of an engine—to the accompaniment of considerable, often expensive, mechanical noise! What happens is that the valve remains off its seat more or less permanently in a state of vibration, being kicked up again by the rocker arm or the cam lobe before the spring has fully returned it to its seat. This state of affairs can lead to serious engine damage.

The usual remedy for valve float when tuning an engine for the attainment of higher rpm than standard is to increase the spring tension and, in the case of pushrod engines, to lighten the valve gear. A simple switch to stronger valve springs will not always solve the problem because of another design limit that cannot be exceeded: one cannot use springs made from wire that is so thick that the spring coils come into contact as the valve opens, thus preventing the valve from being opened farther. Consequently, double or triple springs, arranged concentrically, are a better solution (Fig. 4-2). In cases where valve spring resonance causes the spring pressure to vary at certain harmonic rpm, a supplementary damping spring (Fig. 4-3) can be fitted, which prevents the unwanted longitudinal oscillations of the main spring. The use of washers to reduce the assembled height of the springs has its place but, if too thick a washer is used, spring coil interference is a distinct probability with high-performance camshafts.

Fig. 4-2. Dual valve springs separated. By using more than one spring, each coil can be softer and of lighter-gauge wire.

Fig. 4-3. Stock spring, spring with damper coil, and assembled dual spring. Flat-steel damper prevents main spring vibrations.

Rapid opening and closing speeds, which are desirable from the standpoint of volumetric efficiency, can cause acceleration loads and inertia problems. Supertuned pushrod engines sometimes make use of roller tappets to achieve the rapid opening and closing rates that would otherwise require an overhead camshaft. In some cases, the valves open so quickly that, at high rpm, the inertia of the valve gear keeps the valve opening beyond the point that it would be lifted to by cam action alone. There is then a distinct danger that the valve will strike the piston. When a valve closes quickly, it may strike the seat with sufficient velocity to bounce open again—leading to possible valve/piston contact.

The potential of valve float for causing damage and its undeniable limiting effect on performance might lead one to assume that, up to the actual rpm where mischief begins, the valve gear and the spring perform their job efficiently—following exactly the valve timing laid down by the designer. Actually the movement, even at normal highway speeds, can be far removed from the "open 10° before bdc, close 60° after bdc" of the specification book.

When the valve is lifted by the cam, either directly as in an overhead camshaft engine or through the medium of tappet, pushrod, rocker arm, and so on, it is accelerated very rapidly against increasing spring resistance as the spring closes its coils. Nevertheless, the acceleration of the valve may be such that after about half-lift, despite spring pressure, the cam does very little work—the spring barely controlling the opening movement of the valve, which is under its own momentum as a result of the initial kick imparted to it.

When the valve is closing, the spring, in its closed-up (and thus most powerful) condition, will exert its full pressure, and the designated lowering action of the cam contour will be followed. As the valve approaches its seat, however, the spring, having opened out, will be at its weakest. It is usually at this point that good valve gear and cam design shows up, because it is possible—given adequate spring tension—for the valve to be lowered right onto its seat with comparative gentleness by a well-designed layout, thus with a minimum of the tendency to bounce off again. Incidentally this tendency of the valve to bounce off its seat (after it should have closed once and for all) has nothing to do with our previous remarks concerning high-rpm float, and it is important to appreciate this fact. The valve gear rarely operates so that the valve stays closed at the first attempt. What happens is that the valve bounces several times to an ever-diminishing degree before finally coming to rest. It is important to limit the height of the first bounce, and the acceptable amount of bounce off the seat is microscopic. It is, however, usually taking place, and can readily be detected by the use of a stroboscope—a light that can have its flashing frequency adjusted so that the movements of the valve seem to be slowed down to the eye.

Effect on Power

Evidently several undesirable effects will arise from the foregoing conditions: the potential for gas leakage, the unpredictable stresses imposed on moving parts, and the mechanical noise. A further important factor concerns the power necessary to drive the valve train. When the spring strength is increased to give effective control over the valve, the power absorbed must be deducted from the power delivered to the flywheel. Furthermore, the whole of the operating gear is placed under additional load, which prejudices reliability.

It says much for the design of modern competition camshafts that the undesirable features of spring-closed valves have been nullified to such a remarkable degree. Desmodromic valves have scarcely been heard from for almost twenty years. It is, after all, the profile of the cam that makes or mars the operation, particularly in attaining such essentials as rapid opening and closing, maximum duration of full opening, acceleration and deceleration of the valve gear to give the least stress and noise, and so on.

Camshafts and Drives

High-quality competition camshafts are generally machined from a forged steel billet and casehardened. On the other hand, most production engines now have cast nodular iron camshafts. Typically the cast material is an alloy that contains nickel, chromium, molybdenum, or some other wear-resistant element. This makes it possible to machine driving gears for the distributor or the oil pump directly into the camshaft material. Camshafts of this kind are universally used in American V8s and are sometimes reground for better performance. Nevertheless the substitution of a billet cam is normal when tuning engines for serious competition.

Because the cam surfaces are very heavily loaded, a good depth of hardening is necessary. For production cars, the chill-casting method is often used. In this process, iron chills or insertions of the required shape are placed next to the concerned surfaces. These chills cause a rapid cooling of the casting at the lobes and the journals, producing the formation of a very hard carbide-of-iron layer, which is then finish-ground. On alloy steel shafts, the induction hardening process is widely employed. The principle is that of rapidly heating the surface layer of the material to a high temperature and then quenching it before the main body of metal has had time to absorb too much heat.

The heating is accomplished by passing a high-frequency electrical current through a muffle that surrounds the surface of the camshaft; clearance between the muffle and the camshaft allows quenching water to be sprayed in. Electrical current induced in the camshaft surface as a result of its proximity to the muffle causes rapid heating for a few seconds, followed by a water quench. The process is repeated as often as required to obtain the necessary hardness depth. Obviously the heating and quenching can be controlled with great precision, and therefore induction hardening is used frequently—not only for camshafts but also for many other alloy steel parts that are manufactured in quantity.

The power imparted to the crankshaft is in the form of impulses. Thus, the turning of the crankshaft is not uniform; the degree of irregularity depends on the number of power strokes (that is, the number of cylinders). Similarly the camshaft loading is not uniform; the fewer cam lobes there are on the shaft, the more irregular will be the turning effort. Thus, even such a simple-appearing assembly as a camshaft drive can be quite a problem for the design engineer, involving as it does the coupling of two shafts—one running at half the speed of the other and neither rotating with uniform effort. The resulting torsional vibrations can place far greater loads on the camshaft drive than would be the case if loading and revolution speed were uniform.

The normal location for the camshaft (s) of all recently designed engines is atop the cylinder head. On older designs, and particularly on large American V8s, the camshaft is usually located in the valley between the cylinder banks or, especially on inline engines, in the crankcase. In the latter location, the lubricating problems are simplified because the cam lobes are kept drenched by oil thrown off the crankshaft. Apart from the bearings, a good flow of oil is necessary to the highly loaded faces of the cam lobes. However, in the case of V-type and OHC engines, an excessive flow of oil at high rpm can cause trouble by filling up the cam covers or the valve lifter gallery and thus isolating a considerable quantity of oil from the pump pickup, where it is sorely needed to keep the crankshaft bearings supplied.

Gear Drives

With the camshaft located in the crankcase or in the valley between the cylinder banks of a V-type engine, the simplest kind of drive is a pair of gears—one on the crankshaft and another, with twice as many teeth, on the camshaft. This is the system used on VW air-cooled engines, the Ford V6 (Fig. 4-4), and on quite a lot of production car engines built during the early and middle decades of this century. Extreme accuracy of the gear mesh and a good gear tooth design are essential if noise is to be avoided. But gears are always difficult to silence, and the use of a camshaft gear of nonmetallic material is sometimes resorted to on this account. However, gears do provide a very positive form of drive, an important consideration; a slight variation in the relative positions of the crankshaft and the camshaft can adversely affect the timing because of the flexibility in the camshaft drive.

Fig. 4-4. Gear drive for camshaft of Ford V6, showing simplicity of obtaining correct timing during engine assembly.

Gear drives thus are frequently used on competition engines. Several proprietary gear drive units are available for installation on supertuned American V8 engines, replacing the chain drives that are used in production. These units, which incorporate one or two idler gears, have straight-cut teeth for zero end-thrust, accuracy, and low friction losses; consequently they howl so loudly that they can often be heard shrieking like sirens above the bellow of the exhausts on the more potent oval track and drag racing engines. Overhead camshaft racing engines use gear drives almost universally, often with as many as three idler gears interposed between the gear on the crankshaft and the gear (s) on the camshaft (s). Bevel gear drives, with shafts—used on the Porsche Carrera four-cam four-cylinder engines of the 1950s and 1960s—are light, simple, and compact, but they lack the accuracy of timing that is associated with spur gears.

The difficulty of silencing gears comes mainly from the necessity arising from space limitations of using small gears with fine tooth pitching. These, in consequence, are extremely sensitive to any vibration in the shaft center distance, which naturally alters the meshing of the teeth. The shaft centers are bound to change because of the heat expansion of the engine block. Thus, apart from the irregularities unavoidable in quantity production, the fact that the engine operates over a widely varying temperature range is inimical to the maintenance of a constant center distance. Regardless of the number of teeth there are on a gear, the inescapable fact remains that only one tooth at a time is transmitting the load. It is the impact of transferring the load from one tooth to the next in rotation that sets up the whine, which is increased by the impulsive conditions of load reversals across the backlash between the teeth.

Chain Drives

Aside from the VW and Ford engines, nearly all contemporary pushrod OHV engines use chain-driven camshafts. On imported engines, single-row or double-row roller chains are usually employed; the most common example is the Ford Cortina engine used in Formula Ford racing and in many of the overhead camshaft derivatives of this engine. Roller chains (Fig. 4-5), which are also used to drive overhead camshafts on the Datsun, Toyota, and Mercedes Benz engines, three popular marques, offer accurate valve timing and good resistance to breakage at high rpm. On American engines other than high-power units such as the Chrysler "Hemi", a link belt "silent" chain is commonly used in the interest of quiet operation. This is usually replaced by double-row roller chains when the engine is supertuned for competition.

Fig. 4-5. Double-row roller chain camshaft drive of Daimler 2.5-liter V8 engine.

Gears bear all of the camshaft drive load on a single tooth at a time and consequently tend to wear more at some points of rotation than at others because of the impulsive movements of the shafts. On the other hand, a chain starts with two important advantages. First, in wrapping around the sprockets, it engages a large proportion of the teeth simultaneously, thus spreading the load. Second, the chain itself possesses a degree of weight, which acts to some extent as a damping medium on the impulsiveness of the drive.

Roller chains, the kind normally used in competition engines, are by no means as silent running as link chains. But so long as sprockets of reasonable diameter are used (generally the case on production engines) and the amount of chain slack is kept within limits, little or no noise will be heard above the general noise level of the average sports engine. The use of excessively small sprockets will increase the impact force between the first tooth and the chain; this is accentuated by chain whip caused by excessive slack, and, in extreme cases, where there is not much clearance, a worn chain will rattle against the timing cover.

When the distance between the crankshaft center and the camshaft center is short, as on most American V8 engines, the chain tension does not vary much with wear. Providing that the initial amount of chain slack is at the minimum specified (to allow a free-running drive and to compensate for center distance variations caused by temperature changes, there must be some degree of slack), it is quite possible to design a satisfactory drive that has no provision whatever for chain tensioning or adjustment. However, if the camshaft center is at some distance from the crankshaft center—as on all overhead camshaft engines—some form of tensioning device is an absolute necessity. In fact, tensioning devices will be found on nearly all production engines that have roller chain drives.

Gilmer (Toothed) Belt Drives

If any one factor can be named as having made possible the present rapid trend toward overhead camshaft engines in production cars, it is the perfection of the Gilmer belt drive (Fig. 4-6). Though these toothed, reinforced belts had been used for a number of years previously to drive the superchargers on big blown drag racing engines, it was not until 1964 that the little German-made Glas sports coupe was introduced with a belt-driven overhead camshaft. At first the innovation was viewed with suspicion—even by the now-defunct Glas Company, which recommended that the belt be replaced each 25,000 miles.

The second production engine with a belt-driven overhead camshaft was a six-cylinder inline engine made by the Pontiac

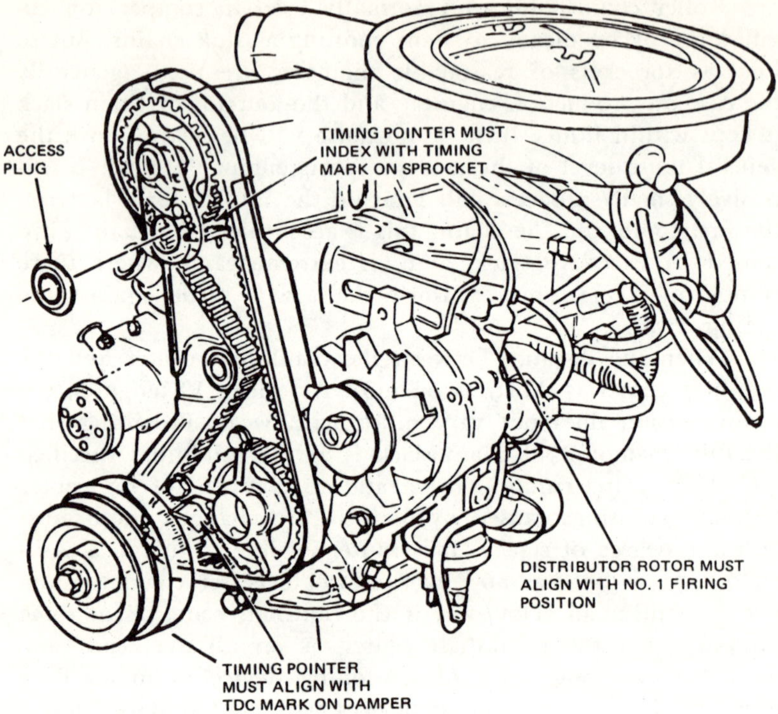

Fig. 4-6. Gilmer (toothed) camshaft drive belt installation on Ford 2300 OHC engine. Rather complex timing instructions are given here, which must be carried out with complete accuracy during belt installation.

Division of General Motors. The Pontiac's belt was reinforced by fiberglass strands, whereas the Glas had used a steel-reinforced belt. Steel-reinforced belts are more widely used today, and most car companies have stopped recommending the routine replacement of the belt. It is not uncommon to hear of Capri and Pinto 2000 SOHC engines that have been driven well over 100,000 miles without belt replacement. In addition to greatly simplifying engine design, the belt has two other good points to recommend it: it is as near silent as any drive system could possibly be, and it requires no lubrication.

Another attractive feature of the toothed-belt camshaft

drive is that radical modifications can be made to the engine without making necessary a major reengineering of the camshaft drive. During development the DOHC Lotus engine was tested with a number of different cylinder heads. The new heads could be bolted onto the original block with no more concern for camshaft center changes than if the engine had been a valveless two-stroke. With gear-driven camshafts, and even with most chain drive systems, the drive modifications alone could have been overwhelmingly inconvenient and expensive.

Chain and Belt Tensioning Devices

Gilmer belts generally have an automatic tensioner that consists of a spring-loaded idler pulley. There is, in addition, some form of manual adjustment that is made when the belt is removed and installed. The idler pulley is often plastic, and despite the obvious attention given to cost engineering, the tensioners are so lightly stressed that they never fail. A simpler, less expensive, and more reliable camshaft drive system for production cars is not likely to come along for many decades—if ever.

Chain tensioning devices are another thing altogether; chain inertia at high speeds can cause the phenomena of "whip" and "wave surge" that tend to drive designers prematurely gray. It should be mentioned here that normal wear at the chain joints has little or no effect on the gearing action of the chain on the sprocket teeth in spreading the load. The chain simply takes up automatically a larger pitch circle higher up the teeth. Between the sprockets, however, wear inescapably means more slack, and with it comes the possibility that the chain will begin to strike the cover. Also, with the chain running high on the sprockets, the sprocket teeth tend to wear to points, and this can have a detrimental influence on accurate timing. Consequently an efficient tensioner—or tensioners—can be a decided boon on engines that will be used in competition.

As long ago as 1912, chain drive pioneer Hans Renold, in a treatise directed at engine manufacturers, stressed the desirability of incorporating some means of adjusting the timing chain so that the tension could be maintained at a correct and uniform level. Because the chain-driven accessories in those days frequently included a magneto, it was not too difficult to incorporate a tension adjustment into an adjustable magneto mounting. With the passage of time, the general simplification of engine auxiliary drives, and the advent of quantity production, nonadjustable chains became accepted so long as the centers were kept short. With longer centers, where excessive whipping could cause actual chain breakage, spring-blade tensioners of an automatic type or manually set idler sprockets were incorporated. With the latter, of course, there was the danger that too tight a setting (such as one done by inexpert hands) could be even more detrimental to the drive than too much slack.

Hence, apart from those engine makers who fairly successfully have taken the view that no adjustment at all is better than further complication and the possibility of faulty servicing, there are others who have attempted to find a solution. Particularly on British overhead camshaft designs, this often took the form of a flat spring-steel strip, bearing on the back of the chain at all times. Superficially this device might pass muster as an aid to quieting a worn drive, but it falls short of what is required in several respects. If the spring strength is sufficient to exercise any real control over chain whip, the frictional loss is considerable, and wear takes place very quickly on both the spring face and the chain-plate edges. If too tight, the spring is quite capable of uncontrolled oscillations in unison with the chain (wave surge); in fact this phenomenon can take place at certain engine speeds however strong a spring is used.

Attempts to incorporate in chain tensioning devices some method of damping the spring oscillations, while often successful, often result in a bulky and relatively costly assembly—particularly when compared to the inexpensive tension adjuster that handles the job so well on Gilmer belt drives. Further, there is the inescapable fact that a greater movement of the tensioner spring is permitted as the chain wears. Obviously, then, the more the chain wears, the less tension on the spring and the less pressure exerted on the chain.

The most successful solution to the problem has been to use the engine's lubricating oil pressure to operate the chain tensioner. The tensioner's spring exerts only light pressure, thus maintaining the necessary pressure for keeping the chain tight until the engine has started. One device of this kind is shown in Fig. 4-7; it should be familiar to all who have worked with MG and other British sports car engines.

A. Slipper pad
B. Detent sleeve
C. Cylinder spring
D. Cylinder body
E. Backing plate
F. Bolt
G. Binding strap
H. Screen trap
I. Gasket

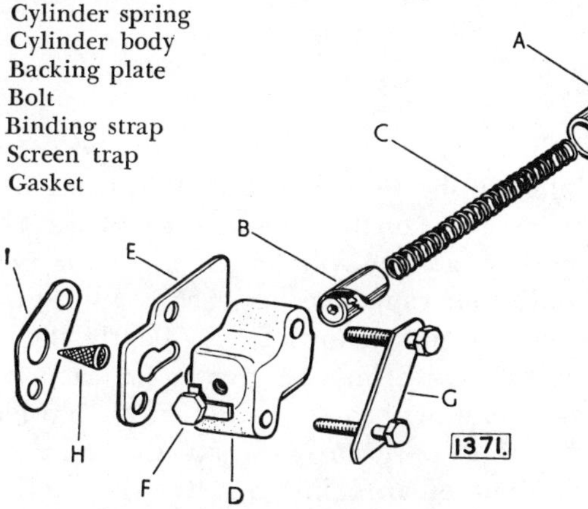

Fig. 4-7. Components of Renold automatic chain adjuster, as used on DOHC Jaguar engine.

The principle is that of a rubber-faced slipper pad carried on a plunger that protrudes from a small cylinder. The cylinder body is bolted to the engine, with a backing plate that keeps the slipper pad aligned with the chain and provides a sliding surface for the pad to move in and out against. The tensioning device is mounted in close proximity to the outside of the non-driving side of the chain. The slipper pad is in constant contact with the chain, urged against it by the pressure of a light compression spring. When the engine is running, the spring pressure is augmented by lubricating oil pressure; the oil emerges finally onto the chain from a hole in the slipper pad.

The essential feature that enables control of the chain

slack to be maintained at a constant value irrespective of the amount of chain wear is embodied in what is termed the *detent mechanism*. This consists of a ratchet system that allows the slipper to move against the chain without restriction yet prevents any return movement beyond the small amount necessary for free chain operation at all times. Another, though considerably bulkier, oil pressure assisted tensioner is used in the Porsche 911 "flat-six" engines. Instead of a slipper pad, however, the Porsche tensioner has an idler sprocket that is engaged with the chain under pressure.

Camshaft Problems

Although enthusiasts naturally expound the virtues of DOHC engines, the duplication of camshafts brings in a few mechanical problems that further complicate the working of the camshaft drive. In a four-cylinder engine with two camshafts, there are only four cam lobes on each shaft. This means an extremely uneven turning effort. Consequently the loading placed on the camshaft drive may be greater for each camshaft than on a similar engine with a single overhead camshaft.

Loadings are more uniform with all the cams on one shaft, and the camshaft of an inline SOHC "six", such as the Datsun "Z-car" (with twelve lobes), provides quite a smooth turning effort. It is understandable, therefore, that most production cars have engines with single camshafts because the smoother turning—especially on sixteen-lobe V8 camshafts—results in very quiet operation and minimum loading of the camshaft drive.

Of course overhead camshafts are the rule today; no pushrod OHV engine has been newly designed in at least ten years. It is probable that, in America at least, any all-new V6 or V8 engines will be designed with pushrod-operated overhead valves; this is the least expensive method of valve operation when there is more than one bank of cylinders. Neverthless, the only all-new American engine designs of the past decade have been four-cylinder inline engines, and all have single overhead camshafts. The engines are the Ford 2300 Capri/Pinto/Comet/Mustang engine, the Chevrolet Vega engine, the Chevrolet

Chevette engine, and the American Motors 2.0-liter engine introduced in the 1977 Gremlin.

Previous objections to the use of overhead camshafts centered mainly on the cost and on increased drive complexity and noise. The use of Gilmer belt drives has solved all three of these one-time problems. If one includes the imported cars, over half of the car models sold in the United States are now equipped with overhead camshaft engines, and the vast majority of those pushrod engines still in production are developments of designs that originated fifteen to twenty-five years ago. Insofar as inline engines are concerned, a belt-driven SOHC design is less costly to manufacture than a pushrod OHV design.

The Overhead Camshaft Engine

The possibility of greatly improved volumetric efficiency, obtained by angling the valves relative to the cylinder bore axis, thus straightening the ports and increasing the potential for larger valves, has traditionally been the main reason for using overhead camshafts. While this efficiency remains the principal virtue today for racing engines, it is but one of the attractive features offered by overhead camshafts when they are used in production car engines. Some examples are:

1. Because the valve gear is comparatively lightweight and consists of few parts, the engine can be made to rev higher without encountering valve float.

2. Because of the small number of valve gear components and because the valve clearance is subject to less change as the engine temperature varies, an OHC engine can be made to operate more quietly.

3. Accessibility to the camshaft is improved, and it is therefore usually possible to change camshafts with relatively little engine disassembly (particularly welcome on engines that are modified for competition use).

4. The camshaft bearings are more easily replaced (if

there are any camshaft bearings).

5. With a Gilmer belt drive, overhead camshafts can represent a cost saving for the manufacturer.
6. Rapid valve opening and closing can be obtained with more accuracy and less mechanical stress.

A majority of today's production car overhead camshaft engines have rocker arms of one kind or another interposed between the camshaft and the valves. Mainly this is so that the exhaust valves can be set at a different angle to the intake valves or so that the exhaust valves and the intake valves can be positioned at opposite sides of the cylinder head. Though the use of rocker arms increases the valve gear mass and thus increases the possibility of high-rpm valve float, there is virtually nothing lost in terms of valve train rigidity, and the added capability for easy valve adjustment is an important side benefit.

The matter of valve train rigidity is of paramount importance to the competition engine builder. Easily half the problems associated with valve, valve gear, and camshaft design on pushrod OHV engines would disappear if the valve train were perfectly rigid and incompressible. Most of the springiness that makes the pushrod system compressible is in the pushrods themselves, and it is this springiness that makes it impossible to get the valve movements to follow the cam profiles. Consequently the camshaft designer must devote a great deal of time, effort, and experimentation to developing a cam grind that will get the valve to open and close correctly when it *does not* accurately follow the cam profile.

The design and manufacture of special camshafts is a huge industry in the United States—largely because all of the big V8s have pushrod-operated valves and therefore need extremely complex cam designs to obtain the power that is potentially available. Overhead camshafts solve the problem completely. But, aside from the short-lived SOHC Ford 427 V8 of the 1960s, no American manufacturer has approached the problem head-on.

DOHC Engines

The double overhead camshaft engine, though famous as a producer of power since the earliest days, was, until a few years after World War II, an exclusive feature of Grand Prix or Indy-type racing cars and a few exotic GTs. Then came the Jaguar DOHC "six" and the Porsche Carrera "four", which brought twin-cam engines within the reach of many enthusiasts. More recently the Cortina Lotus reached the market, making possible twin-cam motoring for the masses—though both Alfa Romeo and MG had relatively low-priced twin-cam "fours" on the market by the time the Lotus was introduced. Nevertheless, it was not until the Fiat DOHC came on the market with its belt-driven camshafts that DOHC engines became commonplace.

A very simple valve mechanism can be designed (Fig. 4-8) by suitably mounting the two camshafts, one above each row of valve stems. The need for rocker arms and rocker arm shafts is obviated, and extreme lightness can be obtained in the only reciprocating part—the tappet or cam follower—by making this a simple, one-piece component. Still, some kind of adjustment system is required so that valve clearances can be set accurately to the required specifications.

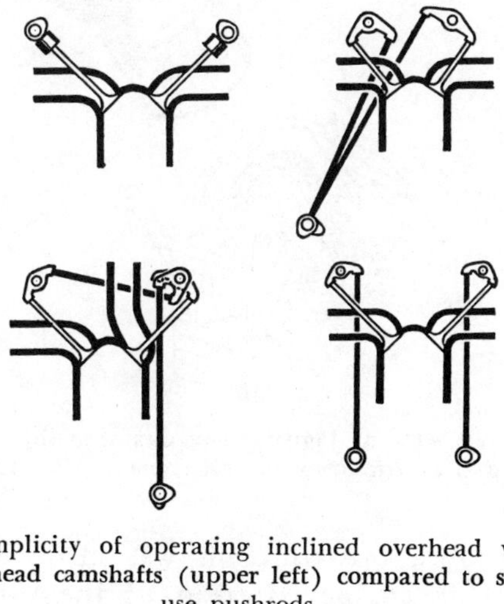

Fig. 4-8. Simplicity of operating inclined overhead valves with double overhead camshafts (upper left) compared to systems that use pushrods.

Some early racing engine designs did not, in fact, have any provision for adjusting the valve clearance. A small amount of metal had to be carefully ground off each valve stem during assembly to obtain the correct gap. Alfa Romeo, and later other carmakers, used mushroom-type tappet/spring retainers with internal threads that could be screwed onto the large-diameter threaded valve stems. Adjustments were then easily made by moving the position of the mushroom tappets on the valves.

More commonly DOHC valve clearance adjustments are made with shims of various thicknesses selected from a range of available sizes. On most racing engines, the Jaguar "six", and the various Fort Cortina-based twin-cam engines, the shim is placed atop the valve stem (Fig. 4-9). After installation of the shims, the bucket-type tappets are installed, and finally the camshaft is bolted on. The shims in this case are extremely small and lightweight. Nevertheless, adjusting the valve clearances is complicated and time-consuming work.

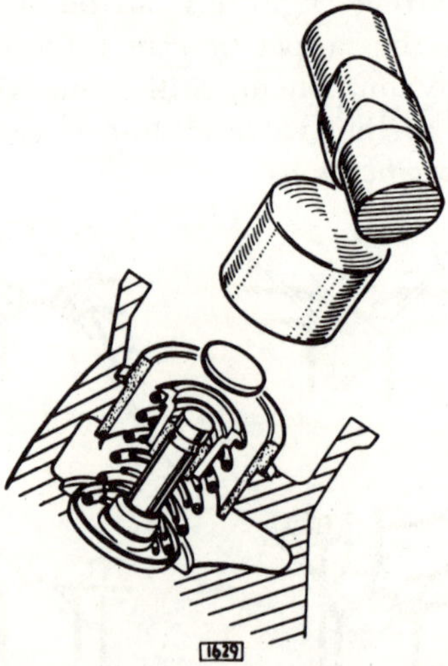

Fig. 4-9. Detail view of Jaguar valve clearance shim installation. Camshaft and cam follower must be removed to change shim.

Something of a breakthrough was made by Fiat on its overhead camshaft engines. Instead of the valve adjustment shim's being placed between the valve stem and the underside of the tappet, it is placed atop the tappet in a recess *where it in contact with the cam lobe itself.* The shim, of course, is large—about the size of a 25¢ piece. Still, the system has many advantages—demonstrated by the fact that Volkswagen and Audi pay royalty fees to Fiat for the use of this system on their SOHC engines.

The foremost advantage of the Fiat system is that no disassembly is required to adjust the valves. The tappets or cam followers can be pressed down by using a special lever against valve spring tension and the adjusting shims lifted out with special pliers. A second advantage is that most tappet wear is confined to the shim (Fig. 4-10), which is easily and inexpensively replaced. There is a related virtue: the cam follower need not have such an elaborate and expensive heat treatment.

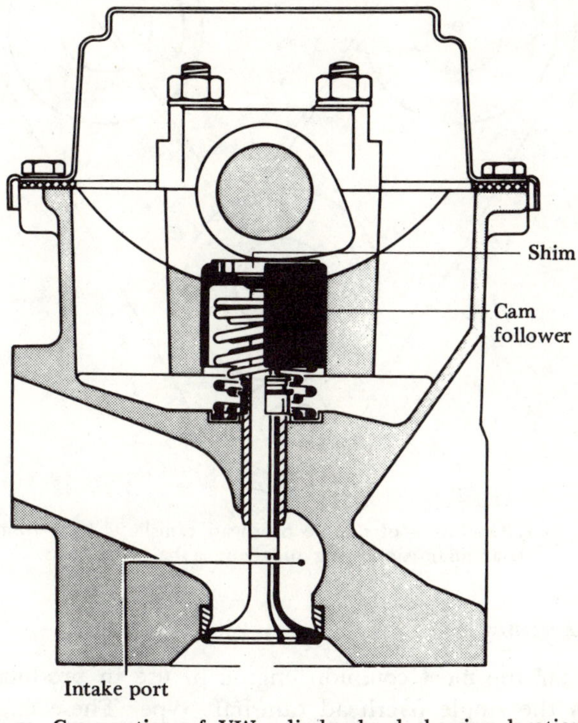

Fig. 4-10. Cross-section of VW cylinder head showing location of valve clearance shim atop cam follower.

In addition to the Gilmer belts used on the Fiat twin cam, the Lotus LV 240, the Ford-Cosworth BDA, and the Cosworth-Vega, there have been many unique forms of camshaft drive used on DOHC engines. One of the more unorthodox schemes of years gone by was the Y-frame system shown in Fig. 4-11. In this arrangement, each extremity of the Y is connected to a crankpin, one on a half-time gear driven from the crankshaft and the other two on wheels bolted to the two camshafts. Rotation of the half-time crank is thus transmitted to the other two cranks with little complexity and considerable reciprocating mass.

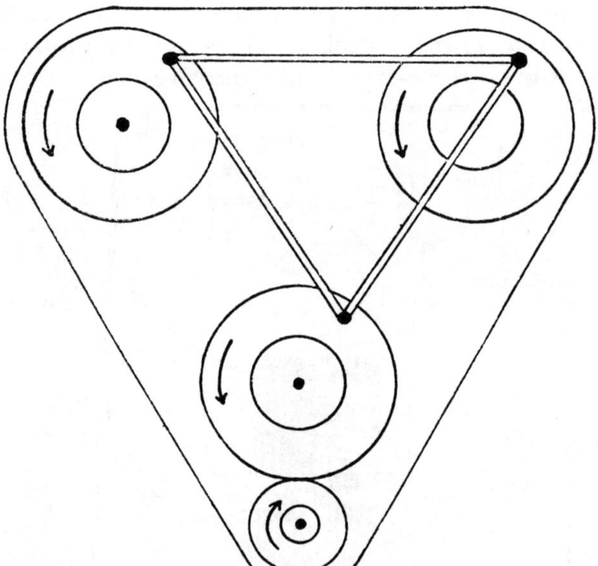

Fig. 4-11. Operation of double overhead camshafts by Y-frame or triangular connecting rod from a half-time gear.

SOHC Engines

By far the most common engine in use in production cars today is the single overhead camshaft type. These engines are found on everything from inexpensive Pintos and Hondas to such rare and costly exotica as V-12 Jaguars and various Ferraris.

There are many reasons why the SOHC engine, which faded from popularity during the 1940s and 1950s, has made an overwhelming rebirth. Probably the most outstanding of the new generation of SOHC designs is the Porsche 911. The competition successes of this engine are too numerous to catalog in a book of this kind and are already so well known that any such listing would be presumptuous. The Datsun and the BMW SOHC engines have also accrued an outstanding competition record, and in the future we can expect to see more and more belt-driven SOHC engines in competition—particularly the Volkswagen Dasher/Rabbit/Scirocco powerplant used in the Formula Super Vee class.

The merits of the SOHC engine as an alternative to the pushrod OHV engine deserve further examination. Certain components of the conventional engine are inevitably subject to reciprocating motion—notably the pistons and connecting rods. Designers, however, have usually felt that the more reciprocating parts that can be eliminated, the better. By moving the camshaft to a location above the valves, it is possible to do away with the pushrods, the rocker arms, and the valve lifters, thus eliminating two-thirds of the reciprocating parts for each valve.

Advantages previously cited were the more predictable valve timing, obtained by eliminating the springiness of the pushrods, and reduced weight in the valve train, which helps eliminate high-rpm valve float. In addition, the weight reduction removes considerable inertia losses, with a consequent gain in mechanical efficiency. Finally, notwithstanding the general reliability of many pushrod engines, their components include those that are most likely to fail when worked hard at high rpm.

The possibility of achieving the absolute minimum of components by contacting the top of the valve stem more or less directly by the cam lobe has exercised the imaginations of designers for a long time. Most of this mental exertion has gone into camshaft drives and—perhaps to an even greater extent—into the design of valve clearance adjusting arrangements. All of the adjusting systems previously discussed in connection with DOHC engines have also been applied to SOHC engines.

In addition to the Fiat/Volkswagen/Audi system of valve adjustment, there are the rocker arm adjustment systems used on SOHC engines by Datsun, Toyota, Honda, Ford, BMW, Mitsubishi, and others. On the Ford 2000 engines, which are made in Germany, the valve clearance is adjustable by varying the height of the rocker arm's ball-joint pivot. This is done by first loosening a locknut and then turning the ball-joint pivot to screw it farther in or farther out of the cylinder head casting. The locknut is then tightened to keep the ball-joint pivot in place. The American-made Ford 2300 SOHC engine (Fig. 4-12) requires no valve adjustments. Though the valve gear is superficially similar to that of the 2000 engine, the rocker arm ball joint is on the upper end of a hydraulic piston. Hydraulic pressure is supplied by the engine's lubrication system so that the effect is identical to that of the hydraulic valve lifters commonly used on American V8 engines.

The various Japanese-made SOHC engines have valve adjustments that are not unlike those of a pushrod OHV engine (Datsun is an exception; its engines use an adjusting system exactly the same as that of the Ford 2000). There are screws and locknuts in the ends of the rocker arms that contact the valves. There is also the system used on BMW engines, and it is undoubtedly the most convenient arrangement for mechanics—especially for those who in the past have limited their practice to the tuning and repair of pushrod powerplants.

A unique system of valve adjustments is used on the Chevrolet Vega engine. This engine has inline valves, with the single overhead camshaft acting directly on the bucket-type cam followers. Instead of shims, however, the GM engines have a tapered screw inside the cam follower (Fig. 4-13). One side of the screw is flat and contacts the valve stem. If an Allen-type wrench is inserted, the adjuster can be turned in increments of one full turn in either direction. Each turn alters the valve clearance by .003 in. This system is undoubtedly more convenient than those used with rocker arms and, once the mechanic has learned the trick, adjusting the valves on these engines becomes a simple task indeed.

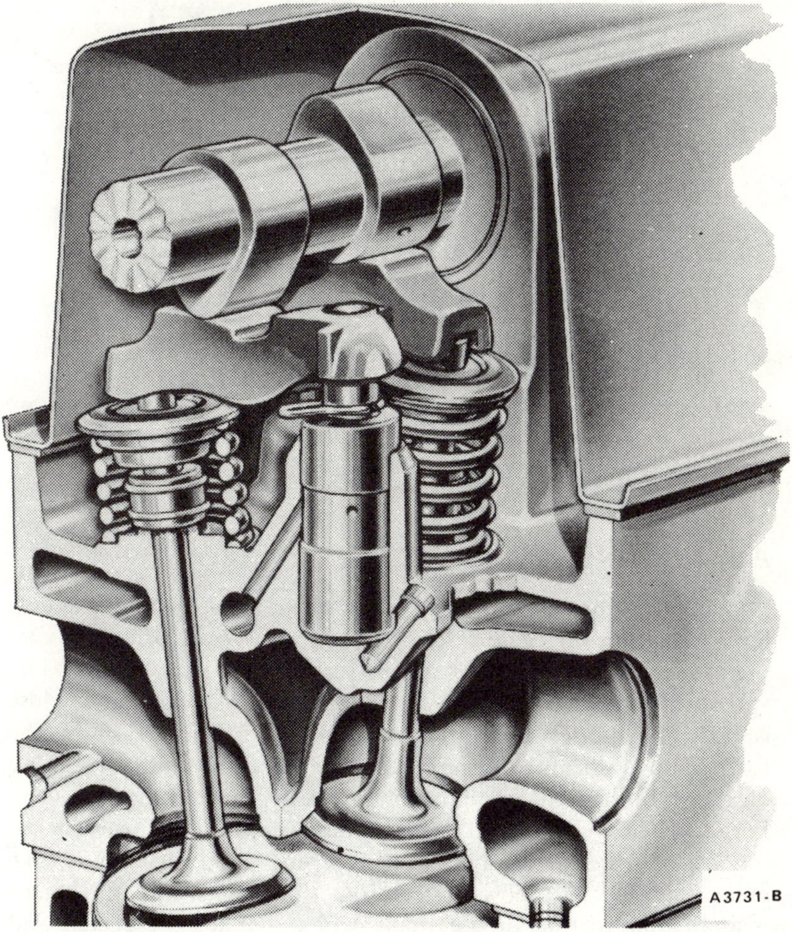

Fig. 4-12. Ford 2300 SOHC engine, showing stationary end of rocker arm supported atop hydraulic piston device.

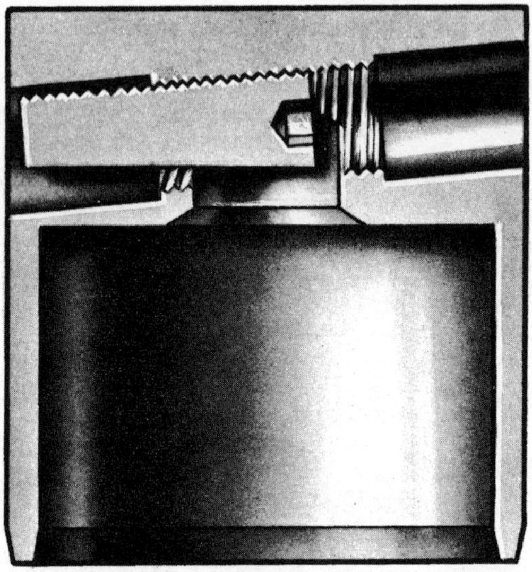

Fig. 4-13. Cross-section of Chevrolet Vega cam follower showing adjusting screw. Screw has one flat, tapered side that contacts valve stem. One turn of screw changes clearance .003 in.

Pushrod OHV Engines

The overhead camshaft system of valve operation presents few difficulties in regard to obtaining precision of valve timing. The pushrod engine is not so fortunate in this respect. The valves of modern high-speed engines must open and close rapidly, have high lift, and have quite long duration in the fully open position to ensure complete charging and scavenging of the cylinders throughout the necessary wide range of rpm. Perhaps because pushrod engines have enjoyed an overwhelming superiority in numbers in the recent past, we do not take greater notice of their shortcomings. The situation at present is analogous to that of the 1950s, when that long-respected and revered powerhouse—the Ford flathead V8—began to be overshadowed by the newly introduced OHVs. So long as there were no OHV engines around, the old flatheads seemed to be very good engines. But it was not long before the faithful Ford began to seem only slightly less archaic than Newcomen's steam engine.

In production car racing classes that are defined on the basis of piston displacement, pushrod engines are barely competitive when there are OHC engines available. In large-displacement classes, where the big V8s hold sway, their handicap is not so apparent because of the dearth of "big inch" competition-developed OHC engines. Typically, in a class such as Formula 5000, the pushrod engines are permitted a displacement of 5 liters in fully supertuned form; overhead camshaft racing engines in the same class are limited to 3 liters.

Noisy operation, long a criticism of the pushrod engine, has been eliminated on most American engines through the use of hydraulic valve lifters. These devices, however, do not work well in competition; they tend to "pump up" at high rpm and hold the valves open. Therefore, pushrod competition engines and all production pushrod sports car engines use solid lifters —and noisily beg the driver to be tolerant of the racket.

The biggest culprit is the requirement of large valve clearances, having little or no bearing on engine temperature. In the distant past it was assumed that clearance between the rocker arm and the valve stem was needed to ensure that the valve would seat properly under high working temperatures. Enthusiasts who delighted in a silent-running engine would drive as hard as possible to obtain the maximum working engine temperature and then quickly set the clearances almost to zero. This is a practice that is continued today by many Formula Vee racers, who are striving to get as much lift and duration as possible from their stock Volkswagen camshafts. The adjustment, of course, must be repeated quite frequently to make sure that the valves are not being held open.

With the cars of forty years ago (and with stock VW 1200s) there is little chance for obtaining high rpm. Slow engine speeds mean a relatively leisurely lifting and lowering of the valve, and the timing under these conditions follows the cam profile more or less exactly. The faster-turning engines, however, need some clearance so that the inertia of the weighty valve gear is partially overcome before the additional load of opening the valve commences. But the use of very large clearances on modern pushrod engines, which do not vary by more than one or two "thou" hot or cold, has obviously little or nothing to do with compensation for temperature variations and is more than adequate to limit valve train loadings. The relatively wide clearances are, in fact, used to obtain the final working valve timing with which the best overall engine performance is provided.

With a large clearance, it will be evident that quite a lot of cam rotation will take place before any opening movement is eventually conveyed to the valve stem itself. There are clearances between the cam lobe and the tappet, between the tappet and the pushrod, and between the pushrod and the rocker arm to be considered. All of these components, of course, remain in more or less intimate contact with each other, but they still represent clearance.

The use of rocker arm return springs, designed to limit clearance in the valve train to the gap between the rocker arm and the valve stem, has been tried, but it offered little noise-reducing advantage while placing an additional loading on the valve gear. The load is increased because the shock of taking up the considerable clearance comes in one impact instead of being spread over three additional oil-cushioned clearances.

On production pushrod engines that have factory valve clearance specifications in excess of .010 in.—particularly if the specifications are between .020 and .025 in.—it often pays to experiment with different valve settings when the engine is used in competition. As a rule of thumb, wide clearances improve low-rpm power, and narrow clearances may improve the output at high rpm. Thus, on a course where acceleration from low speeds is of importance, the wide clearances may be the best possible setting. But if maximum speed down a straightaway is a determining factor, reducing the clearances by .002 to .010 in. may bring a benefit of 200 to 500 rpm in terminal engine speeds.

It is obviously important that the weight of pushrod valve train components be kept within reasonable bounds to reduce inertia loading. The valve spring strength has to some extent to be gauged in respect to its duty in moving the valve train against inertia, as well as to its primary duty of seating the valve and keeping it closed. In fact, the top limit of rpm available with a pushrod engine is quite frequently determined by the weight of these parts, as tuners of this type of engine are well aware.

Light alloy pushrods and rocker arms are commonly installed on most large supertuned engines. Machined aluminum rocker arms with needle bearings and hardened steel rollers to contact the valve stems (Fig. 4-14) are almost universally used in the V8 engines found in drag racing. With proprietary components such as these, both inertia loadings and mechanical losses are reduced. Generally the lightweight rocker arms are accompanied by tubular aluminum pushrods with hardened steel tips and lightweight solid tappets in place of the stock hydraulic lifters.

Fig. 4-14. Aluminum rocker arm, showing hardened steel roller that contacts valve stem. Hardened steel pushrod socket and needle bearing pivot are other features. Stud screws into cylinder head.

Further weight reductions are possible by removing metal from the valves and by using machined aluminum alloy valve spring retainers in place of the stock stamped steel retainers. Many of these stamped steel retainers, incidentally, are prone to failure when heavier valve springs and high rpm operation place loads on them that are far in excess of those encountered in highway driving. If the rules of the racing class permit the use of roller tappets, it is possible to use a camshaft in a pushrod engine that approximates the rapid opening and closing and the long valve-open periods that would otherwise require an overhead camshaft to obtain. However, the possibility of loss of power through valve gear failure remains a constant problem in supertuned pushrod engines.

Crankshafts

The crankshaft speeds of competition engines have tripled or quadrupled during the past fifty years. This has obviously not been achieved without some very remarkable advances in the design of crankshafts and, particularly, their bearings. Because of the great strides made in metallurgy, the cast crankshafts in most of today's production cars are superior in strength and wearing ability to even the finest machine-forged steel crankshafts of the 1920s and 1930s. Cast crankshafts are also less expensive to manufacture.

Though the number of cast shafts is on the increase in competition, racing engines and supertuned production-based competition engines usually have crankshafts constructed of high-quality case-hardened steel. Interestingly, crankshafts for four-cylinder inline engines were once made with only two main bearing journals; today a three main bearing "four" is considered marginal even for highway service. The majority of modern production crankshafts have a main bearing journal on either side of each crankthrow. Thus, the five main bearing inline "four" and V8 and the seven main bearing inline "six" and V12 are pretty much the norm. In 1976 Ferrari reduced the number of main bearings in their flat-12 grand prix engine —then won the world championship! Thus, it seems in order to mention here that pancake engines generally have lighter bearing loadings because there are opposing forces provided by the other bank of cylinders, and all stresses are concentrated in the middle of the engine instead of being off at one end.

The materials used in the crankshaft and its bearings have a great deal to do with how many bearings the crankshaft requires. But with increasing demands for greater smoothness as well as more power, it has become evident that there is a decided merit in having each crankthrow supported on both sides. This has been made practical in part by modern bearing shells, which do not need to be very wide to supply good support. Thus, despite an increase in the number of main bearings, today's engines are no longer than the two and three-bearing engines of yesteryear.

Bearing Surfaces

Casehardening really refers to the traditional method of obtaining a hard exterior on a low-carbon steel by carburizing the surface layer, the process resulting in a "case" of high-carbon steel. Another interpretation of this term is that, in early times, gunsmiths and other craftsmen placed parts in iron "cases" that contained bone charcoal. The cases were then heated in the forge until the part reached high heat and acquired a high-carbon surface from the surrounding charcoal. But whichever etymological interpretation you prefer, the fact is that today the term *casehardening* covers many modern processes, such as chill casting, cyaniding, and nitriding.

The prime requirements of a good crankshaft are a tough core and an adequate depth of hardening; the first property gives strength and the second gives wear resistance. Carburizing, the process most similar to the original casehardening process, requires a fairly lengthy exposure to high temperatures to obtain a good depth of hardening. This process can lead to distortion. The nitriding process differs somewhat; the composition of the surface is altered much as it is by carburizing but by the introduction of hard iron nitrides, plus alloying material. These substances are introduced in combination with ammonia gas, and there is no quenching process after hardening.

Nitriding provides an excellent degree of hardness and does not need a high overall temperature; thus the structure of the metal is not likely to be affected, even when the material is alloy steel. Consequently the use of a high-grade basic material, plus nitriding, is generally the hallmark of a high-grade crankshaft. Crankshafts of this kind are widely available as proprietary components for supertuned production-based engines—particularly for American V8s.

Also available are "stroker" crankshafts that increase the engine's piston displacement and crankshafts with chromed journals for use in nitro-burning drag racing engines. Stroker crankshafts can be either newly manufactured competition components, or they can be modified production components. If a production crankshaft is taken as the starting point, steel is added to the crankpins by welding; then the crankshaft is nitrided and reground to a longer stroke. The chrome surfaces applied to the crankshafts of "fuel burners" are mainly to prevent corrosion caused by caustic substances released during the combustion of the exotic nitro fuel mixture.

Cast Crankshafts

Because of the wide use of the common cast crankshaft in all forms of production, strictly stock, and showroom stock racing, we shall look closely for a moment at it. The casting technique for crankshafts was probably pioneered, at least on a large scale, by the Ford Motor Company, and it is now used by several specialist manufacturers and foundries; the production advantages compared with those of forging are obvious. The material used should really be referred to as cast-iron (though it is a highly special brand), even though the designation *cast steel* is often given to it by auto manufacturers. The accuracy of the definition depends on what is meant by "steel".

Taking cast-iron as a basic material, its qualities for the purpose can be summarized as resistance to fatigue failure and, of course, its excellent wearing properties as a bearing surface. The virtue of elasticity also means that small machining or casting errors in associated components, such as the crankcase or the bearings, are not likely to induce extra internal stresses in the crankshaft itself—an advantage that is especially desirable for mass-produced engines. Further, the self-damping properties of cast-iron reduce the stress caused by torsional vibration to a greater degree than does a steel shaft.

Specialized heat treatment may well result in a cast crankshaft that is quite comparable in every respect with a high-quality shaft of medium alloy steel. Though there are various alloys and added elements used in manufacturing these nodular (or ductile) irons, the material originally used by Ford—and still used by them for the most part—consists of copper-chromium iron, the alloys representing about 6 percent of the total.

The cast shaft is made in one piece. Most forged steel crankshafts are also made in one piece now, though in the past bolt-on balance weights were relatively common. At present, built-up crankshafts, consisting of as many or more pieces as there are cylinders, are seldom seen. Before the development of modern, thinwall bearing shells, however, built-up crankshafts were common both in racing engines and in many high-rpm sports car engines. The reason was that the plain bearing materials available at the time were not adequate for high-speed use; consequently ball bearings and roller bearings had to be employed on the crankshaft's main journals and crankpins.

The method of assembling built-up crankshafts ranged from the Bugatti practice of pressing together crankshaft sections that had zero-tolerance clearances between parts to the Porsche method of using Hirth threaded fittings (which are sometimes used in aircraft engines). In all cases, the need for these built-up crankshafts was dictated by the necessity for using one-piece ball or roller bearing races.

In addition to making one-piece crankshafts with counterweights a simple and relatively inexpensive piece to form, a further advantage of casting is that the shaft (or parts of it, such as the crankpins) can be made hollow—without drilling—thus reducing weight and internal stressing. The extent to which this can be done depends on the loads imposed by engine operation. Fig. 4-15 gives a good view of a lightweight cast-iron crankshaft.

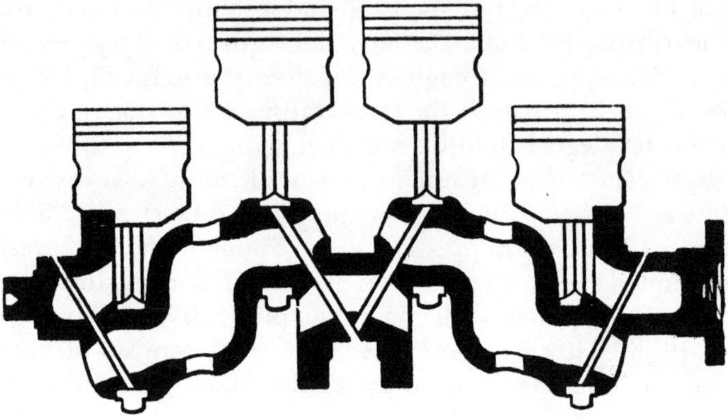

Fig. 4-15. Cast hollow crankshaft in section showing internal oil pipes to big ends (Ford of England).

As received from the foundry, the cast crankshaft is already quite smoothly finished, and subsequent machining operations are confined to surfacing the bearing journals, drilling the oilways, and final balancing. This last operation is usually done by either drilling or plugging "correction" holes, and sometimes by grinding the appropriate crankweb or balance weight.

Apart from the increasing popularity of cast shafts as alternatives to steel forgings, notable advances have taken place in recent years in production line balancing techniques. Nevertheless, vibration dampers are common on the fronts of the crankshafts of American passenger car engines. These are designed to limit torsional vibrations. The intention, however, is not to protect the crankshaft from stresses that could lead to abnormal main bearing wear—as in former times—but to increase the smoothness of the engine for purposes of comfort and silence.

Bearings and Materials

Today, with the exception of those few engines equipped with ball or roller bearings, such as the Porsche 904, all modern engines are fitted with bearings of the thinwall type. Thick babbit metal bearings—whether sprayed, cast, or bonded to thick backing shells—are for all practical purposes as obsolete as the hand crank and bulb horn. This change, which took place over the space of a few years in the 1940s and 1950s, resulted from the development and high-speed production of a new type of bearing by the Vandervell organization in England. These bearings, of perhaps more importance than any engine design improvement of the century, were made possible by adopting an entirely fresh approach to bearing technique.

The universally applied engineering bearing of previous tradition was a brass or bronze sleeve. This was of hefty section and bored to fit the shaft, which was pushed into the bearing during assembly. This operation required using built-up crankshafts in some racing engines. In production sports car engines and other engines that did not have built-up crankshafts, the sleeve had to be split diametrically into two halves to allow for assembly.

In either case, the bearing was located rigidly in a suitable pedestal or housing, which, in the case of the split bearing, was also split in a similar plane. The bearing material actually in contact with the shaft, and which carried the load, had to be chosen in relation not only to the loading but also to the speed of rotation of the shaft. The bronze or brass that formed these bearings was entirely suitable when it was used in conjunction with a steel shaft, moderate loading, and medium speed, and lasted indefinitely. But with increases in rpm and the use of special alloy steel shafts, hardened finishes, and so on, it was necessary to provide the bearing surface with a lining of another metal that was, in effect, cast onto the original surface in a thin layer.

The white metal of tradition, used on early "high-speed" engines, is called Babbit metal (named after the original 1839 patentee). It contains a large proportion of lead. For higher speed applications, the lead is replaced by tin. Thus, a modern white metal bearing alloy might be composed of 88 percent tin, 8 percent antimony, and 4 percent copper. This composition is still widely used in heavy-duty engines, such as diesels, that operate at lower crankshaft speeds than are usual in sports car or racing powerplants. The heavier loading is taken care of by incorporating such substances as copper, nickel, and silver.

One feature remained inherent in the designs of these older bearings and set them quite apart from modern thinwall bearing shells. The thick bearing shell itself was depended on to resist distortion, which could cause variations in the clearance between the bearing shell and the shaft. The hole in the connecting rod big end or the main bearing web was not always machined accurately; accuracy was a feature reserved for the manufacture of the bearing itself. The sloppy clearance between the main bearing web and the shell also allowed lubricating oil to escape. One quaint method of keeping the oil in, aiding heatflow, and supporting the bearing shell more uniformly was to assemble the engine with beef tallow between the bearing shell and the bearing housing. The tallow carbonized at high temperature when the engine was put in operation, thus sealing the gaps with a hard, black carbon. Needless to say, these engines did not take well to high-detergent oils!

Quite apart from the inaccurate machining affecting the bearing fit on its shaft, there is another, more important, point to consider. In a modern engine, a lot of heat is produced, and much of it has to be transmitted through the bearings to the crankcase casting, where it can be dissipated to the atmosphere. No matter how accurately any machined surfaces fit together, they still constitute a barrier to heatflow. Obviously a thick shell bearing that maintains its shape faithfully in relation to the shaft but is an indifferent fit in its housing can prejudice heat flow quite a lot.

The modern approach to bearing design, exemplified by the thinwall type shell, lies in the fact that thinwall bearings are deliberately flexible. No attempt is made to keep the bearing "round" by incorporating it with a thick base. The accuracy of the actual bearing housings (Fig. 4-16) is depended on to maintain the required clearances between the bearings and the shaft, and the bearing itself can be considered as a semiflexible construction, since its wall or backing consists of a low-carbon steel strip produced in continuous ribbon form with the surface cast onto it.

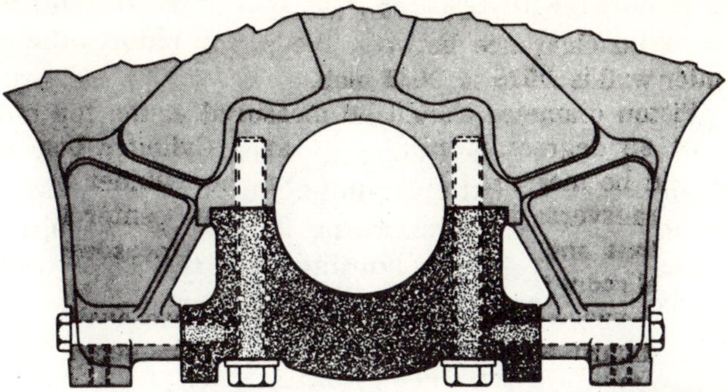

Fig. 4-16. Cross-bolted main bearing cap of Chrysler "Hemi" V8. Horizontal tie-bolts prevent bearing cap distortion that could spoil support for bearing shell.

Strip Formation

The thickness of the steel strip varies with the duty and the journal dimensions. Usually, however, it is in the region of .050 to .080 in. After the bearing surface has been cast thereto, the strip is cut to the necessary lengths and shaped to conform to the required bearing housing contour. During this operation the oil holes, the oil grooves, and the locating lugs are cut and formed. Some high-performance production engines and most supertuned production engines have pinned bearings; that is, a small hole is cut in the shell so that a pin in the bearing housing can be used to prevent "spinning" during competition use.

The bearing surface may be of white metal; its thickness will depend on the operating loading, but generally it is between .004 and .007 in. For current high compression engines, one heavy-duty lining has been much in the news in the auto enthusiast periodicals. This is the lead-bronze bearing, which consists of a thin overlay of pure lead plated onto a bronze layer that is cast onto the steel backing strip. On the lead, an indium deposit is electrolytically and thermally infused to a depth that almost reaches the steel.

This mineral, indium, is becoming associated in the mind

of the motorist with bearings that will "take it" and for good reason; it has good antifrictional qualities, reduces the susceptibility of the bronze to corrosive attack, and toughens up the lead (which is, in its normal state, fairly soft). No hand fitting is required or even desirable; the main necessity is that the "shut" limit (that is, the circumferential length when clamped in the housing) be correct. Being flexible, the strip accommodates itself exactly to the housing, and this close contact ensures maximum heat flow between the components.

Fitting Clearances

It is essential that a high standard of accuracy be achieved in both the housing bore and the shaft journal and that, in the case of the housing, the accuracy is maintained with the bearing clamped up tightly. This is one reason why align boring, with the bearing cap bolts torqued to specifications, is important in the preparation of any stock-block competition engine. In production engines, it is normal to allow .001 in. minimum diametrical clearance between a crankshaft main journal and the bearing surface, on journals up to 2 in. in diameter, and a further .0005 in. per inch of diameter above this. For big ends, the figures are .0005 in. up to 2 in. and a further .0005 in. per inch of additional diameter.

In competition engines, it is usual to use considerably wider clearances to ensure free running and minimal mechanical losses to friction. Typically a small-block Chevrolet V8 prepared for drag racing will have .0025 in. clearance at both the main and the rod bearings. Even little English Ford "fours" are sometimes set up with .003-in. main bearing clearances, when free revving to 6000 to 8000 rpm is important for road racing.

The wider clearances common in racing engines demand a good oil film. Hence, increased oil pressures and modern lubricants, perhaps with a viscosity-improving additive, are necessary. Some tuners lightly roughen the bearing shells with 400-grit sandpaper for better oil film retention. Another necessity when using wider clearances is the previously mentioned lead-indium overlay.

Normally the maximum clearances are determined by the machining limits of the houring bores and the crankshaft journals. The bearing strip thickness tolerance is always held to less than .00025 in.—regardless of size. In the event that the crankshaft is reground to a repair undersize, a correspondingly thicker bearing must be used. However, undersize grinding is quite rare in racing engine preparation. Generally the tuner prefers to have as great a bearing surface area as possible and will replace damaged crankshafts rather than regrind them to a smaller diameter.

No matter how carefully the micrometer measurements and the machining work are carried out, it is exceedingly important to check the actual bearing clearances with Plastigage.® Plastigage® is a thin plastic "string" that is placed on the journal or crankpin during engine assembly. The bearing shell and the bearing cap are then installed and the bolts torqued to specifications. This flattens out the Plastigage.® After the bearing cap and the bearing are removed, the width of the flattened Plastigage® strip can be measured (Fig. 4-17). The width corresponds accurately to the actual bearing clearance. Thus, it is possible to detect faulty bearings and make certain that an error in arithmetic will not be responsible for the destruction of an engine.

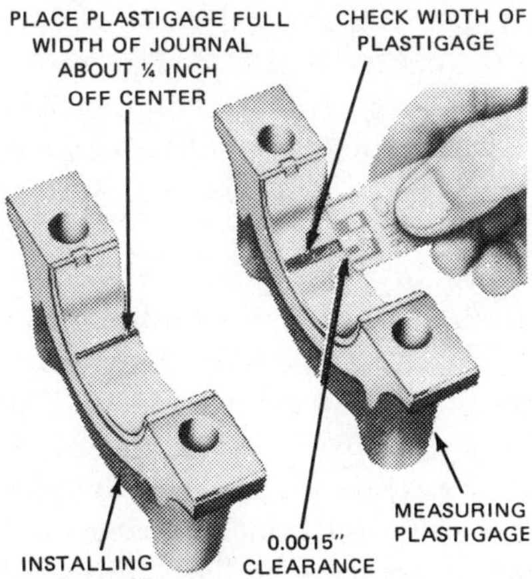

Fig. 4-17. Bearing clearance being checked with Plastigage.® At right, width of flattened Plastigage® strip is being measured.

Of course the ring-shaped, one-piece bearing shells used in air-cooled VW and Porsche engines cannot be checked with Plastigage.® Here, careful micrometer work is a necessity. It is possible, however, to check the clearance fairly well with a dial indicator, moving the bearing from side to side before installation of the crankshaft. However, because most of these engines depend on a predetermined amount of bearing "crush" to obtain the final working clearance, the dial indicator method has limited reliability. Mainly it will prevent the accidental installation of an undersize bearing or a damaged bearing. If the crankshaft cannot be easily hand turned after the crankcase halves are bolted together, something is definitely wrong, and it will pay to disassemble the engine and check the "crushed" bearings with a dial indicator to determine where the tight spot is.

Bushings

Aside from engines that have piston pins that are an interference fit in the connecting rod little ends, there is a tubular bushing installed at this point. Similar bushings are also sometimes used in rocker arm bores and as camshaft bearings. Though these bushings were once seamless pieces of solid metal that had been drilled and reamed to size, the strip process is being increasingly applied here; in this case the bushing is known as a "wrapped" or "rolled" bushing. The abutting ends of the strip are so close that it is difficult to see the seam. The bearing surface may be of white metal, lead-bronze, or other composition as required.

The importance of fitting these bushings accurately cannot be overemphasized. Piston pin bushings must be honed to precise sizes on a machine specially designed for the purpose. Haphazard fitting can be disastrous and may produce either a seized pin or a broken piston. When the piston pin is of the full-floating kind, the clearance between the pin and the piston bore must also be precision honed during engine assembly to achieve an exact clearance. Similarly the camshaft bearings must have precision clearances. If the clearance is too great, the camshaft will deflect, causing valve timing irregularities; if the clearances are too small, there is danger of seizure or galling.

Bearing Surfaces in Major Castings

If the cylinder head casting or the crankcase casting is of aluminum or magnesium alloy, with good wearing characteristics, it is possible to build an engine without camshaft bearing shells or bushings. The overhead camshafts of Volkswagen water-cooled engines have no bearing shells nor do the Datsun OHC engines.

At one time, Volkswagen air-cooled engines were made without bearing shells. However, after long service, the crankcases often wore to such a degree that the camshaft could shift from side to side and cause valve clearance and valve timing problems. Subsequently bearing inserts were incorporated in these engines.

During supertuning work, it is often desirable to remachine the castings to accept bearing shells or bushings. However, this work need be considered only if it will save an otherwise good casting or if the engine, in supertuned form, produces wear at points where wear is not a problem in the stock powerplant.

Water-cooling Jackets

Only fairly recently has it been fully realized that even a large volume of liquid, adequately cooled and circulated at a high rate, is not necessarily the most efficient way to cool an engine. First of all, the coolant must be directed to the right places so that locally heated parts of the intricate casting passages are not bypassed. Second, if the flow is too great, the coolant may not remain in contact with hot spots for a long enough period of time to absorb the heat. Nor are thick water jacket passages desirable; the flow can stagnate, boiling sometimes commencing before the heated water can be pumped out of the engine and into the radiator.

Any person who has looked closely at a cylinder head gasket will have remarked that the holes provided for the passage of water are often considerably smaller than the corresponding holes in the cylinder head and block castings. Under no circumstances should these holes in the gasket be enlarged with the intention of improving engine cooling. Quite the opposite result is likely. The restricted holes in the gasket act very much like the immersed jets of a carburetor, accurately limiting the flow of coolant to certain parts of the engine so that greater flow is available at other points.

The heat given up to the cylinder block and the head on the expansion and exhaust strokes must be of widely varying degrees; at the bottom of the cylinder bore, heating will be uniform and moderate, whereas, for example, in the neighborhood of the exhaust valve and port there is a lot of heat to be got rid of in a compact, localized area. This sort of unevenness goes on all over the casting, and when it causes distortion, rapid bore wear may occur.

The lower part of the cylinder block presents little difficulty because the shape of the affected parts is symmetrical and uncomplicated. Uniform cooling merely demands the presence of a correctly sized cooling jacket going all the way around each bore. From this point of view, the practice of joining adjacent cylinders in pairs on a line parallel to their bores (in other words, "siamesing" them), while making for rigidity, tends to block the coolant circulation. Probably this form of construction is no worse than having an inadequate space for water—especially if it is thin enough to become rust- or scale-clogged. However, the provision of a sizeable space means a longer engine; thus a sensible compromise has to be arrived at.

The real problems of avoiding cylinder distortion start at the tops of the bores. The almost-universally accepted detachable head demands, particularly on high-output engines, a large number of head bolts or studs (Fig. 4-18). These fasteners have to be anchored in the block by means of tapped bosses of large size. In the past, the temptation has been in some instances to make the water jacket outer walls as thin as possible, both for lightness and economy of material, and this often meant that the bosses had to be carried largely on the cylinder bore castings. The cylinders were thus forced to assume more than their share of the pull of the stud or bolt, with consequent cylinder distortion. Uneven spacing of the bosses and their studs sometimes made the effect even worse. To avoid this situation, most modern engines with thin-walled iron blocks have bosses that are placed farther out so that they are on the *exterior* of the block. Undeniably this gives modern blocks a somewhat "lumpy" appearance. In former times, designers seemed to be offended aesthetically by this, striving at all costs to keep the outside of the engine smooth—even if it meant compromises inside.

Fig. 4-18. One of the strongest competition engines of all time, the Jaguar "six". Head studs are evenly spaced, ample block thickness for anchorages. Though this kept the outside of the engine smooth and cylinders free of distortion, the powerplant is heavy by today's standards.

Cooling and Wear

The cylinder walls should be maintained at temperatures that will not adversely affect their strength and shape. Also, to reduce the danger of piston seizure or friction losses further, the lubricant must maintain its film strength without being so viscous that it causes undue drag. These requirements are quite simple to meet, and thus it might be questioned why, with accurately dimensioned and adequately lubricated bores, renovation of the piston/cylinder department still constitutes a major item in overhaul work.

It is frequently argued that the current high-revving, high-output engine cannot be expected to maintain dimensional accuracy in these major components for such long periods as did the older power units. If the engine has adequate lubrication and cooling, however, running wear is very little influenced by either load or piston speed. The modern competition engine operates over a wider range of rpm than of old—to the disadvantage of consistent lubrication, since at low speeds the amount of oil thrown up into the bores from the big ends may be barely sufficient, and at high speed there may be too much. The excess has to be closely controlled by the pistons' scraper rings. Only over a relatively small portion of the rpm range will be lubrication be generally correct. Not only did the old engines run for long stretches with little variation in speed, but when such changes did occur they were neither so frequent nor so rapid as today when very quick acceleration is a prime competition requirement. We are fortunate that developments in lubricating oils have managed to keep pace with other technical advances in internal combustion engine design.

Engine Block Materials

There is one other marked difference between the old and the new that has undoubtedly had a great effect on wear: materials. Though we have already touched on this matter, it seems in order to take a closer look at some of the factors involved.

It is evident that the old irons were better from the point of view of wear resistance than the modern varieties; their disadvantages lay in the necessity for thick castings (because of reluctance of the molten metal to flow freely as well as the large number of rejects from the machine shop as a result of blowholes and distortion). Inevitably this produced a weight penalty that, though it was merely a disadvantage in production cars, was completely unacceptable in racing cars.

To quite a considerable extent, the modern engine owes its light weight and compactness to the use of a cast-iron that is very free flowing and can thus be patterned in thin sections. These castings are also satisfyingly free from flaws and defects, both because of the metal's properties and because of advances in foundry techniques. Today's production demands also make it impractical for castings to be aged, or weathered, as was customary in the old days; the "green" castings sometimes were buried in sand in the foundry yard for a period of months or even years. This technique was aimed at letting the internal stresses in the metal become relieved gradually by the passage of time. Modern castings have relatively less internal stressing, and because of the metallurgy, even this could not be relaxed significantly using the old-fashioned "seasoning" method.

Modern cylinder block iron thus combines the features of light weight, ready casting, and machinability without rejects with a rate of bore wear that is generally acceptable. The wear tends to decrease with increased mileage (especially after the initial indication of wear, increased oil consumption, has taken place). This is probably because larger amounts of oil are passing the pistons.

An excellent method of ensuring long-lived and accurate cylinder bores is to make them of material selected to meet the specific requirements of the bore itself. The material does not have to fulfill the needs for cylinder block construction; usually it takes the form of a tubular liner, a centrifugal casting of an iron alloy that has been developed for its properties as a bearing surface.

A typical alloy has a pearlitic structure combined with a hard phosphide "network." The latter actually forms the bearing surface, while the graphite content in the interstices of the alloy constitutes a minute honeycomb of oil reservoirs that assists lubrication by absorbing the engine oil into the surface. The surface is tough rather than excessively hard. The material, somewhat malleable, soon takes on naturally a working "skin" or "glaze" more effective than any hard-deposited finish at resisting wear.

Liners are of two basic kinds: wet and dry. The latter has a thin wall, usually about .100 in., so that the bore finished to receive it must be machined with a high degree of accuracy in order to avoid distorting the liner. As in the case of a bearing, heat flow will be adversely affected by any lack of uniformity in the fit between the back of the liner and the bore.

The liners are in some engines a press-fit, in which case they become virtually part of the block. A pressure of from 2 to 5 tons is usually employed for assembly, and the liner is given its final finish by boring, grinding, or honing after it is in position. When renovation is subsequently called for, the worn liner is pressed out, and a new one is fitted as before. Thus, though the process is fairly exacting, it probably represents the most efficient assembly for dry liners in regard to rigidity, heat flow, and bore accuracy.

Dry liners can also be of the "slip-fit" kind, requiring no press for installation or removal. These obviously make for ease of servicing, but their disadvantage compared with wet liners will be evident. In any case, dry liners are rarely used in present-day competition engines.

Wet liners, which are common in racing engines with light alloy blocks, take the place of the complete cylinder wall; the cooling water circulates around the outside of the liner, which is of thick-walled construction. There are very definite advantages in this kind of liner, notably the ease of renewal, facility of access to the water jackets, and a machined liner surface in contact with the water, which makes for efficient and uniform cooling. Disadvantages are mainly the fact that cylinder head support must come wholly from the block walls (the bores no longer help in this respect) and that a bottom water joint is required on the liner—failure of which allows coolant to enter the crankcase.

Another weakness has made itself known in modern grand

prix engines that attain very high rpm. This is the tendency for the cylinder liners to follow the pistons. The problem shows up as coolant leakage. Thus it is often mistakenly diagnosed as a sealing problem. The sealing arrangement may be redesigned or modified several times—with no resulting cure for the illness—before the designer discovers that the real problem is inadequate support to keep the cylinders from moving up and down together with the pistons.

Wet liners are prefinished (Fig. 4-19) and made of similar materials to the dry liners. At the top, the method of sealing is usually similar to that used with slip-fit dry liners: a flange located in a counterbore, sometimes incorporating an additional sealing washer at the lower face of the flange. If this lower seal is too easily compressed, it can allow cylinder end-travel. In racing engines, the Cosworth-Ford DFV, for example, there is a groove machined around the top end of each liner. This groove accepts a metallic O-ring, which serves as the head gasket for the cylinder.

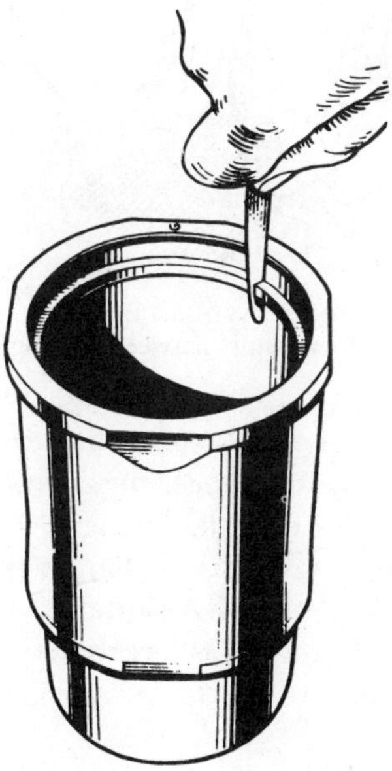

Fig. 4-19. Piston ring fit being checked in prefinished wet-type cylinder liner. Liner shown has flange for sealing gasket near lower end. Upper flange fits in recess in engine block.

There are variations in the method of locating and sealing the bottom end of a wet liner. When the liner lengthens from heat expansion, it must be free to slide. Thus, the well-tried method of annular grooves, shown in Fig. 4-20, is probably most often used. In some cases, however, the grooves are machined into the block instead of into the cylinder liner. In these grooves, rubber or synthetic rings are fitted, which act somewhat in the manner of piston rings, forming the seal between the water jacket and the crankcase.

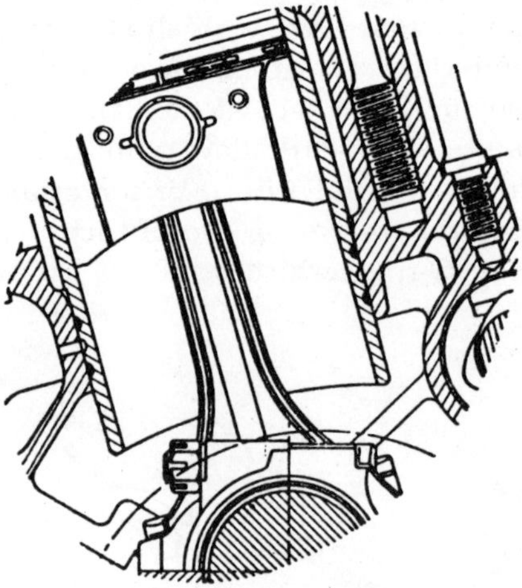

Fig. 4-20. Cylinder liner bottom seal on BMW engine. Notice drain hole to outside between sealing rings.

Where relatively short strokes are concerned, with fairly short liners being a consequence, heat expansion is not a problem, and in this case, a completely rigid location is provided at the bottom. This may comprise a flange that seats on a corresponding flange in the block. A suitable composition or metal washer is interposed; one washer sometimes serves a pair of cylinders and is called a "spectacle" washer (after its characteristic shape).

Both slip-fit and wet liners are easily renewable and are an undoubted aid to quick overhauls and restoration to full efficiency of the engine; however, they also constitute another "loose" part. Consequently they are used almost exclusively in light-alloy blocks today, iron blocks generally having the cylinders as an integral part. Integral cylinders can more easily be machined with accuracy *vis-à-vis* all other block dimensions, such as the deck height and the crankshaft centerline. Thus, to obtain equal precision in an engine with liners, it is essential that high standards of accuracy be maintained—especially to avoid alignment errors in relation to the crankshaft. This is just as important as ensuring that coolant and gas sealing are done properly.

Needless to say, all of this sealing complexity and machining precision adds up to greater manufacturing costs, which explains why cylinder liners are now reserved almost exclusively for use in expensive racing engines. Also, the water jacket is a notoriously messy location for precision-machined surfaces, which are liable to collect much rust and hard deposit in normal highway service. Any such defects or corrosion can affect the accuracy of the fit and must quickly be remedied to avoid extensive engine damage.

Surface Finishes

Apart from the development of special irons with the object of reducing wear from the various causes stated earlier, some manufacturers have introduced special surface finishes on the bore. Chromium plating has been tried extensively, both on iron surfaces and, in the case of some Porsche engines, on aluminum surfaces. Chrome has been applied both as an overall finish and also as a finish for the piston ring contact area only, on the premise that most wear takes place in this region. Chromium is extremely hard and in addition strongly resists any tendency for the shedding of small particles of metal by the action of the piston rings. This characteristic is not uncommon in orthodox bores and results in continuous exposure of fresh metal (shorn of the hard skin or glaze formed by usage), which is readily corroded.

It is not possible to use a normal polished chromium-plated finish, however; it is virtually nonporous and would thus resist oil retention; the final surface must be etched for this purpose, and it is then still possible for corrosive influences to get through the top surface to the iron or the aluminum underneath. Corrosive undermining of this kind can cause the chrome plate to flake off, spoiling the bore. Nevertheless, there is little doubt that chromium plating implies a very high-class design and is the subject of continuous investigation in relation to this application.

A unique, and controversial, surface finish is that first introduced in production cars on the Chevrolet Vega—though it had previously been used in aluminum Chevrolet blocks used as the basis for supertuned engines in Can-Am racing. With this design, the aluminum block has integral cylinders—that is, the bores are aluminum without the customary iron liners or chrome plating. The aluminum alloy, however, has a high silicon content, similar to alloys used for pistons. After the cylinders are machined, the surfaces of the cylinder bores are etched to remove aluminum particles—leaving the glass-hard silicon particles exposed. Some contact with aluminum is, of course, unavoidable, so the pistons are iron plated to prevent the galling that would be caused by aluminum-to-aluminum contact.

Because the iron plating on the pistons and the microscopically thick silicon layer on the bore can easily be destroyed by poor lubrication or metal-to-metal contact induced by overheating, there have been servicing problems. These center around the difficulty of restoring the silicon surfaces to the bores in an ordinary repair shop. Nevertheless, the system is unsurpassed for light weight and good heat transfer and will undoubtedly be used in some future competition engines.

Cast-iron bores have microscopic, graphite-filled pores in them that absorb oil for good lubrication and wearing characteristics. This should be kept in mind when cylinders are being bored and honed for racing. The pores in the iron can also absorb the kerosene or other lubricant used on the hone, together with a considerable amount of fine abrasive dust from the stones. If the abrasive is not removed, the piston rings will quickly be ruined when the engine is placed in service. Most tuners have found that nothing surpasses strong detergent and water for this purpose, and these are therefore the cleaning agents most commonly used.

Pistons and Rings

Aluminum alloy is the universal piston material for high-efficiency automobile engines; the main requirements of the material are a low expansion rate with heat and the minimum loss of strength from the same cause. It is interesting to note that with few exceptions—high speed racing engines—weight is not a very important consideration because there is a continuous tendency in operation for the piston inertia to be (to a greater or lesser extent) balanced by the various gas pressures. Obviously an alloy piston is usually lighter than the equivalent type made from iron, but the great advantage is in heat flow.

Very high temperatures are reached at the piston crown centers, and the heat must be conducted away entirely through the piston body. Some is transferred, via the rings and the piston skirt, to the cylinder walls and thence to the coolant or, in an air-cooled engine, directly to the atmosphere. The remaining heat is dissipated to the underside of the piston and thence to the lubricant or through the piston pin bosses and the pin to the connecting rod.

In all cases, rapid conduction through the piston material is of paramount importance. The modern "low-expansion" alloys have an advantage apart from the one indicated by their designation; the material lends itself to rapid production by inexpensive diecasting. These castings are clean and uniform in both weight and dimensions. But for most competition purposes, forged aluminum pistons are used whenever the rules permit it and the stresses demand it.

The technique of surface finish for pistons varies. Some pistons are finished with fine turning done with a diamond-tipped tool, which cuts through the silicon particles contained in the alloy instead of gouging them out. Other makers prefer grinding. It is evident, however, that in either case a highly polished appearance is no guide to the degree of actual "roughness", that is, the good-wearing silicon particles exposed. Most forged racing pistons are, in fact, highly polished, with grooves turned every half-inch or so along the skirt for oil retention.

To assist break-in, the piston skirts of production engines are sometimes tin plated by a dipping process. Another technique is anodizing to achieve a hard skin on the skirt surface; the resulting skin has a matte finish, which retains lubricant and, if applied also to the ring grooves, reduces wear of the grooves and the sides of the rings.

Ring Wear

Very little wear takes place on the piston skirt; the rings and the bores wear because of the high bearing pressure of the former against the cylinder walls. The bearing pressure is produced largely by gas pressure, and where there is considerable clearance between the rings and their grooves, gas can easily pass behind the rings and thus expand them. This gas-induced pressure is added to the radial pressure produced by the springiness of the ring. It is neither possible nor desirable to eliminate this extra pressure fully since it aids the sealing of the ring. The designer's aim is to maintain a uniform average pressure over the four strokes of the operating cycle, thereby avoiding ring flutter or vibration. The use of special ring section profiles enables a balance to be struck between the pressure exerted by the bearing face of the ring on the cylinder wall and the gas pressure in the cylinder in a manner that is not normally feasible with a conventional square or oblong rectangular-sectioned ring—hence, the popularity of tapered or L-sections on modern rings.

The head land compression ring and the Dykes ring are often seen in racing engines. These rings have a deep L-section that, in the case of the head land ring, extends upward all the way to the top of the piston, thus replacing completely the piston's head land. Expanding gases force the ring outward against the cylinder wall. At the same time the ring prevents nearly all blow-by and thus protects against localized heating and melting, which often begins at the head land of a conventional piston. Both head land rings and Dykes rings (which do not extend all the way up to the piston top) prevent piston ring flutter, a condition induced by piston deceleration and acceleration in combination with gas ingress behind the ring.

Ring drag can be reduced by the application of a low-friction material to the rings' bearing surfaces. Chrome plating has been widely used for many years. It offers a low coefficient of friction and good ring life but sometimes accelerates bore wear. More recently, Teflon-faced piston rings have been tried. At present, the verdict is not in on these when used in competition engines. However, speed tuners have been using them for certain applications with good results.

The quiet running of modern production car engines—freedom from what is commonly called "piston slap"—results not only from improved materials but also from ingenuity in design. The split skirt has, of course, been used for a long time; the vertical split, which is located on the nonthrust side, allows a very close clearance to be maintained with the cylinder wall without risk of seizure.

Another popular method of reducing piston slap is to offset the piston pin bosses by a small amount toward the side of the bore that is "behind" the direction of rotation of the engine (Fig. 4-21). The result of this is to make the piston change sides at the dead centers in two movements instead of one, thus cutting down both noise and wear. However, speed tuners sometimes install these pistons the wrong way around, thus creating pronounced piston slap noises but giving the connecting rod a more direct "push" against the crankpin. In any case, the piston clearances in competition engines are usually so wide—about .001 in. for every inch of bore diameter—that piston slap noise is normal, at least while the engine is cold.

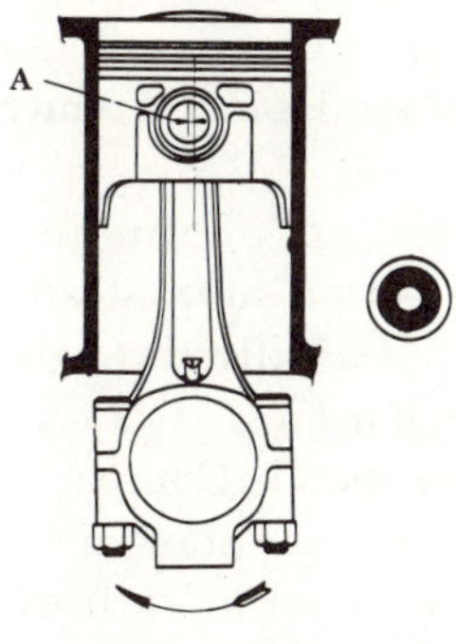

Fig. 4-21. Offset piston pin (A). Offset is arranged to keep connecting rod straighter on the upstroke, thus minimizing sidethrust during compression but increasing sidethrust under power.

Heat Flow

The piston skirt has always formed a major element in heat conduction, but a properly dimensioned piston of modern type can also pass heat from a point much nearer the crown (that is, from the lands between the rings). Even the topmost land, or head land, can be made to conduct some heat away, though this is a notorious danger point for overheating. The use of the lands as bearing surfaces in this way has the added advantage of preventing the buildup of deposits and "varnish", which formerly tended toward more rapid wear and in bad cases promoted seizures.

Of course the necessary piston clearance is very slight if much heat is to be dissipated through the ring lands; consequently competition engines cannot profit very much in this way because they demand wide clearances for free running. Nevertheless, a further investigation of the subject is in order because it does apply to showroom stock and strictly stock racing and can be employed in other kinds of engines if piston overheating is a weak spot in the design.

Obviously, when the designer intends to use virtually the whole of the piston exterior in contact with the cylinder wall, the dimensions have to be calculated most carefully in relation to the differing temperatures. There are many ingenious methods for doing this. One is to insert small plugs of metal, which melt at predetermined temperatures, into the piston body at various points. These are only suitable for measuring surface temperatures, and the available increments of melting points are limited.

A better method consists of running the piston in an engine for an extended period at a steady load. The piston is then removed and split vertically through the piston pin bosses, and a large number of hardness tests can be made all over the surface of the sectioned metal. Comparison of the actual measured hardness with a known standard for the metal enables the local temperature attained at each of the measuring points to be assessed. Apart from the complete picture obtained of the temperature gradient throughout the piston, there is also the advantage that this applies to the full depth of the metal.

This is much more important than knowing what is happening on the surface. Experts consider a 50-hour run the shortest period in which the piston can obtain a thorough heat "soaking" in making this test.

While piston manufacturers have made enormous strides in ensuring good heat dissipation without distortion, still the heat applied in running is not transmitted uniformly to the piston crown, as might be thought at first consideration. Immediately below the exhaust valve, for example, the temperature is about 15°C higher than the average; below the intake valve there will be some cooling effect. Thus, the piston crown must be designed to even out these temperatures as much as possible and should be of such thickness that the overall temperature is kept within bounds. Both of these objectives can be achieved by the presence of sufficient material in the right place.

There are other ways of controlling the piston expansion to bring it more or less into uniform close contact with the cylinder bore. One of these, which is applied to air-cooled motorcycle engines, consists of several turns of steel wire that are wound in a groove located below the bottom ring and above the piston pin bosses. The wire winding prevents diametrical increase beyond a definite amount.

High-performance Engines

Piston noise in sports car and racing engines is secondary in importance to freedom from friction and reliability under the most strenuous conditions of sustained high power output. Thus, solid skirt pistons are always used, with the weight reduced as much as possible. In some designs, crown rigidity and increased heat conductivity are achieved by the provisions of ribs or webs across the underside. Though the diecast pistons used in many sports car and high-performance passenger car engines are entirely capable of competition service when the powerplant is left relatively stock, forged pistons are generally considered essential if the engine is supertuned.

The use of multigrade oils has called for further development of the oil control arrangements. Too high a radial pressure on the oil scraper (or oil control) ring is undesirable since the extra friction would nullify the advantages gained by using this kind of lubricant. Here again, special ring sections have been the answer; the actual method of transferring the oil from the ring to the inside of the piston, via drain holes, is still being used. Sometimes the lowest compression ring is also called on to assist in the oil scraping action. The old practice of fitting an additional scraper ring below the piston pin is virtually obsolete, though pistons of this kind work well as replacements in engines with worn bores.

The hollow casehardened steel piston pin is universal, but several means of fitting and retaining this vital part are common. Full-floating pins, which are free to oscillate in both the little end of the connecting rod and in the piston bosses, are almost invariably retained by circlips in the latter. So long as these are a precision fit and correctly installed, the method is completely safe. Nevertheless, circlips have come loose, causing serious bore damage. Hence, some full-floating pins have plugs of aluminum or a low-friction material (such as Teflon) inserted into their ends to keep the pins from contact with the bore. The plugs have been used both with and without accompanying circlips.

The old design whereby the piston pin is clamped firmly in the connecting rod by a bolt has disappeared. However, some engines have interference-fit piston pins that are a press-fit in the connecting rod's little end. One outstanding merit of the interference-fit pin is that its performance as a bearing that oscillates only in the piston bosses is predictable. In contrast the precise movement of a full-floating pin is difficult to ascertain. Obviously the danger of circlip failure is also eliminated. However, care must be taken during engine assembly to ensure that the pin has an adequately tight fit in the connecting rod bore, which, of course, has no bushing.

As in production engines, the piston rings used in competition engines are of cast-iron. A fine-grained alloy is used. It is recognized as being superior to any other material with regard to wear resistance, heat conductivity, and accurate seating.

Two or three compression rings are usually employed; chromium-plating is applied to the top one or two rings. Though chrome rings have reputedly caused bore wear in some engines, they generally act to reduce bore wear—providing lubrication is adequate in both quantity and quality, as well as in viscosity. The chromium-plated ring cuts down wear through its noncorrosion properties when the ring is at rest—an especially important consideration when nitro fuels are used. Furthermore, because the plating is too hard to pick up abrasive particles, chrome rings are not so likely to "lap" the cylinders oversize. With suitable cast-iron used in the cylinders, the chrome will actually help to burnish the bore into an extremely wear-resistant surface.

Ring Pressure

The distribution of an even pressure against the cylinder bore all around the ring periphery was formerly obtained by a hammering operation applied to the inside, or back, of the ring. Evidence of the process could be seen on the finished article. This method is still used but through the application of a more modern technique. It involves the use of a precision former, which contacts the ring while it is undergoing heat treatment. This method eliminates any possibility of local weaknesses of the kind that sometimes resulted from old-fashioned hammering.

Taking advantage of experience gained with heavy-duty supercharged aircraft engines, the rings now available for competition engines differ in several respects from the normal article so it is essential to choose carefully. A material known as HG 22, which contains chromium and vanadium, is employed by one maker; the rings are centrifugally cast, resulting in a chilled condition. Therefore, a subsequent annealing treatment is applied. The particular properties of the finished rings, which make them suitable for strenuous work, are extra wear resistance, high strength against breakage, and increased elasticity that ensures good sealing properties.

Connecting Rods

Aside from their use in very high-revving racing engines, forged aluminum connecting rods are more likely to be used in drag racing engines and sprint racers than in engines that will be used in long-distance events. In the tuner's mind, there is always the possibility of fatigue failure. Consequently the manufacturers of the forged aluminum rods build in as great an overstrength margin as possible, and speed tuners discard any rod that shows the least sign of weakening.

The normal check is to measure the length of each rod. If any rod has "stretched" by .001 in. or more, it is discarded—even if it means discarding an entire matched set of rods that has cost nearly $300. Of course, there is not much danger of failure in a drag racing powerplant because the race times are measured in seconds, and the more highly modified engines are routinely torn down for inspection after each event. Keeping an eye on forged aluminum rods is just one of many reasons why competition engines are disassembled after every race or after a certain number of hours in competition.

At one time, aluminum alloy rods were favored for another reason. They could be run without any big end bearing, provided that the alloy was suitably wear resistant. Eliminating the big end bearing was a much more attractive design feature in the days that preceded the development of the thinwall bearing. Bearingless aluminum alloy rods are not seen much today except in small engines for Karts, in outboard motors, and in some motorcycle engines—the latter two powerplants being frequently used in SCCA D Sports Racing cars.

Production car engines generally have steel rods forged by a hot stamping process—although a few engines now have rods cast from nodular (or ductile) iron. Apart from machining the bores at the big and the little ends, it is not usual to finish the body of the rod—except in the case of a few expensive and powerful GT cars. When steel rods are used in racing engines, they are always highly polished. If the rules permit it, the polishing of steel rods is also practiced in the preparation of blueprinted and supertuned competition engines. This polishing helps to reduce the danger of fatigue failure by eliminating stress-raising imperfections.

Connecting rod big end bolts are one of the most critical points in an engine (Fig. 4-22). The bolts must be of extremely high tensile strength and highly resistant to weakening caused by fatigue or crystallization. The tightening torque applied to the connecting rod bolt or nut during engine assembly must place a tension on the bolt that is at least equal to the loads imposed on it by engine operation. The bolt must have adequate tensile strength to resist stretching at this torque.

Fig. 4-22. **Damage to piston and connecting rod as a consequence of big end bolt failure.**

Let us assume that the tension placed on the big end bolts during engine operation is 300 psi. If the tightening torque puts a tension of only 250 psi on the bolt, the bolt will be subjected to a fluctuating load of 50 psi with every revolution of the crankshaft. This fluctuating load will eventually cause fatigue failure in the bolt. If, however, torquing places a tension of 310 psi on the bolt, then the tension will remain constant during engine operation, because it exceeds the tension placed on the bolt by running the engine.

This does not mean that it is good to overtorque the connecting rod big end bolts routinely. A classic example is the Type 2/Type 4 VW air-cooled engine that is now an almost unbeatable powerplant for use in oval track midget racers. When these engines were first tried, there were many connecting rod failures. Subsequently it was discovered that the tuners were overtorquing the big end bolts with the intention of ensuring greater resistance to the extra loading imposed by racing. However, the tension placed on the bolts by overtorquing was in itself sufficient to weaken them. When the bolts were torqued to normal VW specifications, the failures ceased to occur.

Connecting rod big ends have come in for some unusual design treatments in several popular British sports cars. Because these cars are frequently raced, it might be well to touch on these oddities here. For example, to facilitate withdrawing the rods and pistons together (upward out of the cylinders after removing the head), many makers have resorted to splitting the big ends diagonally instead of straight across. The object, of course, is to reduce in effect the overall width of the big end so that it will pass through the small bore of a long-stroke, small displacement engine.

Another minor aberration applied to the big end, in the interests of standardization of engine types and economy of material, is the practice of offsetting the big end as a whole axially, as shown in Fig. 4-23. This allows the cylinder spacing to be reconciled with the crankpin spacing *vis-à-vis* the main bearing and balance weight spacings along the length of the engine. On the Volvo pushrod 4-cylinder engine, for example, the bores were enlarged and the bore centers spaced farther apart for increased displacement; using axially-offset big ends, it was possible to retain the original crankshaft and thus to avoid expensive retooling. Obviously if carried to absurd lengths, such design would be reprehensible because all the loading would be taken by one end of the bearing shell.

Little end construction has already been touched upon in our discussion of piston pins. Rods for use with full-floating piston pins have pressed-in bushings, whereas rods for use with interference-fit piston pins have no bushings but nevertheless must be machined precisely.

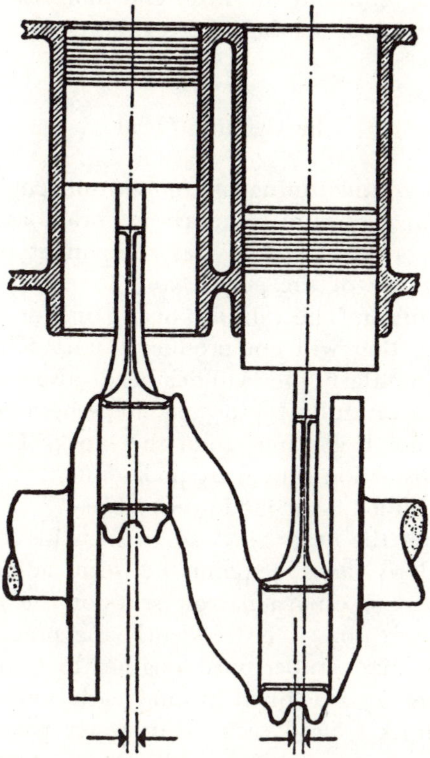

Fig. 4-23. Offset big end bearings to obtain required dimensional proportions on an engine that was originally designed for smaller bore, more closely spaced cylinders.

It is usual to have, on some portion of the connecting rod forging, a point or a block of surplus metal, some of which can be ground off for balancing the connecting rod. This often consists of a small projection above the little end boss and, not uncommonly, a similar projection on the big end bearing cap or on the sides of the big end, above the bosses for the bolts. During competition engine preparation, the connecting rods are balanced not only against one another—so that all of the rods in the engine are each of identical weight—but also so that the weight of the big end and the little end matches the little end and big end weights of all the other connecting rods.

The Cylinder Head

In the design and tuning of competition engines, no component comes in for quite so much attention as the cylinder head. The proportions of this vital component determine the volumetric efficiency of the engine and have a great influence on the power output. The cylinder head studs or bolts must be located in a way that will not produce localized stresses on the cylinder block. Similarly the cylinder head—also a complex casting—must have a carefully engineered shape because it is subject to an even greater heat range than the block. There is always the danger of distortion caused by poor design or by an unwise approach to cylinder head modifications.

Distortion of the head can lead to such ills as head gasket failure—particularly at the point between adjacent cylinder bores where there is only a narrow seat for the gasket to seal against. For this reason, it is frequently the practice on racing engines and on many supertuned engines to O-ring the head. Circular grooves are machined around each combustion chamber and sometimes around each oil or water passage; synthetic or metallic O-rings are installed in the grooves so that they will be compressed into highly pressure-resistant gaskets when the cylinder head is bolted down.

Distortion can also cause warping of the valve seats, particularly the exhaust seats, resulting in leakage and, ultimately, loss of compression as the valve becomes gas-cut or the seat becomes eroded. Local overheating can cause the seats or the head itself to crack. Another problem is distortion of the valve guides, which may cause the valves to stick. Severe damage can occur if a valve sticks open and is subsequently struck by the piston (Fig. 4-24).

Fig. 4-24. Rocker arm broken by piston striking valve. Rocker arm shafts may also break, and obviously valve can be bent or broken.

Liquid-cooled Cylinder Heads

Apart from ensuring a good flow of liquid coolant through the head and ample passages for its circulation, it is important that the flow be directed to the backs of the exhaust valve ports and around the valve guides; a major source of overheating trouble is thus eliminated. With improved exhaust valve cooling, it may be possible to use a higher compression ratio without preignition or detonation taking place, and through this change more power will be obtained.

It can be concluded from examining many old engines that water jacket design was very much of an afterthought. In some cases, a sheet metal tube, with holes bored into it at intervals, had to be inserted into the head casting to direct the coolant to the exhaust valve areas. Today most such considerations are taken care of entirely by the design of the casting itself.

Air-cooled Cylinder Heads

Air-cooled cylinder heads have given their designers many problems involving distortion and localized stresses because it is very difficult to maintain even heat at all parts of the combustion chamber area using air as the cooling medium. The coolant in a liquid-cooled engine tends to even out the temperatures somewhat, making the designer's task a bit easier.

The depth and the spacing of the cooling fins must be finely calculated to carry the heat away in proportion to the heat produced—that is, the fins must generally be deeper near the exhaust valve and the exhaust port and proportionally smaller at cooler parts of the head. In this way, too much heat will not be conducted from "cold" places, which could cause as great localized temperature difference as would the inadequate cooling of the exhaust area.

Aluminum is used for the cylinder heads of all present air-cooled engines. The valve seats must always be shrink-fitted or cast in inserts of some hard material, such as chrome molybdenum steel. The valve guides may be either of brass, bronze, or sintered iron and must be adequately cooled to prevent valve sticking, galling, or seizure. The range of operating temperatures in an air-cooled head is much wider than in a liquid-cooled head; this is because liquid coolants tend to retain heat, whereas air is always being obtained in an unheated state direct from the atmosphere. At idle speeds, an air-cooled engine cools down rapidly, yet at high speeds the temperatures quickly rise above the levels that are commonly encountered in liquid-cooled engines. These wide, and sometimes rapid, changes in heating and cooling can create considerable problems for the designer or tuner.

Valve Seats

The vast majority of automobile engines with cast-iron cylinder heads have the valve seats ground right into the cylinder head material. This design has been made possible by the use of modern alloys and, to a lesser extent, by the "lubricating" qualities of the tetraethyl lead that is commonly used as an agent for increasing fuel octane. Engines with aluminum alloy heads must have valve seat inserts (Fig. 4-25). These are rings of hard material that are either placed in the mold, so that the cylinder head is cast around them, or are precision fitted to the cylinder head at a later stage of manufacture. Inserts can also be installed in iron heads that did not originally have them.

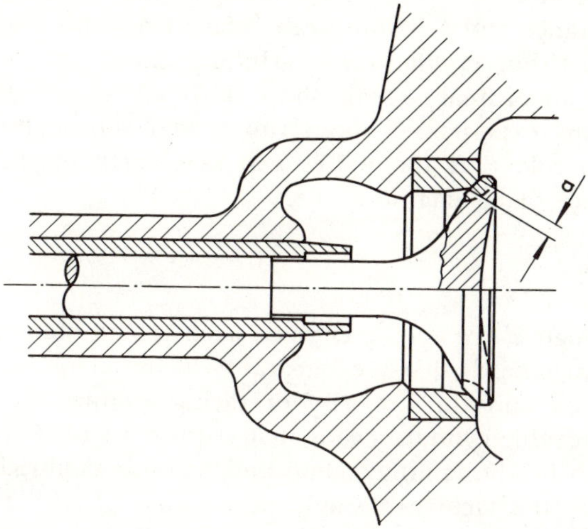

Fig. 4-25. Cross-section of VW 1200 cylinder head, showing valve seat insert that is set into aluminum head casting. Valve contact area is at *a*.

Cast-iron inserts, which are the least expensive kind, should be used for intake valves only—and then exclusively for the repair of cast-iron heads that have developed defective valve seats. Molybdenum alloy inserts are usually used for the exhaust seats in engines not originally equipped with inserts. Chrome nickel seats are installed in many aluminum heads, and chrome cobalt seats are a standard material for heavy-duty truck engines. However, the best seats for use in competition engines are chrome molybdenum inserts.

A chrome molybdenum insert costs at least three times what a common cast-iron insert costs, and these seats will be unaffected by nitro fuels and will show no increased wear when used with fuels that do not contain lead. The chrome content makes the material more corrosion resistant, and molybdenum steels have an exceptionally low coefficient of friction, which minimizes the need for lubrication.

Few factors are more important in competition engine preparation than the precision grinding of the valve seats. A lot of performance and reliability can derive from this work alone, and competition engine valve grinding should never be entrusted to an ordinary repair shop. This job should always be done by an experienced speed tuner who thoroughly understands the role that the valves and seats play in the overall performance of an engine.

Gaskets

Although a few racing engines, such as the immortal Offy, have cylinder heads that are integral with the block, all production engines and the majority of racing engines employ separate, removable cylinder heads that require a gasket seal. Head gaskets can be the familiar metal and asbestos sandwiches often installed at the factory, a single thin embossed steel sheet—the so-called shim gasket in varied form—or O-rings pressed into grooves machined around the cylinders. In any case, the gasket must deform to fill every imperfection in the mating surfaces as the head is bolted down and still retain enough resiliency to expand as the load is relieved—thereby maintaining firm contact with both metal surfaces during engine operation.

By meeting the above requirements, the modern head gasket has made possible not only removable heads on racing engines but lighter engines with thinner, more flexible cylinder heads. When correctly installed, a well-designed, top-quality modern gasket is almost impossible to "blow." O-ringing is resorted to only in racing engines, highly supertuned, production-based engines, and in those few production engines that have more or less chronic head gasket problems when raced because of their fundamental design.

"Sandwich" type gaskets are not all the same. Some are merely copper or soft annealed steel sheets separated by asbestos millboard. These are now outdated, although they were standard equipment on the famous old Ford flathead V8. Gaskets with copper on one side and steel on the other are usually meant to be installed copper-side-up, but there are some exceptions to this (such as the Dodge "Slant Six") so it is best to check the engine maker's specifications—or look for "TOP" or "OBEN" stamped on one side of the gasket.

A careful look at the asbestos "meat" in most of the present-day sandwich gaskets will show that there is metal embedded in the asbestos. This is actually a sheet of perforated steel that has raised "prongs" projecting to the surface of the asbestos itself. The most modern gaskets of this kind have a core composed of rubber and asbestos, in which is embedded the previously mentioned perforated steel sheet. The rubber/asbestos material is recognizable by its dark gray color. A similar type is the "composition" gasket, which has a dimpled steel sheet bonded to another similar sheet by the rubber and asbestos mixture.

Embossed steel gaskets were especially developed for modern high-compression engines. These precision-formed seals rely on the predetermined heights of their embossed ridges to give the proper density for sealing and heat transfer. Asbestos-core sandwich gaskets act as a thermal insulator and interfere with efficient heat exchange between the head and the block. The embossed steel gasket allows such an even transfer of heat that the effect is almost that of a lapped joint—a joint machined so perfectly that no gasket is required. Embossed steel gaskets are used in everything from stock Chevrolet V8s and Ford Cortina GT engines to Grand Prix racers; they are actually a more sophisticated form of the soft copper shim gaskets originated by hot rodders many years ago.

From the heat transfer point of view, O-ringing comes closer to the ideal because it permits the head and the block to be in actual contact. The only engines using a lapped joint, without any gasket or O-rings at all, are the VW and Porsche air-cooled powerplants. The sealing problems are less complex on these engines because there are no liquid coolant passages or oil passages to worry about.

Exhaust Valve Cooling

The exhaust valve has its greatest opportunity to cool off

during the times when it is in contact with its seat, and so heat conduction from the seat is a high-priority objective of cylinder head design. At the same time, however, the valve can advantageously lose quite a lot of heat continuously by conduction from the valve head through its stem, and thence to the valve guide and to the cooling medium, be it air or liquid.

In a majority of engines, the valve guide is pressed into a hole drilled in a cast boss, involving a considerable thickness of metal for the heat path. To shorten the heat path, Ford, in particular, has long made its engines without valve guides; the guides are an integral part of the cylinder head casting. A few relatively expensive engines have the pressed-in guide itself actually in contact with the coolant on its outer surface, which is obviously an excellent scheme from the point of view of getting rid of the heat from the valve stem. But a precise fitting of the guide is necessary, with coolant-sealing arrangements at the points where the guide passes through the walls of the water jacket.

Another approach to exhaust valve cooling, originally derived from aircraft practice, is to make the exhaust valve hollow and fill its stem and a part of the head with sodium. The sodium becomes liquid at operating temperatures and greatly assists in the transfer of heat from the exhaust valve head to the valve guide via the valve stem. These valves are at present found in some Alfa Romeo engines and in some Volkswagen Type 2 and Porsche 914 air-cooled engines.

Air-cooled Cylinders

The problems of design associated with air cooling are quite as formidable as those of the liquid-cooled engine. Most of the problems have long since been solved, with two possible exceptions, neither of which is of any concern in racing. The first of these is the matter of engine noise because there is no muffling effect from a surrounding water jacket. The other problem is with emission control. The previously discussed tendency for air-cooled engines to run cold at low speeds tends to produce an abundance of noxious substances in city traffic.

When heat is transferred from a surface such as an air-cooled cylinder and cylinder head, four times the weight of air is needed for each degree of temperature reduction compared with water. In terms of volume, the air flow has to be about 4000 times as great as the coolant flow in a liquid-cooled engine to obtain the same effect. The problems are somewhat compounded by multiple cylinders, and the motorcycle practice of merely sticking the engine out in the airstream is far from satisfactory—especially considering the result of having the car stand idling in traffic.

When the cooling fins are supplied with air from a blower fan, the depth and spacing have to be carefully dimensioned. Wide spacing will give an easy air flow and thus less back pressure on the fan delivery; closer spacing extracts more heat but requires greater power to force the air through. A nice balance must be struck, bearing in mind at the same time the question of noise caused by the fan, plus the rumble of air through the ducting.

It has long been accepted that complete separation of the cylinder barrels is necessary so that the air can be circulated completely around them. Cylinder heads, however, are amenable to casting *en bloc,* the Volkswagen engines having one-piece heads for each of its cylinder pairs. A possible disadvantage of this arrangement could be in the thermal expansion stresses set up if one cylinder ceases to fire for a long period; this might happen unknown to the driver when the engine has a large number of cylinders. Consequently Porsche has used more or less seperate heads for each cylinder on some of its six-, eight-, twelve-, and sixteen-cylinder air-cooled engines.

Because the cylinders must be separate, air-cooled engines have a design advantage in that the cylinder material can easily be selected for wearing qualities, whereas the crankcase and the cylinder head (s) can be of a lightweight material. Another advantage is that individual cylinder and piston assemblies can be replaced separately if they become damaged. Though Porsche has used aluminum cylinders with chromium-plated bores, the normal practice is to use cast-iron cylinders.

Mechanical details must be closely related to the cooling system requirements, but there is much experience available concerning aircraft and motorcycle engines. Such details as cylinder bolt, cylinder and cylinder head anchorages and stud bosses, valve seat inserts, and so on are liable to differ considerably from the liquid-cooled counterparts; even with careful control of the fan output, temperature changes generally take place more rapidly with changes in load. This must be considered in its effect on the expansion rates of different materials, both to maintain working clearances to the closeness required for efficiency and quiet running and to ensure that high-pressure joints do not work to an extent that renders them liable to "blow".

The problem of piston seizure has been a particular concern for designers who have used air-cooled two-stroke motorcycle engines SCCA D Sports Racing cars. High-revving two-strokes have always been prone to piston seizure; although this trouble has been largely overcome—at least when the bike is kept moving—placing the same engine inside an aerodynamic sports racing car body requires that engine cooling be a prime consideration in the design of the car. It is usually not possible to obtain the same cooling efficiency with a blower fan that the engine obtains when mounted in the open air on a motorcycle. Consequently ducted air is generally used to cool these engines in racing cars.

Assuming that the cylinders and the cylinder heads have been designed and finned to ensure that the airstream can pass adequately over all parts with proper direction and with emphasis on the hotter components such as the exhaust ports, which will be served by extra finning if necessary, the next task is to direct the flow suitably from the blower outlet. There are several methods of doing this, but forced draught with the main stream directed over the exhaust side of the head and then to the intake valve side is general practice on inline engines.

A fairly high but even temperature for cylinder barrels is desirable. Because of the high-speed flow, however, serious temperature variations, far in excess of those obtaining with natural air flow, can be set up if too much air is fed to one point. Hence, it may be necessary to bleed off an air supply from the mainstream at certain places to serve other areas, with the object of obtaining a temperature suited to the part concerned (Fig.

4-26). The Czechoslovakian Tatra engine that is illustrated also shows that air cooling can be effectively applied to a high-efficiency V8.

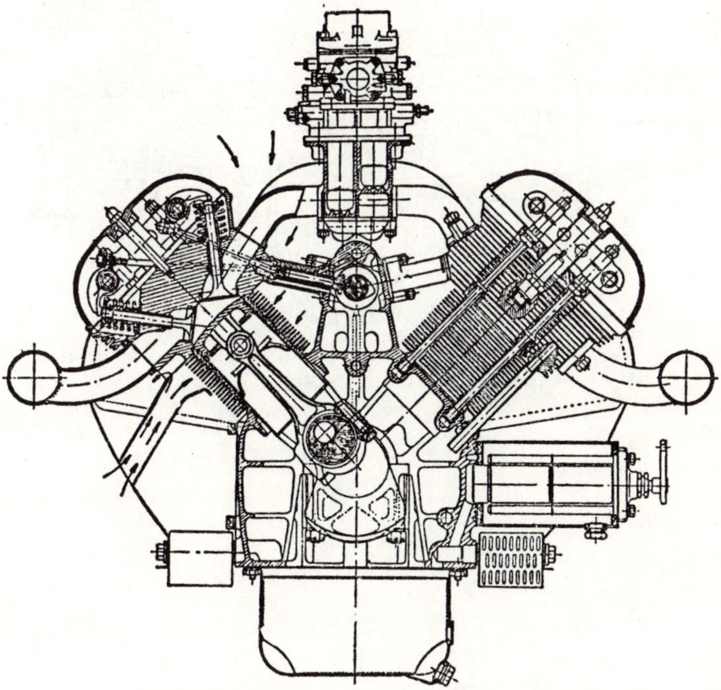

Fig. 4-26. Air flow ducted from fan on Tatra engine. Notice the auxiliary flow (left) to underside of exhaust port.

The Tatra also can maintain uniform temperatures by varying the air flow speed and volume. If the blower fan is driven at crankshaft speed or at a speed that is always proportional to crankshaft speed, the air output will obviously vary widely, depending on rpm. The variable speed hydraulic coupling shown in Fig. 4-27 overcomes this problem on the Tatra engine; the blower fan operates at full output only when the engine is running at high speeds.

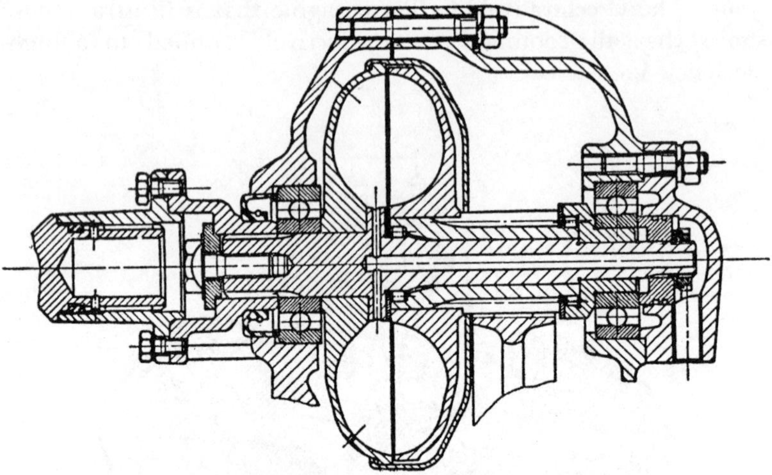

Fig. 4-27. Variable-speed hydraulic coupling controlled by oil from pressure lubricating system (Tatra).

5 / *Crankshaft Design and Engine Balance*

Bearing Location

Considerable unanimity now exists among production car manufacturers on the most desirable number of main bearings and their locations. Twenty years ago, taking four-cylinder engines as an example, expensive high-performance imported cars had five bearings, the common imported sedans and low-priced sports cars had three, and a few primitive cars had only two. Today, five main bearings are normal. The exceptions to this are "pancake" engines, which usually have three or four, and engines that have not been fundamentally redesigned during the past twenty years.

A bearing between each pair of adjacent crankthrows gives a very high degree of rigidity to the shaft when the latter is considered as a beam. Furthermore, there is little or no penalty in terms of increased engine length. Formerly overall length was a serious consideration because the old thick-walled bearings generally needed to be wider to carry the loadings that present-day thinwall bearings carry with half the width. Also, older engines with long strokes and small bores had to be lengthened to make room for additional bearings, with resultant wasted space between the cylinders. Now the bore is usually larger than the stroke so there is ample space for main bearings without separating the cylinders to wasteful lengths.

Another early criticism of separate main bearings between each pair of crankthrows was the liability of excessive torsional vibrations at the main journals—the winding and unwinding effect. This disadvantage has been mitigated in two ways. First, more robust materials and larger-diameter main journals have replaced the small-diameter journals. Second, shorter strokes have caused the crankpin circumferences to overlap those of the main journals to a considerable degree (Fig. 5-1), resulting in an inherently more rigid crankshaft design. It is interesting to recall that, in the early part of the century, main bearing journal and crankpin diameters were deliberately kept small, not merely to conserve material but to cut down the relative velocity at the bearing surface.

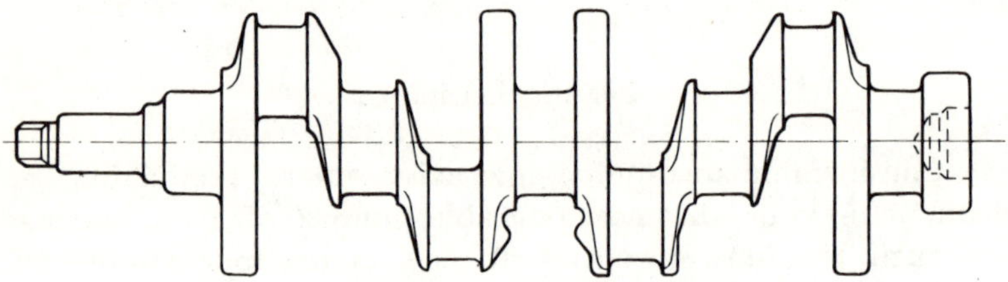

Fig. 5-1. Modern four-cylinder, five-main-bearing crankshaft, showing how diameters of crankpins overlap those of main journals.

The VW and Porsche four-cylinder pancake engines have two crankthrows between main bearings, making them essentially three-bearing engines. However, there is a fourth main bearing—a steady bearing of sorts—at the rear of the crankshaft so that there is a main bearing journal on either side of the timing and distributor drive gears. This practice of an additional steady bearing aft of the camshaft drive is also used on Porsche six-cylinder engines, which have seven other main journals—one between each pair of crankthrows.

It is possible that designers have gone too far in the use of main bearings on each side of every crankthrow. The typical American V8 has only five main bearings with two connecting rods mounted on each crankpin. Bearing and crankshaft failures have not been a chronic problem even though the extent to which some of these engines are supertuned is overwhelming.

Ferrari has recently reduced the number of main bearings in their Grand Prix engine without loss of performance or reliability. Similarly some versions of the Ford Fiesta have a three main bearing crankshaft in their inline four-cylinder engines. This is a return to the original design of the Ford inline "fours" as they were first introduced with three bearings in the 1950s. Certainly in the low-output form that is a consequence of today's fuel economy and emission control concerns, three main bearings are fully sufficient.

Problems of Balancing

The underlying theories of balancing techniques have not changed since the early days of multicylinder gas engines, but their application to modern, lightweight, high-speed automobile engines with various cylinder arrangements and firing orders has called for a good deal of research work. Fortunately electronic balancing equipment is now available that makes possible the full application of these discoveries even in small speed-tuning shops.

When a flexibly mounted engine of modern design is violently accelerated, it rocks on its mountings. Similarly, it tends to wobble when idling if the firing is irregular, the camshaft has considerable overlap, or the flywheel is very light in weight and the cylinders are few in number. These characteristics have little to do with balance but are the result of torque reaction as the engine speed varies. Thus, a wobble that a flexibly mounted engine may betray under some conditions does not indicate that the form of mounting is being used to hide an inferior product that has poor balance.

An engine in complete balance does not vibrate except under the above conditions, whether its mounting is rigid or flexible. Further, the engine as a whole can be balanced without the addition of counterweights to the shaft if the cylinder layout is suitable. For example, an inline six-cylinder engine with three pairs of pistons at 120° firing intervals is completely balanced if it has pistons 1 and 6, 2 and 5, and 3 and 4 up at the same time every 120°. The normal inline four-cylinder engine with pistons 1 and 4 and 2 and 3 up and down, respectively, gives faultless primary balance only.

Vibration is usually caused by deflection of the crankshaft, resulting from the power impulses. Under the influence of the power strokes, the shaft tends to twist and untwist; this is known as *torsional vibration.* It was torsional vibrations in the twin-crankshaft BRM H16 (sixteen cylinders, arranged as two flat-eights, one above the other) that caused the devastating crank coupling gear failures that plagued the design. Eventually development of this powerplant had to be shelved—an expensive and disastrous failure and a monument to the power of little mice to topple great kingdoms (Fig. 5-2). Apparently the main trouble was backlash in the coupling gears, which transformed the violent winding and unwinding of the crankshafts into sledgehammer blows on the gear teeth. Despite the greater crankshaft lengths, no similar problems were encountered in the Porsche flat-sixteen or in the prototype Coventry Climax flat-sixteen.

The phenomenon of torsional vibration can be illustrated by a single diagram, Fig. 5-3, which shows a plain length of shaft that has flywheels of different sizes at each end. If the wheels are imagined as being turned in opposite directions and then released, the wind-up or torsional strain energy in the shaft will cause the shaft to untwist in the other direction, rotating the flywheels accordingly. Because this motion will then be maintained by the kinetic energy in the flywheels, the shaft will overwind in the appropriate direction and repeat, until the inherent frictional damping forces in the shaft have dissipated the energy. The amplitude of the swing will decrease, but the frequency in terms of cycles per second will be constant, just as in the case of a pendulum. This frequency is known as the *natural frequency* of the shaft.

Fig. 5-3 shows what is known as a *two-mass system.* If there are three masses and two interconnecting shafts, as in Fig. 5-4, it is possible to have two natural frequencies. Calculating the natural frequency of either of these systems is fairly simple, but when the number of masses exceeds three it is not unusual to have to assist calculations by practical experiments on a trial and error basis. In Fig. 5-3 and Fig. 5-4 the nodes are indicated; these are the points at which no torsional displacement of the shaft occurs. It is generally considered that the most highly stressed points are the nodes.

Fig. 5-2. The 3-liter H-16 BRM Grand Prix engine, a lovely concept that failed to prosper.

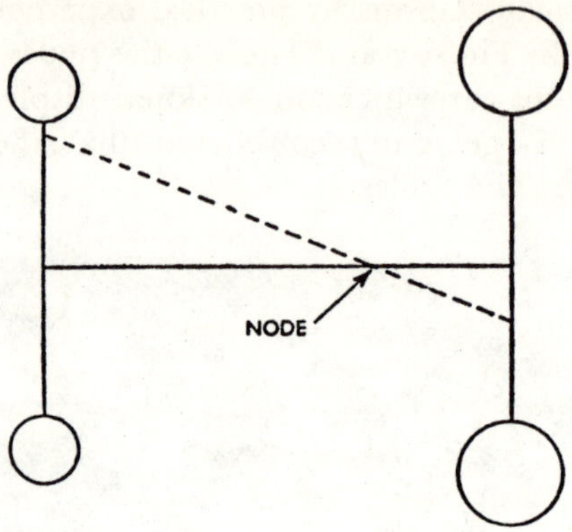

Fig. 5-3. Simple torsional loading in a two-mass system.

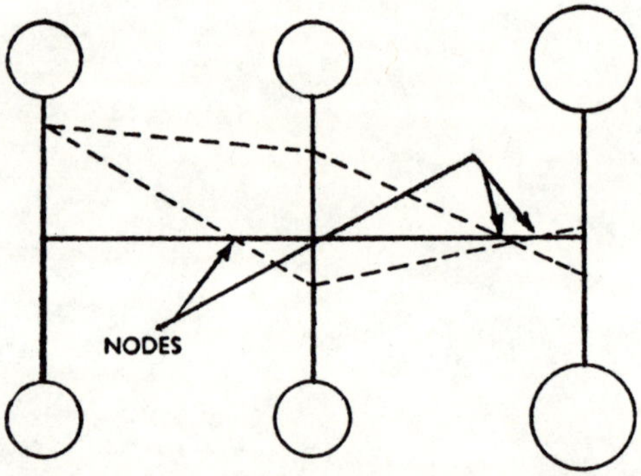

Fig. 5-4. Simple torsional loading in a three-mass system.

Because the torque applied to a crankshaft is impulsive and at intervals along the shaft length, a condition analogous to that illustrated in Fig. 5-3 will be set up since one or more crankthrows will be receiving power while others will be on nonpower strokes and thus will tend to lag behind. This does not occur significantly at low speeds, but at higher rpm, the amplitude of the vibrations increases as does the frequency. Thus, if the latter coincides with the natural frequency of the shaft, the inherent damping forces may not be able to cope, and failure of the shaft is the result.

It is important to arrange that the most-used engine speeds do not necessitate prolonged running at one of the natural frequency periods. It is usually possible to run through such periods without either damage or excessive vibration and bearing loads, but when the periods are near to coinciding, an external vibration damper can be used to alter the natural frequency to give a clear margin of safety. When there are several natural frequencies to run through in operating over a useful speed/torque range, it is not easy to render them all innocuous.

The provision of balance weights opposite the big ends on multicylinder engines certainly assists in smooth running by reducing the loading on the main bearings. The use of such weights adds to the inertia resistance to rotation of the shaft and thus to problems of torsional vibration.

Another factor in this connection is the length of the shaft in the main bearings in relation to its diameter. A long bearing has the virtue of giving a good measure of support to the shaft as a beam, but there is more of the shaft available to twist. This can be kept in check by making the journals of adequate diameter, but in general the shortest possible bearing is preferable.

While balance is certainly very much associated with lack of vibration, there are several other influencing factors, such as the avoidance of shocks in combustion, inequality of cylinder pressures, and the use of a cylinder firing order combining minimum crankshaft stressing with acceptable spacing out of the explosions along the engine to reduce wobble and rock on the flexible mountings. Balancing has been brought to a very fine art in the case of four and six cylinders inline; but the emergence of new engines of V6 and flat-eight type, for example, has called for a good deal of reassessment of the established procedures, and any new conclusions are likely to benefit engine design as a whole. It will be of interest, therefore, to consider a few of the basic requirements in obtaining smooth running.

The Single Cylinder Engine

It is quite impossible to balance a single cylinder engine except by methods impracticable for high-speed competition car applications (Fig. 5-5). In fact no attempt was made to do so on the pioneer industrial types of the last century, which used natural gas, coal gas, or vaporizing oil as fuels. The main goal was to obtain reasonably smooth torque from one power stroke per two revolutions at the very low speeds then obtainable. Thus, a very heavy flywheel was mounted on a crankshaft often formed in precisely the same way as in today's toy steam engines—by putting a U-bend in a bar of round steel. The massive bedplate, combining rigid anchorages for the main bearings and the cylinder barrel, absorbed a good deal of the vibration of the piston, but inertia loading on the main bearings, the big end, and the piston pin was high, and there was a good deal of crankshaft flexing. The addition of balance weights opposite the crankthrow acted at each end of the stroke to relieve the main bearing loading to some extent but did nothing to help the other bearings.

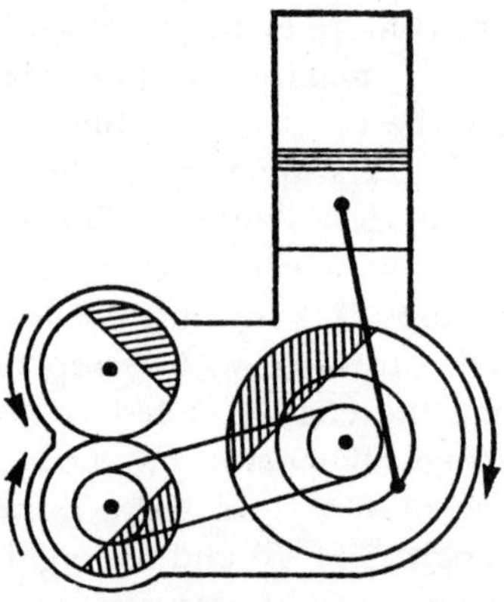

Fig. 5-5. Arrangement of contrarotating auxiliary balance weights for primary balance of single-cylinder engine.

The fact that single-cylinder gasoline engines are now capable of an almost vibrationless 6000 to 12,000 rpm might indicate that in the old days, some fundamental was overlooked, but this is not the case. The success has been the result mainly of dimensioning and proportioning and has not been gained by departure from old principles.

Balancing the Single

With a completely unbalanced crankshaft, a vertical single-cylinder engine will move up and down on its mountings as a whole as the piston reaches its maximum rates of acceleration-deceleration at tdc and bdc, respectively. Considering tdc as an example, as the piston approaches, it is rapidly slowed down; the slowing-down force is applied through the connecting rod and the crank so that the main bearing—and with it the whole engine—is, in effect, "pulled" upwards. On instantaneous reversal over tdc, the piston is accelerated away by a tensile force in the connecting rod, and again the force on the main bearing is pulling upward, combining with the first one. A similar downward force occurs at bdc; at around midstroke when the piston is traveling at nearly maximum speed with little acceleration, these reactions will rapidly decrease. (Loads caused by compression, firing, and so on are ignored for the moment.)

If we provide the crankthrow with a balance weight opposing the piston and connecting rod, and of similar weight, this balance weight will arrive at the opposite dead center simultaneously with the piston reaching top or bottom. The loading on the main bearing will thus be cancelled, and there will be no upward or downward vibratory force transmitted to the engine.

The difficulty now is that when the crankthrow is at about right angles to the cylinder axis (that is, at the point where the piston has little acceleration to oppose), the balance weight is swinging outward with only the weight of the big end to give any countereffect. Thus, the engine will move to the right or left horizontally as the weight changes in direction over the appropriate arc of crank travel in front or at rear of the vertical axis to the horizontal. See Fig. 5-6.

Practical Solutions

The practical solution is to substitute four smaller periods in two directions for two big ones in only one. These are simple terms, though they describe the effect quite well. It will be evident that if the balance mass is made as a smaller proportion

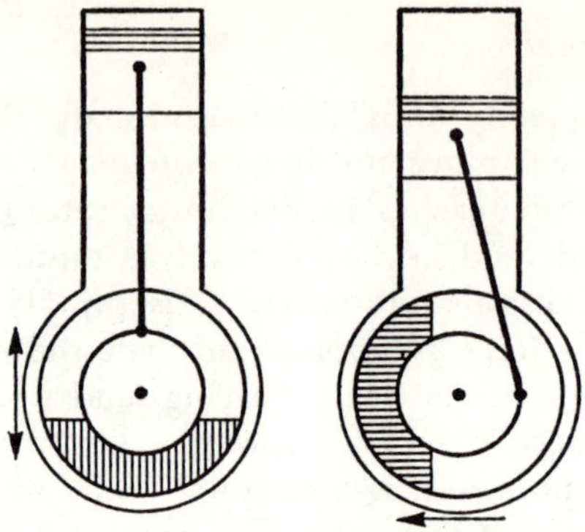

Fig. 5-6. Transfer of primary out-of-balance force in 90° of crankshaft rotation on single-cylinder engine.

of the reciprocating weight instead of completely balancing it, we shall reduce but not eliminate the up-and-down vibration and also generate a fore-and-aft vibration at right angles. The percentage used not only varies with the engine design, power output characteristics, and so on but may also be altered to meet installation requirements. Basically any improvement in the direction of the cylinder axis means more vibration at right angles thereto and vice versa; which of these is the more objectionable may well depend largely on the above-mentioned matters. While not a rigid rule, it is usual for high-revving engines to use less total balance weight than slower ones; the average is about 65 percent of the total of piston, piston pin, connecting rod, and big end. Individual tuners have evolved their own accurate figures for specific engines, which are closely guarded secrets; in a few cases the procedure has been made into formulas enabling balancing to be done with precision.

In using one such formula, the percentage of reciprocating weight arrived at is made up into a weight in the shape of a sleeve to fit the crankpin after the connecting rod has been removed. With the flywheel assembly supported on knife edges, the balance weights are machined until the assembly remains stationary with no tendency to rotate at any point. The sleeve is then removed and the connecting rod assembled. This procedure can be used when altering the balance factor either way on an existing assembly, using the required percentage of weight in the sleeve.

Except for small capacity racing cars and karts, the high-output single-cylinder engine has been defunct for many years as far as its use on cars is concerned. Twin-cylinder engines, however, are fairly popular—particularly in small-capacity and utilitarian vehicles overseas.

Multiple Cylinder Engines

The impossibility of balancing a single set of reciprocating parts except as a compromise no longer applies when there is a second set available; the question then is to decide on the layout of the engine that will give good balance as well as regular firing intervals, the latter being obviously desirable in the interests of smooth delivery of torque. If, however, we consider balance only, it will be evident that when counterweights are added to the crankthrow of a "single", that if a second cylinder is added at right angles to the first—that is, in V-twin formation—the unit will be in balance. That is so if the counterweights are equal to the reciprocating weight of one cylinder's components, because in both the vertical and the horizontal planes there will always be masses moving in opposition (Fig. 5-7). This type has excellent running balance, but unfortunately the cylinder angle means that instead of obtaining a power stroke at every crankshaft revolution, or every 360°, these occur at intervals of 0°, 450°, and 270°.

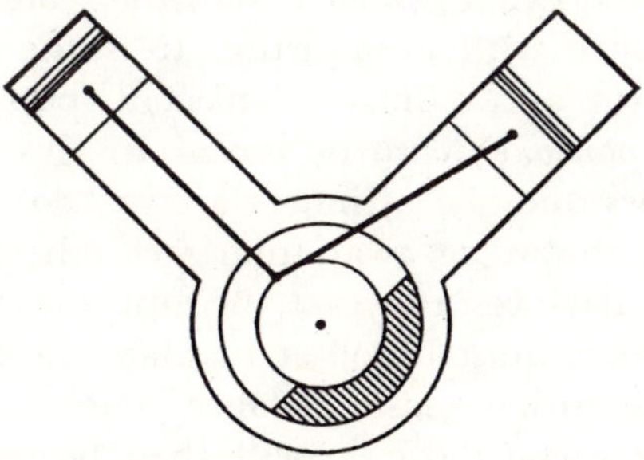

***Fig.* 5-7. Primary balance of 90° V-twin.**

Although this gross irregularity has been acceptable in certain specialized applications—the vintage British G.N. Spider, for example—it is customary with V-twins to compromise again by reducing the cylinder angle to something between that required for meeting the two desirables. For example, with an angle of 60° the firing intervals become 0°, 420° and 300°, but the element of out-of-balance now introduced into both vertical and horizontal planes as in the case of the "single" is again reintroduced, but to a much lesser extent. In this case the counterparts will be of less weight than the total reciprocating weight of one set of components, just as in the case of the single. Many American motorcycles have had the 60° and 45° V-twin cylinder arrangements.

The Parallel Twin

If we again consider the effect of closing up the cylinder angle to obtain even firing, assuming that we keep the two connecting rods on one crankpin (or the equivalent, with a two-throw crankshaft that has both throws in line), we obviously arrive at the parallel twin arrangement with cylinders side by side. In this case one cylinder will fire at each 360°, but the balance will be akin to that of the single. Smoothness will be improved because of the greater frequency of impulses and the shorter stroke for equivalent total capacity, and in fact the best parallel twins reach very high standards. Of sizeable cars, both the Fiat 500 and the NSU Prinz have used this kind of cylinder arrangement.

There is one twin-cylinder design that combines almost unsurpassed balance with even firing. It is the horizontally opposed twin with a two-throw crankshaft—popularly known as the *flat-twin, pancake twin,* or *boxer.* In this case, since both the crankthrows and the cylinders are at 180°, the pistons are always moving toward or away from each other and are in complete primary balance. However, because the crankthrows and thus the cylinders must be offset in plan, the engine will tend to rock in the tilting sense. Balance weights will reduce this effect, but the rocking couple will then be transferred to the up-and-down plane. George Lanchester, a British designer of the early twentieth century, produced an exercise in perfection in his early car engine by eliminating even this minor snag, putting the cylinders in line and duplicating the crankthrows and the connecting rods on one assembly. See Fig. 5-8. The Buick Motor Company earned its early fame with a highly reliable flat-twin, which was sold to a number of other car manufacturers in the United States in addition to powering the early Buick cars. The most famous engine of this type, at least at the present, is the one used in the fine BMW motorcycles.

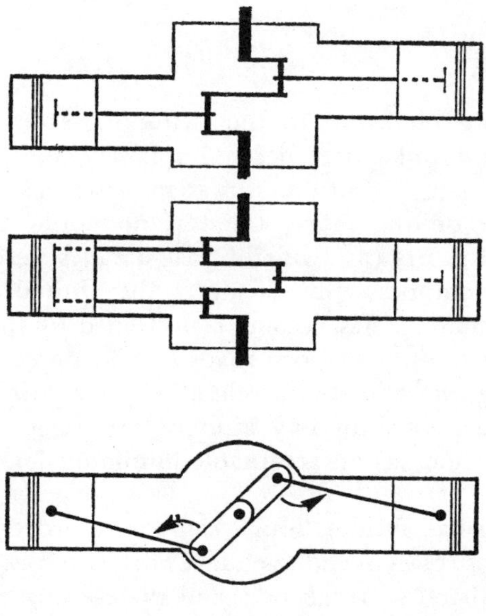

Fig. 5-8. Flat-twin comparisons. Rocking couple caused by cylinder offset (top) can be eliminated by Lanchester design (center). At bottom, side view shows perpetuation of equal crank/connecting rod angles.

Flat-Twin Advantages

The flat-twin substitutes width for length and height in its general dimensions, a far from disadvantageous design in the modern chassis, so that the type lends itself admirably to either front or rear mounting. The famous Citroen 2 CV, and the potent Dyna-Panhard of the mid-1950s, adopted the former position with frontwheel drive; the Steyer-Puch of the same period used the rear position.

With its separate cylinders, the flat-twin calls for rather more separate components than others, but in view of its adoption by European and Japanese makers of highly utilitarian vehicles, it is difficult to justify any objection of high production cost. In the past, carburetion difficulties caused by the long induction tract, as well as lubrication complications resulting from the horizontal cylinders, used to be snags associated with this kind of engine, but these have long ago been overcome.

Four Cylinders

The main objection to the twin-cylinder engine, particularly if its size and power become appreciable, is in the single power stroke per revolution; this is most serious at low engine speeds. Once on the move, there is no doubt that even the highly tuned sports car twins of past decades perform with exhilarating smoothness. But as far as the United States is concerned, the motorist has become conditioned by the sheer necessity of speed limits and poor roads to an appreciation of good slow running and effortless acceleration from low speeds. These are desirable points for any kind of car usage, and thus the four-cylinder remains an acceptable minimum for the American driver.

The normal inline "four" can be regarded as two 180° parallel twins placed end-to-end. Thus, two pistons arrive at tdc as the other two reach bdc, and the rocking couples set up between cylinder pair 1 and 2 and cylinder pair 3 and 4, as a result of the crankthrow spacing, cancel out along the engine and do not tend to rock it from end to end. In addition to this balance advantage, the firing order can also be arranged so that the reactions are nicely spaced, though the torsional load on the shaft is by no means regular. With the normal firing order of 1-3-4-2 or 1-2-4-3, the portion of the shaft at the center bearing receives two twists in half a revolution followed by a quiescent period of the same amount. See Fig. 5-9 and also Fig. 5-13.

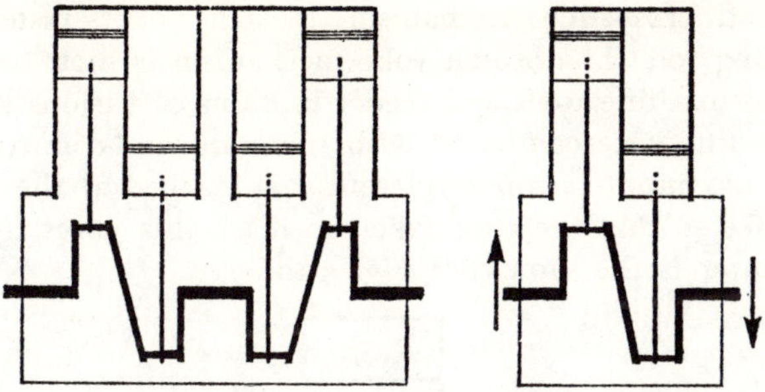

Fig. 5-9. Inline "four" and parallel twin compared. Inline "four" (left) constitutes two parallel twins end to end. This cancels rocking couple present in twin (right).

With the primary balance established using this shape of shaft, there is no need for balance weights opposite the crankthrows to counteract the reciprocating weight, except insofar as they relieve the main bearing loadings. This is particularly important at the center bearing of a three-bearing crankshaft since the crankthrows on either side are in the same plane. At least one famous sports car engine was prone to excessive wear of this bearing only until a counterweighted shaft was fitted.

Secondary Forces

It is possible to obtain primary balance on a multicylinder engine if the directions of piston travel are arranged so that they counteract in respect to their inertia forces. For example, to this is done when traveling in opposite directions and arriving at the dead center simultaneously, as in a flat-twin or an inline four-cylinder (to name only two possibilities). However, because of another geometrical feature of the engine's construction, good primary balance is not the only answer; it is necessary also to consider further out-of-balance forces known as *secondary vibrations.*

To those familiar with reciprocating engine history, it is usual to explain secondary forces by the comment that they are *not* present, and thus cause no trouble, on engines using a crank and cross-slide mechanism (Scotch yoke) instead of a connecting rod. The Scotch yoke mechanism is most often seen today on machine tools and feeder motions of various kinds; its obvious difference compared with the crank-and-connecting rod is that maximum piston acceleration is reached at the point of half-stroke so that the time taken to reach this point from each dead center is the same. See Fig. 5-10.

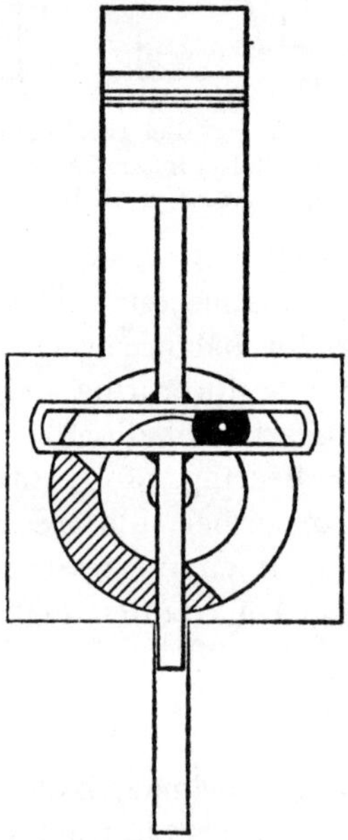

Fig. 5-10. Crank and cross-slide (Scotch yoke) mechanism. This gives harmonic reciprocating motion, eliminating vibrations caused by orthodox connecting rod angularity.

This is, of course, not the case with a connecting rod. It is usual to state that if the latter is infinitely long, the same conditions of equal acceleration will prevail. But with the usual ratio of connecting rod length to crankthrow of around 4:1, the pistons starting from tdc will reach their maximum speed quite a way before half-stroke; those starting from the bottom will correspondingly be a good way past the same point before they stop accelerating.

Obviously, therefore, the piston's acceleration from tdc is much greater than that from bdc. How much greater depends on the dimensional ratio of connecting rod length to stroke length; the shorter the rod in relation to the crankthrow's stroke—actually, half the stroke because it is the difference between the crankthrow's center and the crankshaft centerline that concerns us—the greater will be the difference in the piston acceleration rates. Since this difference has to be resisted by the the same means as in the case of the primary forces—that is, by the engine mass, via the bearings, and so on—it follows that in certain engine layouts a vibration will be set up that cannot be balanced. Also, since the difference in acceleration occurs twice per revolution, it follows that the vibrations caused by this difference will have twice the frequencies of the primary forces. (The latter frequency in a single-cylinder engine is equal to the engine rpm.) The effect of this, combined with the connecting rod angularity, is felt in two ways. At both tdc and bdc, there is an upward vibratory force. At right angles to the cylinder axis, there is a force downward as the crank swings through its right angle with the connecting rod. The intensity of these secondary vibrations is only about one-quarter of the primary value.

As with many other complex vibrations, the secondaries go on to infinity with decreasing magnitude and increasing frequency. But so long as the basic impulse can be made small, the effect of the others is negligible. Rendering secondary effects negligible involves careful selection of the dimensional ratios concerned and adequate rigidity.

It is impossible to cancel out the secondary vibrations on an inline "four". In the case of a flat-twin, however, not only do the pistons move in opposition at all times but the two connecting rods also swing oppositely when making identical angles with their cranks; thus, these forces cancel also. See Fig. 5-8, given earlier.

This perfection of balance is also obtained on a flat-four using a normally disposed four-throw crankshaft—if cylinders 1 and 4 on one side are opposed by 2 and 3 on the other. This would mean, however, simultaneous firing of opposed cylinder pairs so that the torque delivered to the crankshaft would be no better than on a twin. In practice, therefore, with the same crankshaft layout, the cylinder banks are offset as shown in Fig. 5-11, one bank containing pistons 1 and 3 at tdc and bdc, respectively, and the other containing 2 and 4. (This is the arrangement used by VW, Porsche, and Subaru even though the cylinders are numbered differently.) This gives a power stroke at every 180° of rotation, with primary balance, but the secondary forces will set up a rocking couple.

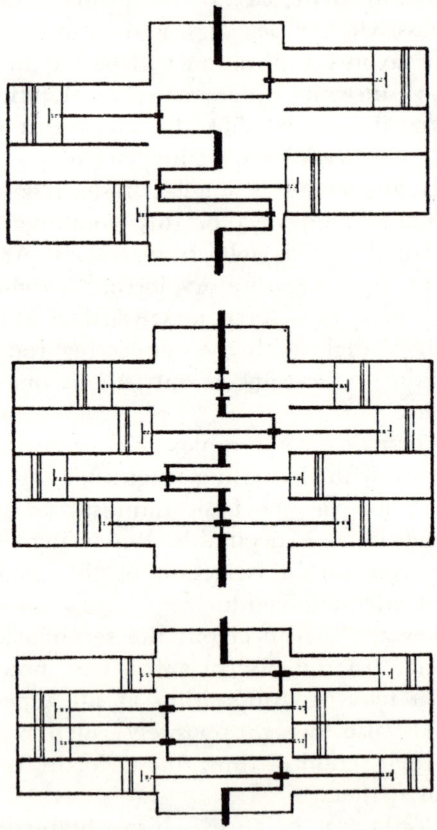

Fig. 5-11. Horizontally opposed cylinder arrangements. (Top) flat-four; (center) flat-eight with 90° two-plane crankshaft; (bottom) British Anderson competition engine of 1920s, a simultaneous-firing flat-eight with 180° single-plane, two-main-bearing crankshaft.

Six Cylinders versus Four

The conventional inline "six" is similar to the corresponding "four" in that it comprises two three-cylinder inline engines in looking-glass formation. The crankthrows 1 and 6, 2 and 5, and 3 and 4 are given the same phasing, with 120° of crank angle between pairs. It is thus possible, as in the "four", to distribute the firing intervals acceptably along the shaft with the usual firing order of 1-5-3-6-2-4. This means that alternate cylinders fire from the center pair of cylinders (3 and 4) to the outers, in turn, at each 120° of crankshaft rotation. See Fig. 5-14, which appears later.

With the extra length of the inline "six", makers normally use seven main bearings, which is highly desirable for maximum rigidity. A few, however, continue to employ four bearings. The Datsun, BMW, Ford, and Chrysler inline "sixes"—the engines of this kind that are most often seen in competition these days—all have seven main bearings. Thus, they are only now catching up with the famous Jaguar DOHC "six" that has achieved so many competition records since its introduction thirty years ago.

The primary forces are fully balanced in the inline "six". As two pistons come to rest at each one-sixth of a revolution, the primary vibratory force has a frequency of three times the rpm and the secondaries have twice this frequency. The latter, however, are also balanced except for a minor unbalanced third-order frequency, which can be disregarded.

It is thus evident that apart from convenience of physical shape and the desirability of having a specified number of power strokes in the interests of good torque characteristics, the cylinder arrangement of the engine has quite a lot of influence on the smoothness of running. For example, in the case of the inline "four" and the flat four, it is not merely a case of whether the latter is a better shape or more amenable to air cooling that makes it preferable to the former. The inline "four", good as it is, cannot be fully balanced without the adoption of artificial balancing mechanisms driven from the crankshaft.

Because of economic and conservation concerns, many Americans who formerly owned six- and eight-cylinder passenger cars have begun to purchase four-cylinder cars. To cater to these pampered drivers who might otherwise rebel at four-cylinder vibrations, Toyota and Mitsubishi (Plymouth Arrow and Dodge Colt) have begun offering U.S. consumers inline "fours" with balancing shafts. These shafts turn at twice the crankshaft rpm and cancel out the secondary imbalance to a degree that makes the inline "fours" equal to other popular engines in smoothness and lack of resonant periods in the rpm range that is most used.

Balancer shafts were pioneered by Ford (Fig. 5-12) and Lancia. Interestingly, these companies no longer build "fours" with balancing shafts, having also abandoned the V4 cylinder configuration; nevertheless, Ford and Lancia now manufacture some of the smoothest inline "fours" built—without balancer shafts. Undoubtedly much has been learned.

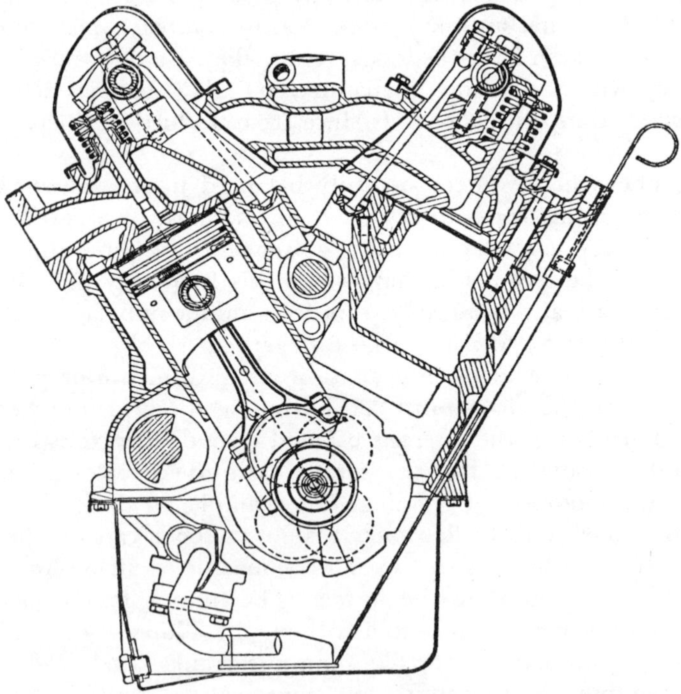

Fig. 5-12. Cross-section of German Ford Taunus V4 engine with auxiliary balancing shaft on left side. Notice phasing of crank-throws, which produces even firing intervals.

In any case, an additional shaft flailing away in the crankcase is no help in a competition engine; it is, in fact, a decided detriment. Thus, we will not go deeply into the influence that such extraneous equipment has on engine balance. It will suffice to say that adopting the flat-four configuration would be a more desirable alternative if the engine is to be a competition "four."

The flat-four in its practical form replaces the out-of-balance secondary of the inline "four" with a minor rocking couple that may be easier to render innocuous. So many other manufacturing and designing points intrude that we are left with the conclusion that you take your choice—both are good.

Straight Eight Engines

Inline eight-cylinder engines are not being built at present. However, at one time it was the predominant type in all forms of racing. The last successful racing straight-eight was the Mercedes Benz Grand Prix engine of the middle 1950s (Fig. 5-13). This engine had a spur gear in the center of the crankshaft that transmitted the power to the transmission via a gear-driven output shaft. Thus, the Mercedes engine was in effect two inline four-cylinder engines attached end to end by a common output gear.

Considering the success of this crank-center power transmission scheme, it is obvious that there is at present only one thing wrong with the straight-eight cylinder configuration; it is too long for use in today's mid-engined Grand Prix cars. Nevertheless, in some drag racing classes where a front-mounted engine is normal—and perhaps even advantageous—the old Buick straight-eight is still a very competitive powerplant. Of all the straight-eights once manufactured in the United States, only the Buick has retained a competitive life; the DOHC Duesenberg has now, sadly, become too valuable as an antique to be risked on the race track.

The normal straight-eight crankshaft has crankpins 1 and 2 up and down respectively, 3 to the left, 4 to the right, 5 to the left, 6 to the right, 7 down, and 8 up. This configuration produces full balance and overlapping power strokes for exceptional smoothness and uniform torque.

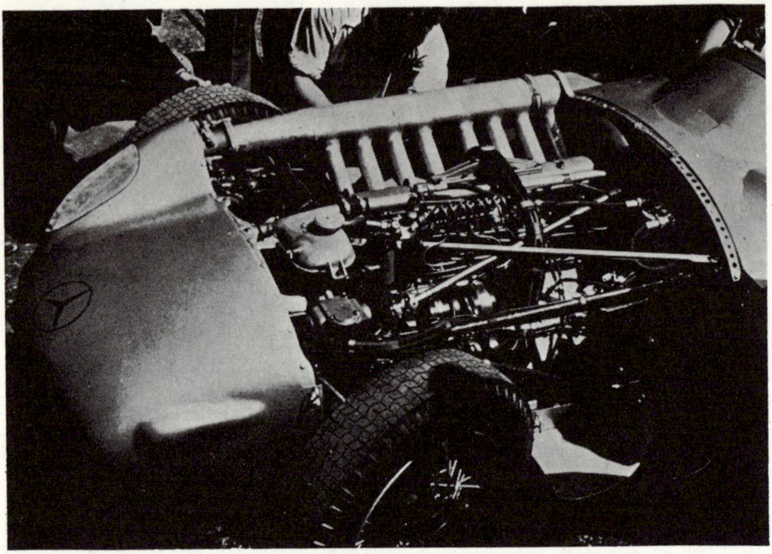

Fig. 5-13. The 2.5-liter Mercedes Grand Prix engine, the last successful straight-eight racing engine.

The V8 Engine

The V8 engine was neither invented in the United States nor was it widely accepted here until the advent in 1932 of the Ford flathead V8. Up to that time, V8 engines were too expensive and complicated to attain major importance in the passenger car market. It is interesting to note that Chevrolet, Ford's arch rival, built a V8—an OHV V8 at that—fifteen years before the Ford flathead V8 came along. (One of these early Chevy V8s can be seen—where else?—in the Ford Museum in Dearborn, Michigan.)

It is wise to start a discussion of the old Ford flathead with an examination of its shortcomings because the success of the engine in competition was largely based on the ingenuity of the speed tuners who overcame them. First and foremost, the old Ford V8 was limited by being an L-head engine. Zora Arkus Duntov (who went on to become the head of Chevrolet's racing department) tackled this problem by designing overhead-valve cylinder heads for the Ford. Nevertheless, most of the Ford flathead's competition successes came to engines that raced in supertuned flathead form.

The main problem with the Ford flathead V8 from a design point of view was its exhaust ports. The block had separate intake ports for each cylinder. Hence, tuners were seldom hindered by difficulties on this side of the engine. But, because the engine was an L-head, the exhaust valves were located beside the intake valves—at the inside of the "vee". Consequently the exhaust ports had to make a U-turn through the water jacket to reach the outside of the block. These long exhaust passages, surrounded by the coolant, were as efficient as the flues in a steam boiler for raising the temperature of the surrounding water. Boiling was common, and the center two exhaust ports of each cylinder bank, being siamesed, caused irregular scavenging and filling of the cylinders.

Nevertheless, Ford V8 engines were cheap and plentiful and thus had the advantage of inexpensive competition development. The engine's rivals were remarkably few in number. The other popular American engines of the day, the Chevy "six" and the Plymouth "six", had even greater shortcomings than did the Ford. The Chevrolet, despite its overhead valves, had three intake ports and four exhaust ports; the intake ports were all siamesed, but only the center two exhaust ports were siamesed. Consequently there were terrible irregularities in cylinder filling and scavenging. Furthermore, the engine lacked a pressure oiling system and relied on the splashing of oil inside the crankcase for much of the bearing lubrication. Like the Chevy, the Plymouth had an irregular pairing of intake and exhaust ports—and an infamously weak, though pressure-lubricated, "bottom end".

As modern high-compression engines began to appear in American passenger cars, it was not surprising that most of them were V8s. Not only was the Ford V8 the only conspicuously successful production engine in competition, which gave it a certain glamor, but many of the younger American designers had served a youthful hot rod apprenticeship on the Ford. In addition the V8 configuration had two distinct advantages for the car industry. First, it was short and compact at a time when hoods were becoming short, and secondly, with two connecting rod big ends on each crankthrow, there were fewer bearing surfaces to machine.

The first American high-compression V8 engines were a stunning experience for speed tuners of the day; they were able to outperform supertuned Ford flatheads without significant modifications. Still, by present-day standards, they were very limited designs. The center two exhaust ports on each bank of cylinders were siamesed on both the 1949 Oldsmobile V8 and the 1949 Cadillac V8. Olds and Caddy engines had some notable competition successes in their day. The Cadillac engine, for example, won many important American road races when installed in the English Allard sports car and even managed a tenth and eleventh place at Le Mans.

Then the Chrysler "Hemi" engines came onto the scene (Fig. 5-14). With their hemispherical combustion chambers and separate intake and exhaust ports, they were an American speed tuner's dream—in every respect other than cost. Derivatives of this engine are still unbeatable in several drag racing classes, despite the fact that this engine is somewhat heavy; their NASCAR successes are also legion in number.

The Chevrolet V8 that was first introduced in 1955 ushered in a new day for America's V8-oriented speed tuners. This engine was soon cheap and plentiful, making possible rapid development for competition purposes. The Chevy V8, known today as the small-block V8, had inherently good breathing and, in addition, was both light and very compact. The scions of this engine form the backbone of a great many racing classes today.

Let us now examine the V8 engine configuration from a technological point of view. If a V8 is built that consists basically of four 90° V-twins coupled end to end and using a straightforward four-cylinder crankshaft, two connecting rods on each throw, and crankpins 1 and 4 and crankpins 2 and 3 coaxial at 180° between pairs, it is possible by suitable arrangement of the firing intervals along the cylinder banks to obtain a power stroke every 90°. The smooth torque is, however, somewhat nullified by the considerable vibrations set up as a consequence of the unbalanced secondary forces, and obviously no arrangement of balance weights can do anything to overcome this.

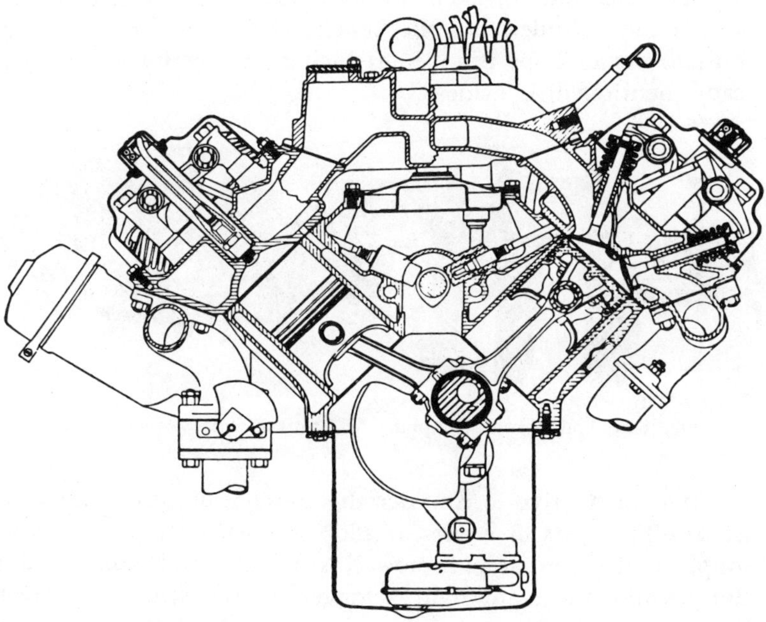

Fig. 5-14. **One of the earliest Chrysler Corporation "Hemi" engines, the 1952 De Soto Firedome V8.**

There is also another objection to the "flat" crankshaft layout: the two center crankthrows, each laden with two connecting rods and pistons, are in the same plane, and the process of operation puts a heavy bending load on the shaft and thus on the engine structure, which eventually has to resist such a force. This, however, is less of a consideration than formerly, because the overlapping diameters of the crankpins and main journals in today's short-stroke engines produces an exceptionally strong crankshaft.

The arrangement used in modern production car V8 engines is the "two-plane" crankshaft shown in Fig. 5-15. As viewed end on, it is clear that the crankpins (each carrying two connecting rods as before) will now appear to have No. 1 pointing upward, No. 2 to the right, No. 3 to the left, and No. 4 downward—each crankthrow at 90° to the next. Obviously the shaft has now forsaken the looking-glass conception of the normal four-cylinder, but the benefit of having the two center crankthrows in opposition, in reducing the bending stresses already mentioned, is evident.

Fig. 5-15. Four-throw, two-plane V8 crankshaft for American V8.

It is instructive to consider the couples, or crankshaft rock, set up. The pairs of cranks at 180° will each set up a rocking couple, and these pairs are now Nos. 1 and 4 and Nos. 2 and 3. But because the crankshaft is longer, in the sense of a lever, between the former pair than between the latter pair, this would normally mean that the two would not cancel each other, and the engine mass would tend to rock on its mountings.

The problem now is to arrange counterweight masses along the shaft to compensate for this "leverage" effect, and it is in doing so effectively that good design really shows up. The old Ford V8 crankshaft had three main bearings, so that the flying web between the pairs of crankpins had the latter offset on them to an extent of 90°. This shaft was one of the pioneer cast ones, a forged shaft being used in 1932 only. As such, it was a creditable achievement in foundry practice. But in arriving at satisfactory balance masses combined with the crankweb arrangement necessary (as outlined above) for three main bearings, the overall shaft weight was quite considerable—another reason why this engine proved disappointing when installed in a light sports car chassis; its power of acceleration was handicapped by the inertia of its crankshaft. In dirt track midget racers, the Ford V8 60 was usually raced without a flywheel.

All modern V8s have five main bearings. This allows much greater finesse in designing the balance masses and is largely necessitated by the high power outputs that—in view of the staggering of crankpins at 90° between cylinders—would now hopelessly overload a three-bearing shaft. The five-bearing shaft has eight webs, each of which can be given the appropriate counterweight for the effect desired. In practice, as any good sectional illustration will show, the general arrangement can be summarized as follows: A heavy weight is fitted on the first and last webs—Nos. 1 and 8; these will be at 180° to each other and in primary opposition. Lighter weights are used on web Nos. 2, 3, 6, and 7, which are at 90° to each other as pairs. The middle webs, 4 and 5, have still lighter balance weights or even none at all.

If all the significant factors are assessed carefully, both primary and secondary balance is obtained. With the ordinary single-plane four-cylinder crankshaft referred to earlier, assuming that two pistons in the left-hand cylinder bank are at tdc, the other two will be at bdc, but all the pistons in the right-hand bank will be about halfway down the stroke; hence the secondary vibrations. With the two-plane arrangement, however, the pistons in each bank must be one up, one down, and two about halfway; that is where the balance improvement lies.

It is fortunate that when a power stroke is available at each 90°, the sequence is perhaps less important than in engines where there are fewer of them because arranging them satisfactorily around the engine is by no means easy. A typical firing order, numbering the right bank front to back as 1-2-3-4 and the left bank front to back as 5-6-7-8, is in the sequence 1-5-4-2-6-3-7-8. With this, equal induction impulses can be obtained on the chokes of two carburetors by coupling them to the outers of one bank and the inners of the opposite bank (Fig. 5-16). The sequence gives irregular bearing loadings, but this also holds good for several alternatives, which in addition are more troublesome from the point of view of mixture distribution.

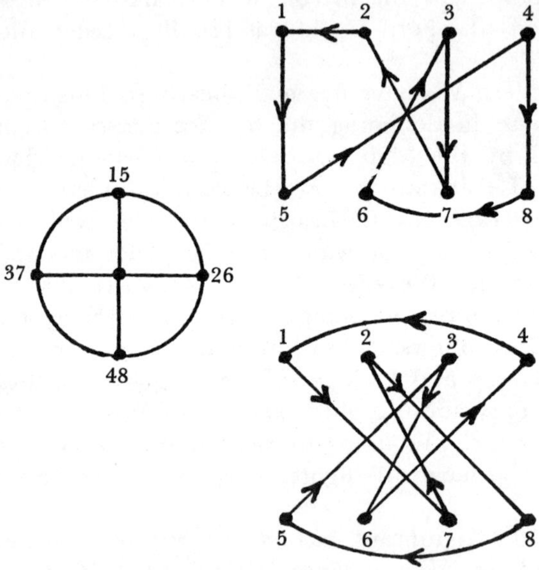

Fig. 5-16. Typical eight-cylinder firing orders using two-plane crankshaft (left). (Top) 90° V8; (bottom) horizontally opposed "eight".

The latter point does not, of course, arise in racing engines where one choke per port is possible (or one injection duct, of course). The difficulty in that case lies in the intricacies of the exhaust system on which so much reliance is placed for obtaining maximum power. It is quite impossible, regardless of how the branch pipes are arranged or their complexity, to obtain an exhausting sequence that will employ pressure-wave action to the fullest advantage. Hence, both the BRM and the Coventry-Climax V8 Grand Prix engines of the recent past and the Cosworth-Ford Grand Prix engine of the present have used single-plane crankshafts, thus enabling each bank of cylinders to use exhaust layouts similar to those used on an inline "four". See Fig. 5-17.

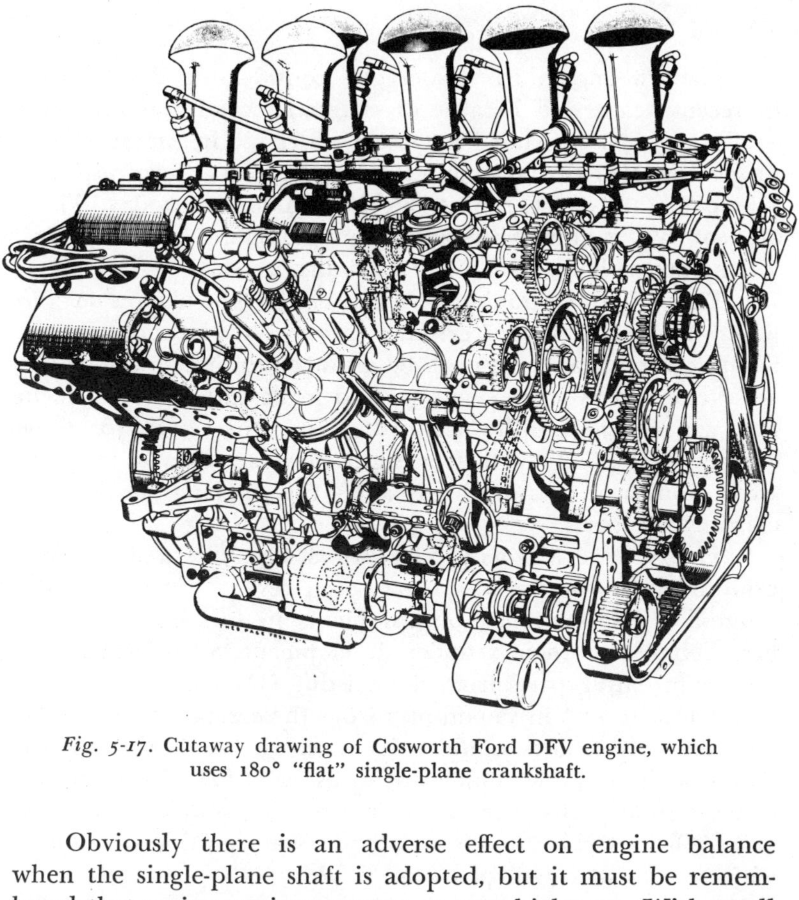

Fig. 5-17. Cutaway drawing of Cosworth Ford DFV engine, which uses 180° "flat" single-plane crankshaft.

Obviously there is an adverse effect on engine balance when the single-plane shaft is adopted, but it must be remembered that racing engines operate at very high rpm. With small cylinders and short strokes, the snags are theoretical rather than practical in that particular context.

The V8 engine, America's first choice for many years now, is seeing ever-greater application overseas. In its normal 90° configuration, the V8 is a very compact engine with a short, stiff crankshaft. Rolls Royce and Mercedes Benz both use V8s in their finest passenger cars; Porsche, in its first move away from the "boxer" or "pancake" engine, has elected to use a V8 in its Type 928.

The V6 Engine

The V6 engine has undergone considerable development in recent years, and because of changing social and economic conditions, it is scheduled for much wider use in American passenger cars than it has enjoyed in former years. Two distinctly different kinds of V6 engines are in common use today. There are 90° V6 engines with three-crankpin crankshafts and 60° V6 engines with six-crankpin crankshafts.

The 90°, three-crankpin V6 can be thought of as a 90° V8 with one V-twin pair of cylinders sawed off. Naturally the remaining three crankpins must be arranged to different angles in an attempt to produce evenly spaced firing intervals—which, in fact, can never be fully attained. The main reason for these engines is implied by their design; they are derived inexpensively from what was originally a V8, thus making use of many existing V8 parts.

The General Motors 90° V6, which first appeared in Buick cars about twenty years ago, has been revived by GM in the interest of badly needed improvements in fuel economy. GM has similar new engines under development, all of which will use reciprocating parts from the existing GM family of V8s. In an attempt to obtain smoothness from these engines, it is necessary not only to use counterweights on the crankshafts to balance out the rotating couple but to use very flexible mountings to disguise an unbalanced secondary rocking couple. The firing order, obviously, is uneven and requires an ignition distributor with an irregularly shaped cam.

Volvo has also accepted these same compromises. In the early 1970s, Volvo had under development a 90° V8 engine for use in the more luxurious passenger cars that were then being planned. However, the energy crisis that surfaced along with the Arab oil embargo of 1973 forced a reconsideration of the plans for an eight-cylinder engine. Consequently the basic Volvo design had two cylinders lopped off, and a V6 was thus obtained, thereby salvaging much of the work that had been done in tooling for the stillborn V8.

From an engineering and competition-oriented point of view, Ford's 60°, six-crankpin V6 engines are a much "neater" design. Ferrari had previously used this approach (and also the 120°, three-crankpin arrangement) in producing a V6 with uniform firing intervals. With the Ford V6, used in America on the Capri, Pinto, Mustang, and Bobcat models, each connecting rod big end has its own crankpin (Fig. 5-18). For the purposes of this book, we will disregard other V6 engines that involve engineering compromises and limit our technical analysis to this kind of 60°, six-crankpin V6.

Fig. 5-18. V6 crankshaft from Ford Capri. Arrows indicate main journals; crankthrows are spaced at 60° to one another.

With a crankshaft that has six crankthrows spaced at 60° to each other along the shaft (Fig. 5-19), the firing intervals are equal at 120°, as in an inline "six"—assuming that the angle of the cylinder banks is 60°. Furthermore, the three pairs of reciprocating components will cancel out to give both primary and secondary balance.

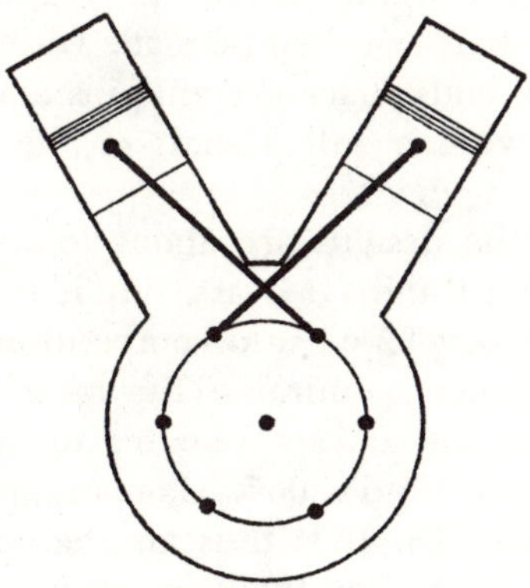

Fig. 5-19. End view of 60° V6 engine with six-throw crankshaft.

Numbering the cylinders front to back as right bank 1-2-3 and left bank 4-5-6 (Fig. 5-20), the usual firing order is 1-4-3-6-2-5, or in the case of the Ford V6 seen in the United States, 1-4-2-5-3-6. This gives alternate power strokes from each bank, as does the 1-6-3-5-2-4 firing order illustrated later in Fig. 5-21. Three additional firing orders are possible but are not used, because they involve consecutive firing of the cylinders in each bank.

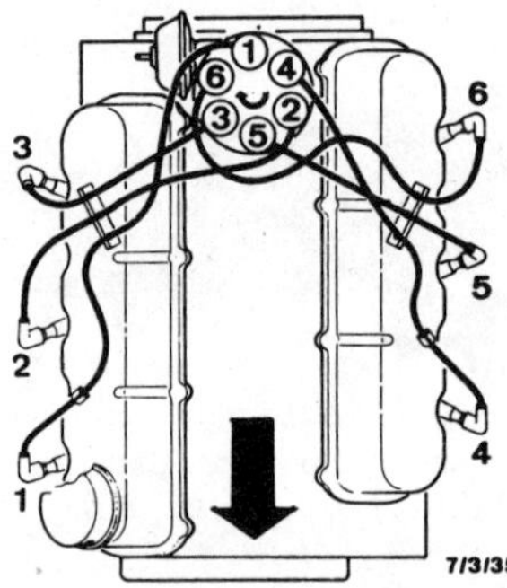

Fig. 5-20. Cylinder numbers of Ford V6 engine. Arrow points toward front of engine.

If the cylinder angle is opened out to 120°, as in the 1960 Ferrari Grand Prix car, and the big ends advance on the crankthrows by 60°, a similar firing sequence will result, and the engine will reduce overall height at the expense of greater width. The crankshaft can then be made with three crankthrows, with opposite big ends sharing crankpins at 120° spacing—that is, half of a six-cylinder inline shaft or the same as used in a 90°, three-crankpin V6.

New V6 engine designs are about to reach the market in American-made Ford and GM cars, but it is uncertain whether these will ever be developed into competition engines. Whether Ford will follow their admirable European practice or revert to the "sawed-off" V8 concept remains to be seen. It is interesting to note that Ford's 60°, six-crankpin V6 was derived from an earlier 60° V4. It is thus an engine that has had two cylinders grafted on rather than an engine that has had two cylinders taken off.

The Flat-six

The flat-six, of course, refers mainly to the Porsche 911 and its derivatives. America's modest entry into this field, the Corvair, is little more than a lingering memory, having suffered an inglorious demise out of all proportion to its shortcomings. There can, in fact, be little said against the flat-six cylinder arrangement, either from the standpoint of balance or from that of firing regularity.

As in the case of the inline "six" and the better sort of V6, the flat-six has six crankpins on its crankshaft. No two crankpins are coaxial, just as in a V6. However, it must not be assumed that a flat-six crankshaft is identical to a V6 crankshaft. In a flat-six engine, adjacent pairs of crankpins are always set at 180° to one another so that the reciprocating pistons make the familiar "boxing" movements toward and away from one another, as in a flat-twin. By contrast, adjacent crankthrows in a 60° V6 are set at 60° to one another. See Fig. 5-21.

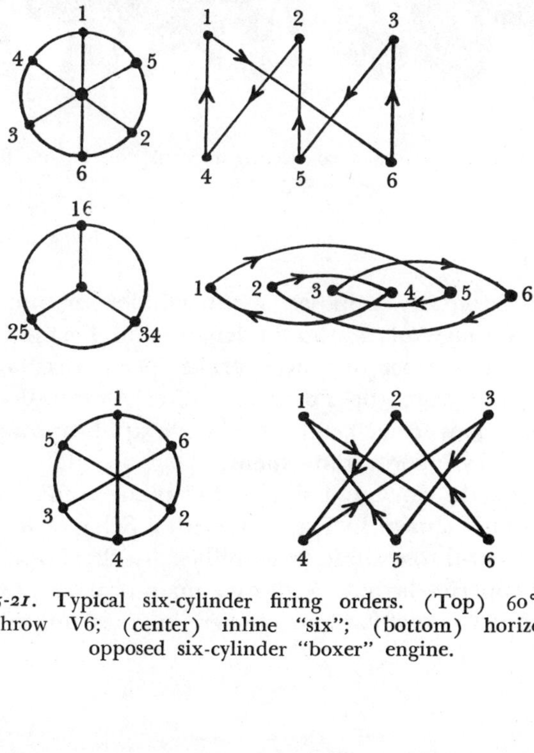

Fig. 5-21. Typical six-cylinder firing orders. (Top) 60°, six-crankthrow V6; (center) inline "six"; (bottom) horizontally opposed six-cylinder "boxer" engine.

The flat-six fires cylinders in alternate banks, as a good engine should, so there are no difficulties involved in making an exhaust system that takes advantage of pressure-wave principles. Because the flat-six is, in effect, three flat-twins spliced end to end, there is no difficulty in balancing such an engine. Consequently, the Porsche 911 crankshaft shown in Fig. 5-22 has very modest counterweights and an amazingly light overall weight.

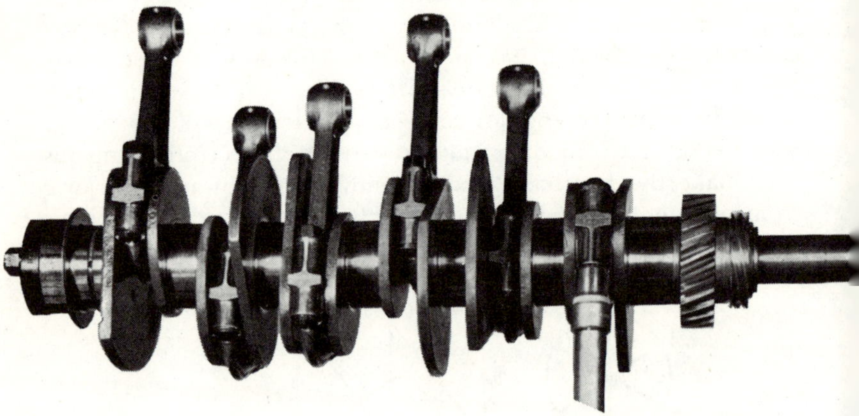

Fig. 5-22. Crankshaft and connecting rods of Porsche 911 flat-six engine.

The Flat-eight

The horizontally opposed eight-cylinder engine has the inherent advantage of minimum length and height combined with an ample number of power strokes per crankshaft revolution. But aside from the Porsche 1.5-liter Grand Prix engine of the early 1960s (Fig. 5-23), very little application has been made of this cylinder arrangement.

The flat-eight can use a similar two-plane crankshaft to the V8, as previously shown in Fig. 5-11, with a firing order 1-7-4-6-5-3-8-2. The eight-throw shaft, exemplified by the Porsche racing engine, is a superior layout. With nine main bearings, this engine can be regarded as two flat-four "boxer" engines end to end with their crankshafts phased at 90° to one another at the center bearing. This gives excellent balance, and the symmetry of firing is well shown in Fig. 5-16 (given earlier), the order being 1-7-2-8-5-3-6-4.

Fig. 5-23. Porsche 1.5-liter flat-eight Grand Prix racing engine.

The Flat-twelve and the Flat-sixteen

At present, flat-twelve "boxer" engines are enjoying considerable success in road racing. Ferrari, long an exponent of the twelve-cylinder engine, has been a dominant force in Grand Prix racing with the Type 312 flat-twelve (Fig. 5-24). Horizontally opposed twelve-cylinder powerplants have also been used with considerable success by Porsche and Alfa Romeo.

Fig. 5-24. The Ferrari flat-twelve Grand Prix racing engine used in the Type 312B chassis. Front of engine, with fuel injection pump and ignition distributor, is in foreground.

A flat-twelve is, in effect, two flat-six engines spliced together, having a crankshaft with twelve crankthrows. Thus, the familiar "boxing" movements of the opposed pistons, first seen in the flat-twin, are retained. Balance and smooth torque are great virtues of this cylinder arrangement. The crankshaft, which resembles two flat-six crankshafts joined in the middle and phased 30° to one another, is rather long but apparently offers no serious problems for skilled and experienced designers.

In recent years, Porsche has also developed a flat-sixteen engine. This is exactly like coupling two flat-eight engines end to end. In turbocharged form, this potent unit developed about 2,000 bhp but was never used as the flat-twelve, which had vanquished all its competitors in international racing, went on in turbocharged form to dominate the Can-Am series, and set the world's closed-course speed record, an honor that for many years had been the sinecure of Indy-type race cars. The "sixteen", it was decided, would simply be a case of overkill.

The V12 and the V16

The V12 has always been an excellent cylinder arrangement, offering extreme smoothness and uncomplicated balancing. The crankshaft itself is identical to that of an inline "six" but with two connecting rod big ends on each crankpin, as in a V8. The Ferrari V12 engines have all been 60° layouts, as are the present Jaguar V12 and the various V12 Grand Prix racing engines of the current 3-liter formula (Fig. 5-25). But there have been exceptions.

Fig. 5-25. BRM V12 engine as used in Type 153 Grand Prix car.

The Cadillac V12 of 1935 through 1937 had a 45° angle between cylinder banks, and for a very important reason; it shared virtually all of its working parts with the Cadillac V16 engine that was built from 1935 through 1940. A 45° angle is, of course, half of a 90° angle, and it is possible to build a 45° V8 with a two-plane crankshaft that has even firing intervals and will fire cylinders on alternate banks. So it is also possible to build a 45° V16 with even firing intervals—eight overlapping power strokes per crankshaft revolution—that fires cylinders on alternate banks. The crankshaft in this case is identical to that of a straight-eight but with two connecting rod big ends on each crankpin.

The General Motors practice has always been to give odd numbers to the left bank cylinders of V-type engines (1-3-5-7-9-11-13-15) and even numbers to the right bank cylinders (2-4-6-8-10-12-14-16). From the firing order of the Cadillac OHV V16 (1-8-9-14-3-6-11-2-15-10-7-4-13-12-5-16) it can easily be seen that the cylinders do indeed fire on alternate banks. The crankshaft of the Cadillac OHV 45° V12, being like that of an inline "six", did not provide uniform firing intervals. Instead, the engine ran as though it were two inline "sixes" that were 45° out of phase with one another. The cylinders did, however, fire on alternate banks (1-4-9-8-5-2-11-10-3-6-7-12), and ignition complications were solved by having a separate six-cylinder distributor for each cylinder bank. Careful analysis will show that this firing order is identical to the one used on today's V12 Jaguar; the Jaguar, fortunately, being a 60° V12, produces evenly spaced firing intervals.

Other Cylinder Arrangements

The V4 engine has had some competition success. Lancia used this layout in the 1950s and, more recently, Ford's German Taunus cars were equipped with a V4. Though the Taunus had some wins in international rallying, the Ford V4 has done most of its racing in SAAB cars, which are especially competitive in ice racing. This engine, shown in Fig. 5-12, had a 60° angle between its two banks of cylinders; one pair of crankpins on its four-throw crankshaft was set at 120° to the other pair, giving absolutely even firing. See Fig. 5-26.

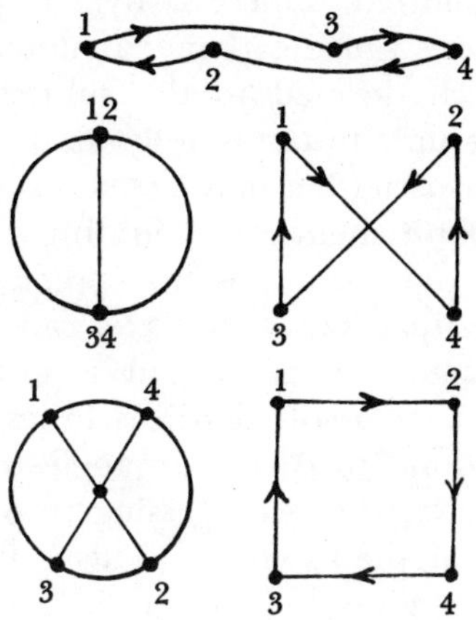

Fig. 5-26. Four-cylinder firing order comparison. (Top) inline "four"; (center) horizontally opposed flat-four "boxer" engine; (bottom) Ford 60° V4.

Five-cylinder inline engines are beginning to appear in passenger cars. Although Ford experimented with this kind of engine in the 1940s, the first inline "five" to reach the public was the Mercedes Benz 300D diesel-powered sedan. Audi has introduced a 2.2-liter five-cylinder inline spark-ignition engine in its 5000-C sedan, making it the first "gas-burner five" on the market.

The main appeal of the five-cylinder inline engine is that it offers greater piston displacement than a "four" and more power impulses per revolution without the added length and complexity of a "six". Four-cylinder engines of above two liters displacement tend to be somewhat rough, because of the large displacements of the individual cylinders. The five-cylinder, with five equally disposed crankpins on its crankshaft, offers a way out. As yet, no such engine has been prepared for competition use.

Engine Mounting

With the engine balanced as well as possible, there still remains the problem of mounting it in the car so that vibration will not be transmitted unnecessarily. In passenger cars, the availability of a wide range of bonded rubber mountings, along with much research devoted to the subject, enables virtually any type of power unit to be securely located; at the same time there is no danger of undue movement.

The basic requirement in mounting the engine satisfactorily is to relate the vibrations that occur in the engine as a whole to the principal axis. This axis can be considered as a line passing through the engine in a fore-and-aft direction, along which the moment of inertia is lowest. In practice, such a line would be found to run at an angle from high up at the front to low down at the rear, passing through the crankshaft centerline at or near the flywheel. (This is in the case of an inline four-cylinder engine.)

Because there are few places that high front engine mountings can be attached on the modern car, most inline "fours" have three-point mountings. The two side mountings, about halfway along the length of the block, are placed near the height of the principal axis. The third mounting is placed at the bottom of the rear end of the transmission near an extension of the principal axis at a point well aft of the flywheel.

Care must be taken in mounting an engine differently from the way that it was designed to be mounted. A mounting that is good from the viewpoint of taking the weight of the engine can cause extra stresses on the block casting at unsuitable points. When the OHV Chevrolet V8 came along in the 1950s, it was not long before hot rodders and dirt-track stock car racers began installing this engine in cars that originally had Ford flathead V8s. The Ford engine used two mountings at the front of the block; the Chevrolet engine had centerpoint mountings of the kind described above for four-cylinder inline engines. The Chevy, when outfitted with front mountings to match the old Ford installation, frequently suffered broken block castings and warped internal parts as a result of the inappropriate distribution of stresses.

Competition cars nearly always use solid engine mountings, and the most solid kind of all are those used with the Cosworth Ford V8 Grand Prix engine. This powerplant has been designed to serve as part of the car's structure. The monocoque "tub" that forms the chassis of the typical Grand Prix car is bolted to the front end of the engine. The engine itself provides the support for the transaxle and the rear suspension and wheels. Thus, the engine must be very rigidly mounted because it is literally part of the chassis.

An engine must be designed for this kind of structural use from the first. It is not possible, say, to use a Chevy small block V8 or a Ford Cortina "four" as a structural member of the chassis because these engines were not designed for that kind of service. The weight reduction made possible by this unique feature of the Cosworth-Ford V8 design should certainly be reason enough for other designers to adopt this approach in the future.

Bore/Stroke Ratio and Other Considerations

The choice of a bore/stroke ratio is one that requires a number of conflicting factors to be taken into consideration. Recently there has been a tendency to oversimplify the question when reviewing designs of engines in which a drastic shortening of the stroke has been carried out. Upon investigation, it will frequently be found that the dimensions concerned have little or no bearing on the power produced or the overall efficiency of the power unit and are adopted almost solely from the viewpoint of costs or to meet installation requirements.

Obviously the significance of the stroke length is that for a given number of rpm, the longer the stroke the higher will be the piston speed, and the latter, in general, represents the limiting factor in the top rpm that it is possible to obtain from any particular engine. Assuming that the same working pressure can be maintained in the cylinders, the horsepower of any engine will increase in direct proportion to the increase in rpm. Thus, if an engine of a specified cubic capacity is redesigned by shortening the stroke, increasing the bore to give the same cubic capacity—and increasing the rpm to give the same piston speed as the original—the engine will produce more power.

A study of current engine specifications reveals a considerable divergence of opinion as to what constitutes the most desirable ratio of bore to stroke, and one might be tempted to inquire whether in fact these dimensions are so important since the power units concerned all seem to perform quite well. First of all, it will be useful to consider the dimensions from the point of view of what is best for power production and sound mechanical design. We can then review other influences that have a bearing on the makers' choices.

Increases in engine output in recent years have come from improved breathing and mixture distribution, higher compression ratios for taking advantage of better fuels, and an increase in certain cases in maximum rpm. The last can be ignored for the moment; it springs from the first item as a result of the engine's ability to "charge" itself adequately because of improved induction characteristics.

These breathing improvements call for the use of adequately sized valves and correct port design—not to overlook the possible use of turbocharging in this regard. In port design, the cylinder head casting is a complex structure, and the ports have to find their ways past headbolt bosses, water passages, and so on. The spark plug, too, must come in correct relationship to the valves and to the general shape of the combustion chamber.

It is obvious that the larger the cross-section of the head the better, since more room will be available. However, the cylinder bore must be sized to match the head, and any decision on this measurement will automatically fix the stroke dimension for a specified cylinder capacity.

Compression ratio is influenced considerably by the length of the stroke. It is not difficult to visualize the impossibility of obtaining the required compression ratio if the cross-section dimension of the cylinder head is carried to absurd limits, remembering that the valves must have room to operate above the piston. However, it might be necessary to use domed piston crowns, designed to "take up space" at tdc, that would mitigate against efficient combustion by restricting flame travel. Thus, to obtain the specified compression ratio and at the same time to allow for the use of a piston of rational shape (so as not to throw away on the power stroke all the advantages gained by the aids to good breathing), a careful choice of stroke dimension is necessary.

Piston Speed

If the crankshaft speed of the engine remains unaltered, shortening the stroke length will reduce the piston speed; this is the virtue most frequently offered by champions of the short-stroke engine. For any reduction to be worthwhile as a virtue in itself, the original piston speed must have been excessively high. Between wide limits, piston speed has little or no effect on engine efficiency, and it is only when the speed exceeds some 3,000 feet per minute on normal passenger car engines that one need be at all concerned; competition engines can go higher than this because of the greater attention paid to precision balance and choice of materials.

There is one benefit of short stroke designs that is undeniable: short strokes mean lighter and more compact engines. Both of these characteristics are highly important advantages in a competition powerplant. We need not look very far to find examples. Despite being of cast-iron, the engines of both the Chevrolet small block V8 and the English Ford Cortina are exceedingly light and compact for their displacements and power outputs. These factors have been responsible for their wide use and extensive development as successful competition engines in formula-type racing cars. Nevertheless, the working parts of such engines must be designed with considerable strength.

The reciprocating action in the conventional engine sets up bearing loads that are very different from those in purely rotary motion, and these are particularly severe at the connecting rod big ends. The loading on crankshaft main bearings can be fairly evenly distributed with a shaft in good balance so that wear is spread around the bearing shell and the crankshaft journal. In the connecting rod assembly, however, the reversal of movement at the top and bottom dead centers produces an alternating "push-and-pull" load in the rod; this is naturally reproduced as a shear stress at the piston pin and piston bosses and communicates a heavy load to the big end eye at its bearing on the crankpin. These loads are independent of the power loading caused by the cylinder explosion pressure, and they increase in proportion to the square of the rpm (that is, the loading at 4000 rpm is four times that at 2000 rpm).

Obviously the weight of the reciprocating parts must influence the loading to a marked degree, and in this respect a big-bore piston will be heavier than a long-stroke/small-bore piston. The same amount of power is available for power production (from equal-capacity cylinders) whatever the bore/stroke ratio. It is simply a question of whether it is produced by fewer psi on a small-bore piston, acting through a greater distance, or by more psi on a big-bore piston, acting through a shorter stroke. In the former case, the power loading on the big ends and so forth will naturally be less than the latter.

More Power

In the 1960s demands for more power from production car engines were met in some cases by the normal processes of increasing the engine speed, more often by an increase of cubic capacity—sometimes both—and by cylinder head and induction system modifications. Thus, in any original design, it would seem best to allow for the possibility of bore (and perhaps stroke) increases and the use of larger valves at some future date. On engines where such items have been ignored or not allowed for in the original layout, the limit of performance is obviously reached sooner than if adequate forethought had been used. For instance, a piston stroke dimension that is excellent at 4200 rpm may result in an undesirably high piston speed at 5200 rpm. Valve port areas that give excellent pulling power for use with a conventional manual transmission may be too small if the engine is later converted to operate with an automatic transmission at generally higher average rpm.

It is thus difficult to state that any particular ratio of bore to stroke is the one that the designer would have chosen for maximum engine efficiency in that particular power unit; there are, of course, cases where this is done, and these are not difficult to recognize. But in the main they are found in the extra-high-quality, limited-production class of cars. In America, during the years of, first, the "horsepower race" and, later, the "cubic inch race", very little thought was given to proportions. The piston displacement was increased almost annually by any means left available.

We can investigate the effect of stroke length on bearing loading, as it applies to the power stroke, in line with this fact, by using a simple example. Suppose we have two 1000-cm^3 four-cylinder engines, each of the same bhp and rpm; the difference is that one of them has a stroke length twice that of the other. Since each cylinder must provide 250 cm^3 in each case, the bores must be dimensioned accordingly; if the long-stroke engine has a stroke length of 100 mm, its bore will be 56 mm. The short-stroke engine, with a 50-mm stroke, will have a bore of 80 mm.

If we assume the peak rpm in both engines is 4500, it will be obvious that the piston speed on the short-stroke engine will be halved in comparison with the other. In figures, the former piston speed will be 1640 feet per minute (fpm), which is very much below even a normal cruising speed figure. On the long-stroke, however, this will become 3280 fpm, which is a shade high for comfort unless the engine has exceptional standards of materials, balance, and workmanship. Note also that the foregoing are the *mean* speeds. The maximum, which occurs momentarily when the crankthrow and the connecting rod are at a right angle, with the big end at the "hinge", would be 2616 and 5232 fpm, respectively. A piston speed for normal fast running of 2500 fpm (mean) is considered a good average value; this is only just exceeded as a maximum by the short-stroke engine.

In connection with increased power, however, supercharging can be used without increasing either the displacement or the rpm needed to obtain the added output. However, other problems are sometimes produced. Nevertheless, both the Offy and the Cosworth-Ford V8—neither of which were originally designed as supercharged engines—have in recent years, along with the Ford/Foyt V8, been adapted to high-boost turbocharging for use in USAC oval-track racing.

Bearing Loadings

The short-stroke engine generally shows up advantageously

in most respects, and the design can be expected to influence cylinder bore wear and smoothness of running; both of these factors depend to some extent on subtleties of manufacture as well. It can also be argued for the alternative, fast-moving piston that there is less time for blowby past the rings to take place.

Now we have to consider the actual development of power. If we imagine that the mean effective pressure available for the power stroke is 120 psi, we can ascertain the total pressure on the respective pistons; the 56-mm piston in the long-stroke engine will have an area of 3.9 square inches, while the 80-mm piston will have an area of 7.8 square inches. This is apparent; having halved the stroke, we must double the pressure available. At 120 psi, the respective pressures will be 468 psi and 936 psi.

Reducing piston speed very much reduces the bearing inertia loads, but it doubles the power-stroke load (this is also true of the compression stroke). However, this considerable extra stress occurs only every two revolutions, over a crank angle of some 120°, and on this count must be preferable to excessive inertia loads from the aspect of bearing life. In considering the reduction of inertia loading by shortening the stroke, the larger-diameter piston on this engine—of nearly 50 percent greater diameter—will be considerably heavier, and this will to some extent counteract the effect of the lower piston speed.

Nevertheless, of the two, reduction in inertia loadings is of greater importance than increase in power loading, if only because of the greater frequency of the former. In the most famous "80-bore" engine, the Ford Cortina, it is interesting to note that in five main bearing, 1600 cm^3 form this tiny powerhouse has a bearing area that is equivalent to that of many V8s. This perhaps explains how these engines can survive at the very high outputs that are obtained in supertuned examples.

Correct Proportions

The question might well be, what then is the best ratio of bore to stroke? It could be argued that if the effect on big end life and reliability is so apparent, an absolute minimum stroke should be sought. Aside from the obvious emission control problem in an ultra-short-stroke passenger car engine, there is another catch that will be apparent: the power loading must be kept to a reasonable figure and must also strike a balance with the other bearing loads so that irregular wear is minimized.

Adequacy of cooling surface can be a problem when increasing the bore of a basic engine, since it is probable that the cylinder spacing is already at a minimum to keep down the engine length. In such a case, there might be recourse to siamesing of adjacent cylinder pairs to enable these to be bored to a larger diameter. This practice, however, is undesirable because it prevents a complete circulation of the coolant around each cylinder barrel.

Valve timing errors are likely to show up rather more with a shortened stroke because the piston movement is small for a considerable amount of angular crank movement. Thus, wear in the camshaft drive and the valve gear can affect engine tune more seriously.

As a basic rule, it seems that an initial "square" ratio—bore and stroke of equal dimensions—is the best compromise, providing that any anticipated capacity increases do not make the bore unduly large, which is hardly likely since there are definite limits to the degree of overboring within the confines of an existing block casting. A condition to be observed when applying this generalization is that the piston speed must not become excessive. When components are produced in vast quantities with, inevitably, fairly generous margins of tolerance on weight and dimensions (as compared with specialized racing components), piston speed should be kept on the low side of normal. If 2500 feet per minute is accepted as something near maximum, all to the good. If it is argued that it is physically impossible to meet these requirements when a large-displacement engine is being considered, the answer is to use more cylinders.

Champions of the short-stroke, "square", or "over-square" engine may well point to the extremely lively performance of this kind of unit in popular vehicles that have recently adopted such dimensions. It should be kept in mind, however, that the engines in question are new designs, replacing engines that have had a long record of faithful service. In these circumstances, any new design should be expected to show a performance commensurate with the general progress that has been made in nearly all aspects of design. In other words, the liveliness must not be put down entirely to the particular bore/stroke ratio that is used. Of course, short-stroke designs do tend to be lighter and more compact.

An additional good point for the big-bore engine is that it leaves the designer ample room for valves. Four valves per cylinder (Fig. 5-27) is the normal number in today's short-stroke racing engines, and this arrangement is used sometimes in exotic GT machinery. Generous valve and port areas greatly extend the power curve upward, permitting very high rpm before the powerplant begins to run out of breath. In supertuning production engines, it is quite normal to enlarge the valves and ports considerably, and modern big-bore cylinders create no great problems in this respect. In fact, valve size is mainly limited on high-displacement V8s by the availability of large valves.

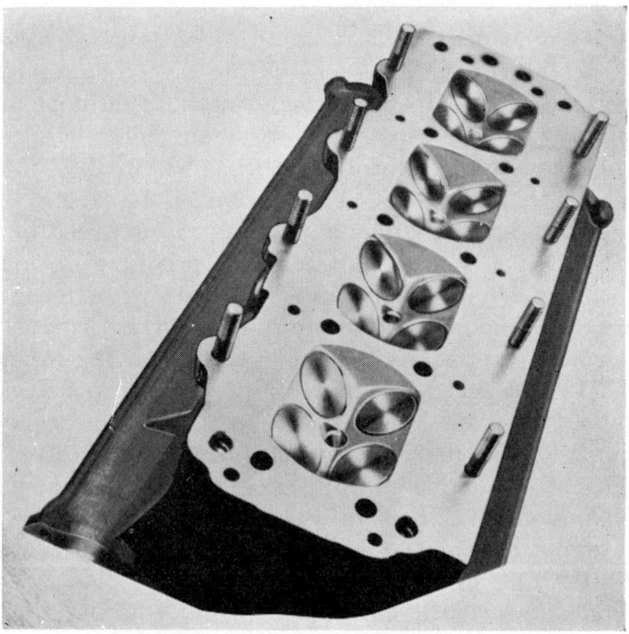

Fig. 5-27. Four-valve-per-cylinder combustion chambers of Ford/Foyt Indy V8.

Hemispherical Combustion Chambers

Where hemispherical combustion chambers are used, any reduction of stroke means a problem. This combustion chamber shape is correctly considered excellent from a thermal efficiency point of view, but it also results in a relatively large space above a flat-topped piston. Even with what might be considered an old-fashioned bore/stroke ratio, it is usually necessary to adopt a domed piston crown to achieve a compression ratio that is suitable for the advantageous use of modern high-octane fuels. A moderate dome on the piston (Fig. 5-28) does little to detract from combustion efficiency, but if the process is carried to extremes, the shape of the combustion space at top dead center might well be likened to half an orange skin.

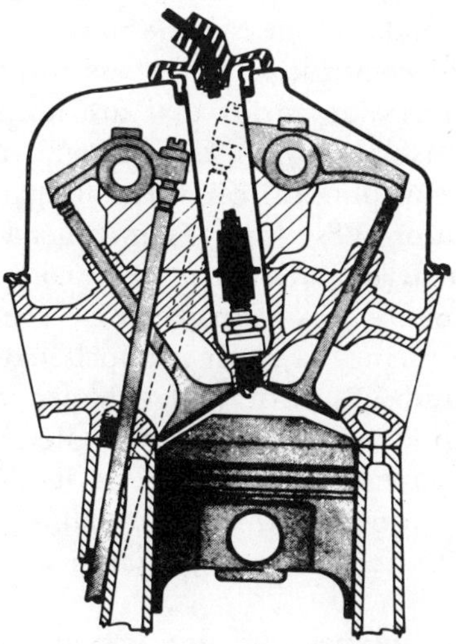

Fig. 5-28. Cross-section of Plymouth "Hemi" cylinder head shows moderately domed piston that is used to achieve a sufficiently high compression ratio.

A study of drawings of high-rpm, very short-stroke racing engines will show that such is often the case. The difficulty of flame propagation in such a space can readily be appreciated, particularly when the spark plug is centrally located (this normally is an almost ideal place for the plug). The use of two spark plugs per cylinder is one way of improving on the situation, but no designer wishes to clutter up his combustion space with a multiplication of uncooled objects if he or she can avoid it and will resort to this solution only when all else fails.

The truth is that hemispherical combustion chambers are somewhat old-fashioned. When strokes were long and bores small, a dome-shaped combustion chamber roof was a means of providing greater room for the valves, in addition to creating a thermally efficient place to burn the mixture. No great problem exists in this regard when the bore is large and the stroke short. Consequently modern racing engines, such as the Cosworth-Ford V8, have combustion chambers that are conspicuously flattened in comparison to a true hemisphere; the pistons are not highly domed, and there are reliefs cut into the piston crowns to make room for the valves.

Hemispherical combustion chambers continue to work well at the lower compression ratios that are normal in production cars. The Dodge Colt, Toyota Corolla, and certain other Japanese cars have hemispherical combustion chambers—as do some Chrysler Corporation V8s (the "Hemi" head having been virtually a Chrysler trademark since that company began using it in high-performance V8s a quarter of a century ago).

In NASCAR racing, Chrysler products have raced with both Hemi-head V8s and V8s with wedge-shaped combustion chambers. However, in drag racing the Chrysler "Hemi" continues to dominate the supercharged classes as it has for many years. There is a good reason for this performance, which deserves to be considered with regard to hemispherical combustion chambers in general.

Supercharged engines do not require high compression ratios; rather, they need combustion chambers of considerable volume to provide more space for the high-boost supercharger to cram mixture into under pressure. For example, though the "mechanical" compression ratio of such a cylinder may be only 5:1, when given the pressure supplied by the supercharger, the pressure on the mixture in the cylinder may be as great as in a normally aspirated cylinder having a 10:1 ratio. In addition, because of the greater space available in the combustion chamber, perhaps twice the quantity of mixture can be placed there and under the same pressure that would be possible with normal aspiration. Consequently the power produced is approximately twice as great, though the compression ratio is only half as high. From this point of view, the Chrysler "Hemi" is an ideal engine.

6 / *Planning for Performance*

Performance Data

In chapter 1, we touched on the definitions of horsepower and torque and the soon-to-be-adopted SI method of rating engine power in kilowatts. Whether given as horsepower or kilowatts, all such figures can be considered as published performance data.

At one time, most car makers were rather reticent on the subject of engine performance figures. Then we passed through the "horsepower race" era of the 1950s and 1960s when manufacturers made extravagant claims based on the most tenuous kind of dynamometer evidence. Now, because of frowns from safety officials (and the shamefully low horsepower outputs of emission-controlled engines), the manufacturers are once again saying very little about horsepower.

Their reticence does not mean a great deal to the competition engine builder or the racing tuner. The professional is mainly interested in two things: how the engine he is working with performs on his own dynamometer and how his dynamometer stacks up against his rivals'. No two dynamometers read the same, so it is always helpful to know that So-and-so's dyno reads seven horsepower higher than yours while What's-his-name's puts out readings that are about four horsepower lower.

Cylinder Head Design

Nowhere is fuel so plentiful that its inefficient use in quantity can be considered a substitute for power obtained by efficient burning. All other factors being equal, the competition engine that burns its fuel more efficiently will produce the greater power. Hence, good cylinder head design is of utmost importance to any engine—be it a racing engine or a low-powered production powerplant.

That a suitable combustion chamber and port arrangement is indispensable is shown vividly in the decades-long struggle to invent a successful rotary engine. Many experimental rotary designs might have functioned very well as rotary air pumps. However, only the Wankel rotary engine has achieved a combustion chamber shape that burns fuel with a high degree of efficiency. Even with the Wankel, despite the high degree of success already attained by various makers' designs, it is in the areas of combustion efficiency and port arrangements that most development work is being concentrated.

Overhead Valve Layouts

For maximum efficiency—the conversion of as much of a given quantity of fuel into power as possible, together with high power output—there is no question about the merits of the OHV engine. It is usual when considering what advantages accrue to the OHV type—and we are speaking not only of pushrod units but also those with double or single overhead camshafts—to refer to the disposition of the valves as an aid to good cylinder filling at the higher end of the rpm range when the time intervals for induction and exhaust strokes are shortening. It can, however, be argued that at average engine speeds this is not so important and, in fact, the old L-head engines were satisfactory on this score as witness the good torque figures obtained with the latter type. This argument is sound, and it is useful to consider other advantages of overhead valves apart from the fairly obvious merits at high rpm.

Compression Ratio and Heat Loss

In order to produce as much power as possible, no matter how much mixture has managed to get into the cylinder, we have to ensure that after ignition the gases are expanded so they produce the maximum pressure on the piston throughout the length of the working stroke. To meet this requirement, the compression ratio and the pressure before ignition should be as high as possible. In addition, the heat loss to the surrounding metal must be the minimum possible.

We showed in chapter 2 that very little power can be gained if all heat loss is stopped. However, in regard to compression ratio, an overhead valve cylinder head with a reasonable mechanical layout will allow up to 12:1 without difficulty. In other words, the compression ratio is governed mainly by the fuel octane, and not, as is the case with the primitive L-head, by mechanical considerations.

The advantage of the overhead valve cylinder head in allowing a relatively unimpeded entry and exit for the gases is evident from a study of Fig. 6-1 and other drawings of engines in this book. Further, with the valves in the head, it is possible to vary the positions of the components within a wide range and also to design the combustion chamber and piston crown to promote turbulence.

Altering the position of the spark plug in relation to one or both valves can effect a considerable change in the engine's characteristics. In addition, the provision of a "squish" area at one or both sides of the head, with the resultant draught directed as desired, can also be advantageous. It is worthwhile to investigate some of the possible arrangements before going more deeply into the technicalities.

In early OHV engines, the combustion chamber consisted of an extension of the cylinder bore, that is, circular in plan. Since it was impracticable to mount the spark plug centrally in the head because of the small amount of room between the valve seats, the spark plug was carried horizontally at one side—usually the side remote from the ports. This form of combustion chamber was devoid of any "squish" but had generous spaces around the valves and a refreshing absence of pockets. Because of the spark plug position, the rate of flame travel was found to be considerably affected by engine rpm and throttle opening, and the best results were obtained when a hand spark advance control was provided for use at lower rpm. Mixture strength somewhat on the rich side was also desirable, for the same reason. Evidently the degree of turbulence set up in this head shape was limited, and although the performance could be improved by experimenting with the plug position, the general arrangement gives little scope for this. The shape has thus been generally superseded by designs that show improved results in the ability to burn weaker mixtures and are less sensitive to ignition timing.

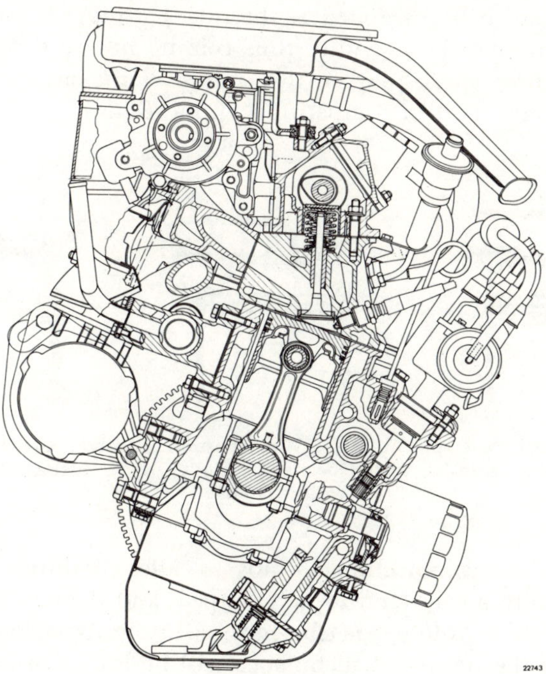

Fig. 6-1. Cross section of Fiat 128 engine showing excellent port shape of modern OHV designs.

Turbulence Features

A simple and effective method of producing a head shape that promotes turbulence, and one that seems to have been a logical step in development, is to make the chamber containing the valves (and forming the combustion chamber) slightly narrower than the bore diameter in the dimension at right angles to the crankshaft (Fig. 6-2). This means that when the flat face of the cylinder head is in position, a ceiling forms on two opposite sides of the bore, which is just clear of the piston at TDC. This "squish" feature, as well as promoting turbulence, reduces the area of the clearance space in comparison with the type previously described, and this enables very high compression ratios to be used if desired—without the need for an abnormally long stroke, which was often resorted to in early engines. The spark plug, though still positioned at the side, has moved inward with the reduction in head width, thus coming nearer the valves. A projected-nose spark plug further improves the situation and is nearly always used in competition applications.

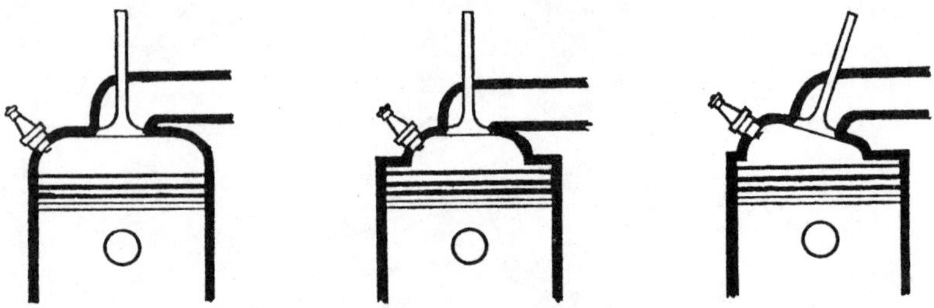

Fig. 6-2. Development of OHV combustion chamber, introducing (center) "squish" turbulence and (right) valve inclination for easier gasflow.

An obvious objection to this so-called bathtub layout is that the valves may tend to be cramped, and if breathing is unduly restricted, volumetric efficiency will suffer, and much of the benefit of the improved turbulence will be lost. It is important, therefore, not to make the valve space too small. As a matter of fact, early engines of this kind did not show the hoped-for advantages over the earlier type, mainly because of breathing restriction. We shall see from later investigations that the chief requirement of valve arrangements is to get the gas in and out with as little restriction as possible; all other aids to efficient combustion, desirable as they are, must not prejudice this.

Without departing from the simplicity of the single row of valves in the head, it is possible to arrive at a combustion chamber that in respect to both high torque figures and good full-power output is not very far behind layouts of much more complicated form. A slight deviation of the valves from the vertical will provide a freer entry and exit to and from the valve ports. At the same time, if the valves are moved slightly away from the center line of the head and the "squish" area concentrated opposite the spark plug, which is situated on the deepest side of the combustion chamber, several desirable conditions will be fulfilled.

First, in this so-called wedge combustion chamber, the all-important requirement of free gas entry and exit has been met. Second, a high degree of turbulence will be set up in the direction of the spark plug, which, together with the deep section of the chamber around the plug, will give an adequately large flame-front area to ensure uniform and complete combustion. Here, again, a projected-nose spark plug usually offers a decided advantage. Last, the shape of the combustion chamber is still of such form that high compression ratios are practicable.

Versatile Design

Of the three combustion chamber shapes shown in Fig. 6-2, the first is hardly ever used now. The bathtub chamber and the wedge chamber are popular in every price class and can be developed to give extremely high power outputs. Two typical combustion chamber shapes, which have been evolved from the classical bathtub and wedge by specialists in combustion techniques, are shown in Fig. 6-3. Also worth mentioning is the Heron combustion chamber, or combustion cell, which consists of a flat cylinder head with a bathtub depression in the piston crown, directly under the valves. There is little danger of the valves being shrouded with this arrangement, and the Cortina and Pinto engines used in Formula Ford racing have this kind of combustion chamber. Its development, however, has been sidetracked because of its tendency toward excessive exhaust emissions.

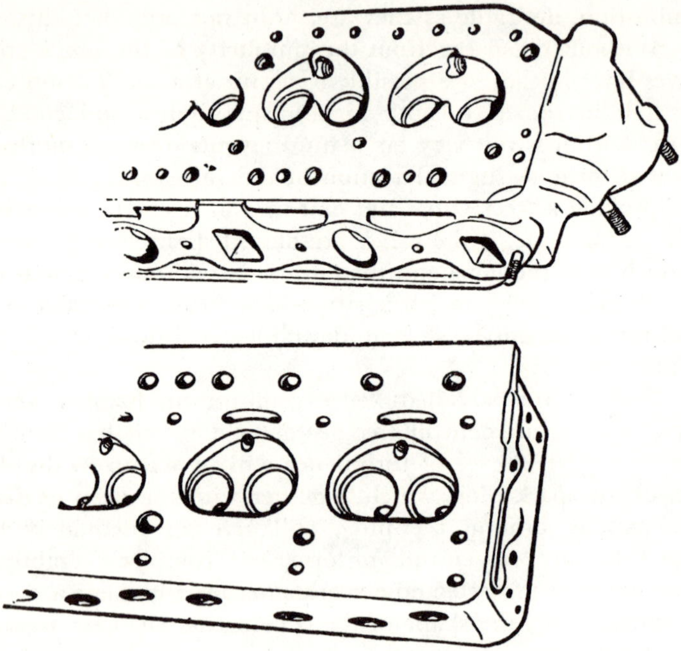

Fig. 6-3. Two typical designs of OHV combustion chamber showing different approaches in regard to gasflow.

With all the ports arranged on one side, there will inevitably be a considerable inequality of temperature. Much thought has therefore been devoted to using plenty of metal at the right places, employing aluminum alloys, and having the coolant flow directed to the hotter spots. Distortion caused by temperature variations does not necessarily lead to cracking of the casting, but it can be troublesome in causing a loss of power through valve leakage, increased friction, piston ring blowby, and so on. The vicinity of the exhaust valve is an obvious danger point, as are any other places where a thickness of metal is not split up by water passages.

At one time, OHV cylinder head designs left something to be desired in these respects—a factor that kept the L-head engine in production for many years longer than it should have been. In current engines, however, there is little cause for complaint, and well-engineered examples, such as the VW Dasher/Rabbit/Scirocco engine, are capable of a performance that is equal to that of other arrangements without suffering at all structurally.

The trend has been away from single port face cylinder heads—those with the intake and the exhaust ports on the same side—to crossflow cylinder heads, which have their intake ports on one side and their exhaust ports on the other. A great deal has been learned about this layout because it is used universally on V8 engines. Consequently, most new inline four-cylinder designs have adopted the crossflow port arrangement from common V8 practice.

Overhead camshafts, now used on virtually all overhead valve inline engines, have also contributed to making the crossflow head practical. In fact, on older, long-stroke/small-bore engines, a crossflow head design was possible only with overhead camshaft operation of the valves because there was insufficient room between the pushrods to accommodate a port of suitable diameter. This is not so great a limiting factor when the bore is large and the pushrods are not crowded closely together, for example, in the crossflow Cortina/Pinto engines that are used in Formula Ford racing. Though a few overhead camshaft engines, such as the VW water-cooled engines and many Datsun engines, have single port face cylinder heads, the crossflow arrangement clearly predominates now.

Considerations of Gasflow

From the standpoint of gasflow, the crossflow cylinder head does not have any inherent advantage, at least in production cars. However, the crossflow layout does make space available for as large ports as any sane designer would choose to draw. The full potential is clearly indicated in four-valve-per-cylinder designs, such as the one shown in Fig. 6-4. Certainly this arrangement would be impossible using a single port face head layout.

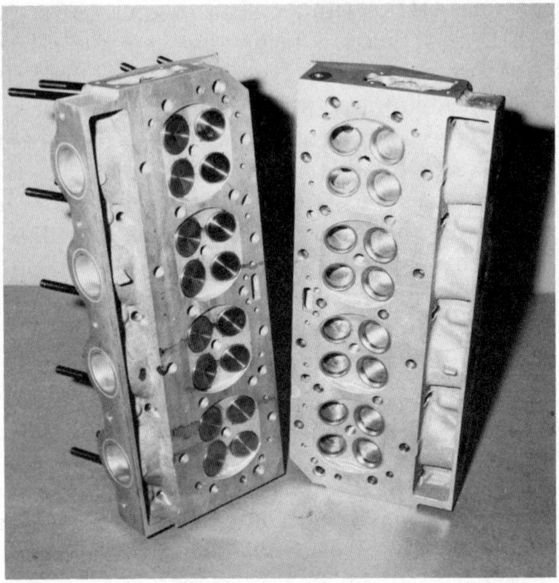

Fig. 6-4. Beautifully made cylinder heads for Moser racing engine. Gasflow is greatly improved by big ports, four valves per cylinder.

On modern single port face cylinder heads, it is the practice to place the centers of the intake ports well above the centers of the exhaust ports (Fig. 6-5). In this way, ample coolant space can be obtained between the ports, the ports can be of larger diameter—overlapping somewhat—and the intake and exhaust systems can be designed with more freedom. The intake ports can also be considerably straightened with this design, further improving the gasflow.

There is another approach to single port face cylinder head design that survives today only in engines that have been in production for twenty years or more. On these venerable powerplants, the ports are siamesed inside the cylinder head (Fig. 6-6). Any person who has tried his hand at dyno-tuning, blueprinting, or supertuning a six-cylinder Chevrolet will recognize the shortcomings of the system quite vividly; it is impossible to get even scavenging and uniform mixture distribution to all of the cylinders.

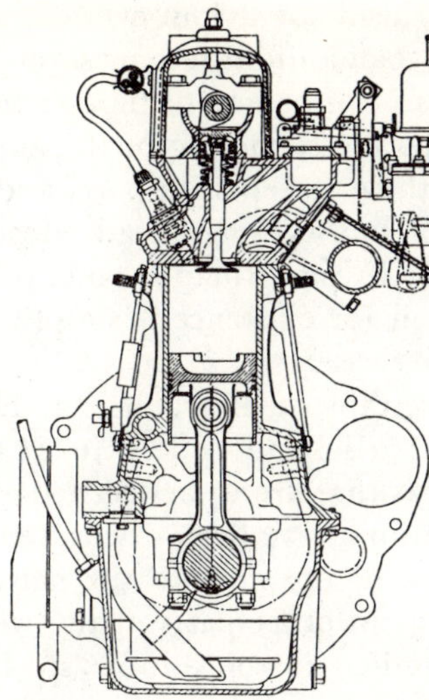

Fig. 6-5. Cross section of Rover 2000 engine. Notice how intake ports are routed upward and thus made more direct, whereas exhaust ports are angled downward.

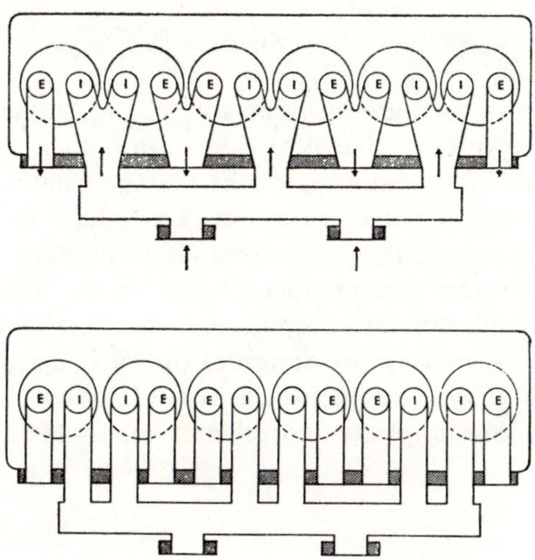

Fig. 6-6. Fully siamesed port cylinder head (top) for six-cylinder inline engine, showing short leakage paths between adjacent valves compared with (bottom) separate-port head.

A fully siamesed head for a four-cylinder engine will have five external ports on its manifold face or seven ports on a "six". This arrangement has been common, though some makers do a little better by deleting the siamesing of the adjacent central exhaust ports, keeping these as separate ports and thus having six ports on the face of a four-cylinder head. Heads with siamesed ports are limited both in performance potential and fuel economy. Their only reason for existence is simplified foundry practice and manufacturing ease.

The MGB and certain other traditional British sports cars with siamesed ports participate widely in SCCA competition. Just about all the tricks that are to be learned about overcoming their breathing limitations have been discovered and widely published. Therefore, we will not go into the matter here. It is interesting, however, that in other parts of the world, where these cars have not been limited to the use of "stock" cylinder heads by racing class rules, it is not uncommon for the factory cylinder heads to be replaced with proprietary units such as the one shown at the bottom of Fig. 6-7.

Port Shape

Engine designers are well aware of the influence of the shape of the valve port, that is, on the "atmospheric" side of the valve. By careful proportioning of the passage from the manifold flange to the actual valve seat, a marked effect on power and economy is obtained. In this respect, it is perhaps remarkable that more advantage has not been taken of the lessons learned by those who develop motorcycle engines. The amount of research that has gone into the design of the valve systems on these engines is one of the reasons for the very high power output and excellent fuel consumption figures for motorcycles.

Intake Ports

In the case of siamesed intake ports, there is no individual port shape. It is possible to do some designing on the passage, starting at the manifold flange and proceeding into the head as far as the bifurcation to the valve ports. The three-zone port evolved at the Weslake Laboratories (Fig. 6-8) is typical of a methodical approach. In this, the first zone aims at building up pressure, hence, its sharply tapered cross-section. Zone 2 is designed to maintain velocity by assisting the gasflow around the obstruction produced by the valve guide and the valve stem. The final zone, at the valve itself, is designed to match the gasflow produced by the preceding zones, that is, it is large enough to accommodate without restriction the gas delivered to the valve but is not so large that the velocity of the gasflow is lost. In short, the port as a whole is designed to provide the maximum possible gasflow.

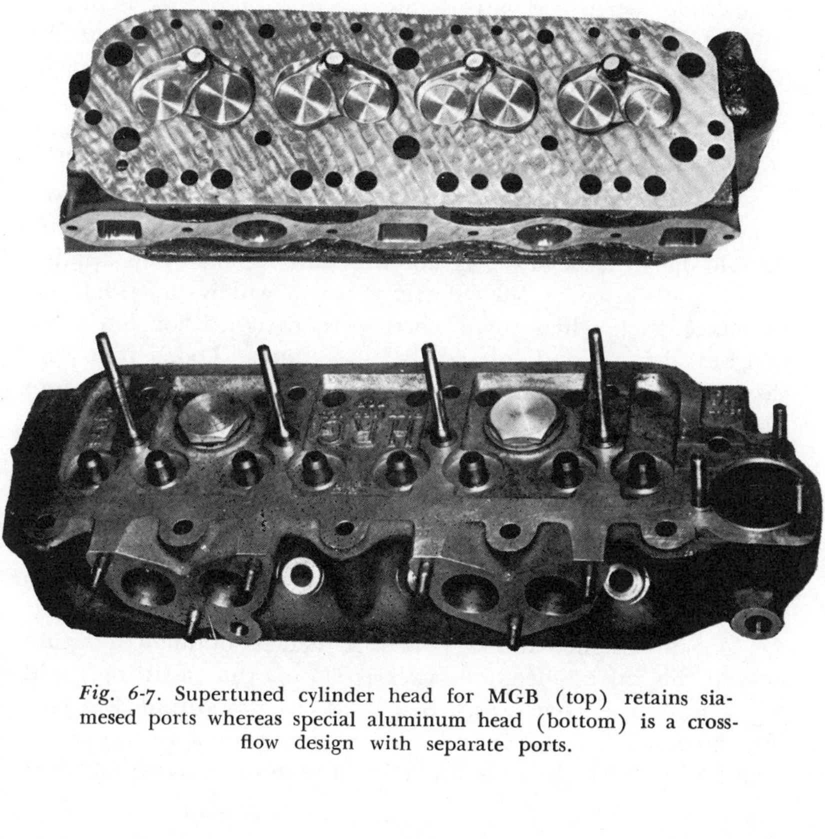

Fig. 6-7. Supertuned cylinder head for MGB (top) retains siamesed ports whereas special aluminum head (bottom) is a crossflow design with separate ports.

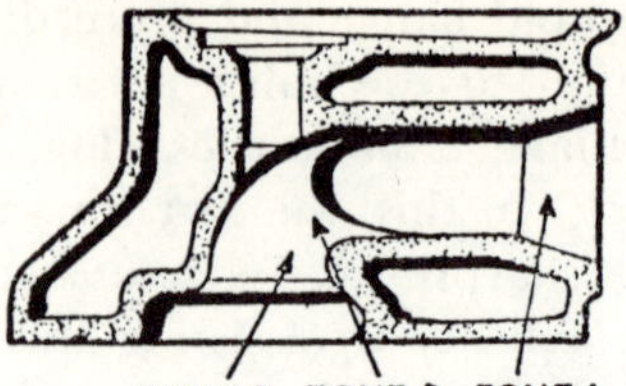

Fig. 6-8. Weslake design of three-zone intake port.

If a single port is used for each valve, several improvements can be achieved. The whole of the passage from manifold flange to valve opening can be proportioned to give a specified velocity to the gasflow; it can also give it some direction in its flow into the cylinder when the valve is open. Speed of gasflow is important—*all important*—since on it depends to a great extent the agitation or turbulence of the mixture, which, persisting right through the compression stroke, influences not only the completeness with which the mixture is burned, but also the nature of the expansion, and thus the feel, of the engine.

In the matter of improving gasflow through improved port shape, it is important to keep in mind that what was good on the old Ford flathead V8 is not necessarily good when applied to a big-block Chevy V8. "Porting" was a widely practiced performance trick when small ports were designed for low-speed highway driving and low compression ratios. Today it is very easy to lose performance if the ports are given anything more than a good polishing. This is true, at least, with a factory high-performance cylinder head. Plainly any increase in port size will cause a loss in gasflow velocity, so what looks good when the head is flow tested might be deceptive insofar as actual power is derived. Really large ports are successful only with supercharging. With normal aspiration, the benefits begin to taper off and then reverse as a certain critical size is reached.

A well-designed intake port will tend to build up pressure behind the valve when the latter is closed, thus assisting rapid filling when the valve opens. It will also be proportioned to take into account the unavoidable obstruction caused by the valve stem and to make its nuisance value a minimum. With adjacent cylinders, there will be far less robbery between pairs with separate intake ports; the path for flow reversal has been greatly lengthened, and the ingoing velocity under pressure is maintained at both intake valves. Robbery would involve a reversal of gasflow in one port back into the manifold. Because of the unidirectional flow at considerable velocity in a well-designed port, it is difficult for this to happen. Thus, with separate intake ports (as opposed to the siamesed variety), more nearly equal cylinder filling, more complete filling, better turbulence—and hence more power from any given volume of mixture—can be obtained.

It is important not to confuse siamesed ports with bifurcated ports. In the case of the former, there is a virtually direct leakage path between the two valve ports. Bifurcated ports, however, while still connecting two valves to one carburetor, manifold flange, or injection duct, do so via individual branches that unite some distance away from the valves. Thus, reversals of gasflow in proximity to the valves cannot take place, because of the individual mixture columns. A typical bifurcated intake port layout is shown in Fig. 6-9.

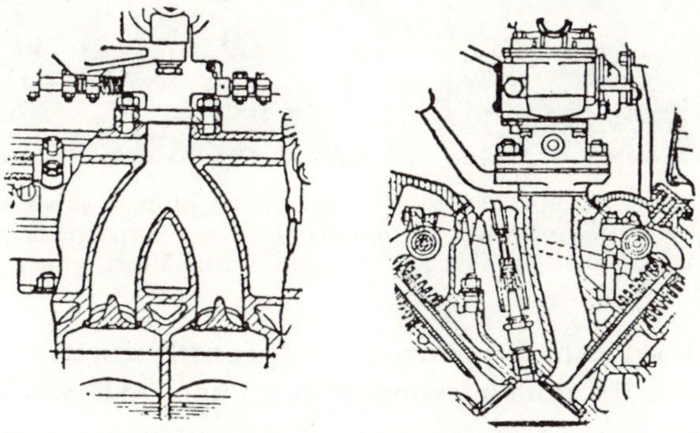

Fig. 6-9. Bristol bifurcated intake port with vertical flow. Volumetric efficiency is high in this design.

Exhaust Ports

Exhaust ports can also be bifurcated. This occurs within the cylinder head itself on the four-valve-per-cylinder Foyt and Offy engines used in USAC oval track racing. It can also occur in the exhaust headers, as on the Moser racing engine shown in Fig. 6-10. As in the case of bifurcated intake ports, reversals in gasflow cannot take place, so there is no penalty involved in the design.

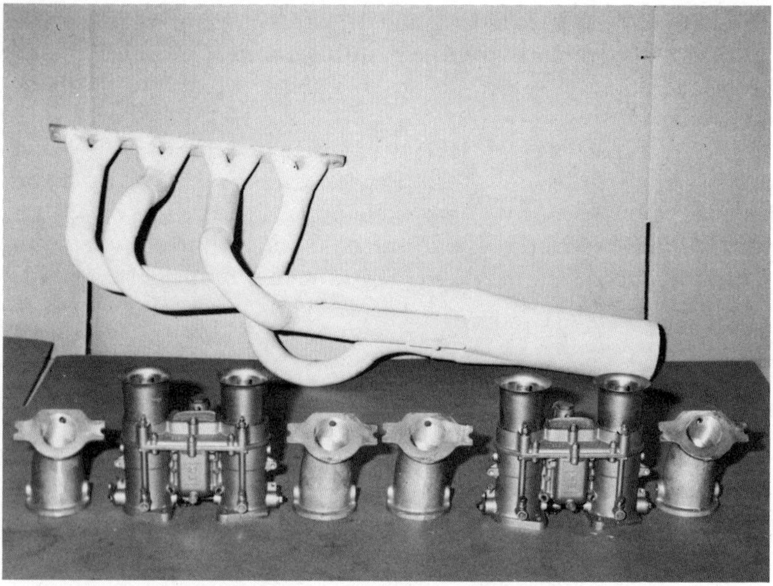

Fig. 6-10. Exhaust headers with bifurcated primary pipes are a unique feature of Moser racing engine. Also shown are short intake pipes and Weber carburetors for Moser engine.

Siamesed exhaust ports are occasionally encountered, particularly on the middle cylinders of a "four". Here again, there is a power loss. Also, the exhaust valve temperatures will be considerably higher on these cylinders because the cooling period available to each valve is curtailed: the cylinder that is exhausting is not only heating up its own valve, but the gases are also allowed to impinge on the stem and back of the adjacent valve. There may also be an uncooled spot in the casting between the combustion chambers.

There is, of course, no actual overlapping of exhaust strokes between cylinders 2 and 3 of a four-cylinder engine, because they do not exhaust consecutively. However, it is essential that the exhaust from all cylinders be cleared away at high speed; it can be done only if the shapes of the exhaust port, pipe, and manifold are correct. Inferior design leads to the possibility of exhaust gas being drawn back into the cylinders at low throttle openings (through the still-open valve) during the overlap period at the start of the intake stroke. A reduction in the exhaust gas velocity, plus a multiplicity of bends and openings at close quarters in the manifold, curtails the useful source of power that is inherent in the energy of the outflow and that assists in scavenging the cylinders.

Exhaust manifolds (of the kinds shown in Fig. 6-11) assist more in fitting the engine into the engine compartment than in assisting the performance of the engine. Great progress has been made in the designs of production manifolds during recent years; however, on V8 engines in particular, the manifold shape is largely determined by the available space.

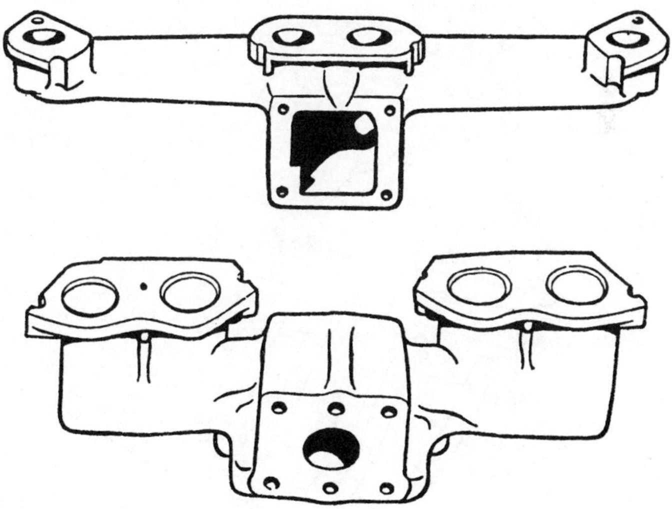

Fig. 6-11. Inferior designs of cast intake and exhaust manifolds.

Special Exhaust Systems

Aside from racing classes that have rules specifying stock exhaust manifolds, special exhaust systems are always used in competition. From a performance point of view, however, there is often a considerable difference between scientifically designed tuned systems developed by racing tuners and the bolt-on "street" headers that are commonly added to production cars by backyard hot rodders. The bolt-on systems from a speed shop or a mail-order catalog often improve a car's performance; however, it is not so much the design of the headers that is responsible for the improvement as the *poor design* of the stock manifold that had previously limited performance.

Unfortunately most "street" headers are designed largely for easy installation and inexpensive manufacture and do not make use of tuning principles that would be beneficial in highway driving. Subdivided systems, such as those shown in Fig. 6-12, are normally best for improving mid-range power; the four-into-one headers, commonly sold for inline "fours" and V8s, are effective mainly at wide-open throttle and high speeds.

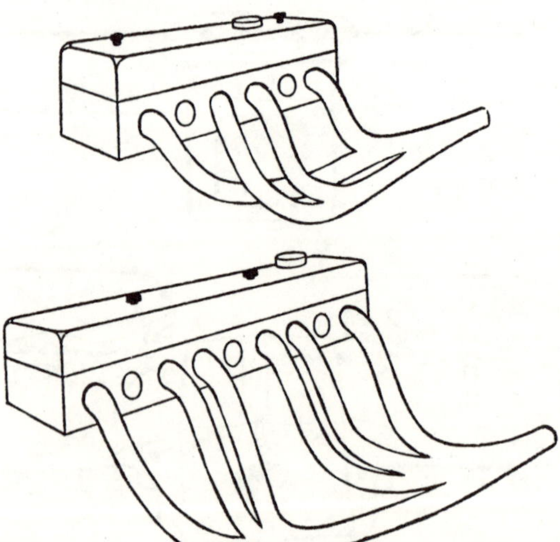

Fig. 6-12. Principle of exhaust manifold subdivision on four- and six-cylinder engines.

In general, these proprietary systems aim at keeping exhausts from individual cylinders separate, using a length of piping that will ensure a minimum intercylinder interference. In addition, the diameter of piping is selected to obtain the correct velocity of gas outflow for average performance. Finally, the individual tubes are led into the common tailpipe in a manner aimed at inducing the cylinders to assist each other, to some extent, in exhausting (by increasing the negative pressure waves in the pipe).

Thus, by using separate ports throughout, there is much more scope for the use of components that will give a really worthwhile increase in performance without complication and with little or no extra expense.

The Combustion Chamber

Reshaping the combustion chamber is now a commonplace method of supertuning, but the results are not always successful. Successful power increases are invariably the outcome of much experiment with selected shapes, though a good deal of preliminary work can be done in observing gasflow in mockup assemblies, obtaining measurements of the flow resistance, and so on.

Research and development on these lines has been responsible for the truly remarkable power outputs now being obtained in all forms of racing from ordinary overhead valve engines with pushrods. Common Ford and Chevrolet engines vie with Chrysler Corporation and American Motors powerplants on the drag strip, the dirt tracks, and the high-banked NASCAR superspeedways—all giving power outputs that thirty years ago would have been unobtainable from the most sophisticated double overhead camshaft designs. Even in road racing, supertuned pushrod V8s, with radically redesigned combustion chambers, commonly rule the big-engine classes—until confronted by something truly exotic and expensive, such as a turbocharged twelve-cylinder Porsche.

It is usually assumed that, for sheer high power, the hemispherical combustion chamber, which, with inclined valves, requires a somewhat complicated mode of pushrod operation, is unsurpassed. This is true, but the normal wedge or bathtub combustion chambers are almost as good when they are equipped in a similar manner—that is, with equivalent carburetion and exhausting arrangements and a comparable compression ratio. Regardless of its merits, the bowl-in-piston, combustion cell, or Heron combustion chamber, which seemed so promising a decade ago, is not likely to be seen in any new engine designs because of its tendency to produce high exhaust emissions.

Hemispherical Combustion Chambers

The OHV engine with hemispherical combustion chambers and both valves inclined in the head is almost as old as the four-stroke cycle, and it has been the natural choice of racing engine designers for over seventy years. After World War II, the type began making inroads not only in the popular sports car field but also as a unit for propelling luxurious sedans. We have previously discussed the "hemi" head from the standpoints of valve operation, compression ratio, and bore/stroke ratio; now we will investigate the various good and bad points that arise from the shape itself.

The general layout will be familiar from the illustrations given in earlier chapters. The valves are mounted in the head, making an equal—or almost equal—angle on either side of the center line of the cylinder bore. Usually the included angle is not less than 60° or more than 90°. The ports are on opposite sides of the head in a crossflow pattern; the intake ports are on one side and the exhaust ports on the other. To allow for adequately large valves, it is sometimes necessary to mount the spark plug in a location that is off to one side instead of in its theoretically desirable position at the approximate center of the hemisphere. In modern "big-bore" engines and in four-valve combustion chambers, however, there is usually little to prevent the central spark plug location.

The ability of this kind of cylinder head to maintain good power outputs at high rpm is largely the result of the extremely good breathing afforded by the easy flow in and out of the valve ports and the unrestricted entry and exit—the circumferences of the valves being free from the proximity of cylinder or combustion chamber walls. If the valve angle, particularly of the intake valve, is designed to take operating conditions into consideration, it is possible to direct the flow of ingoing mixture into the cylinder bore. With this feature and if the gas flow can be maintained at high velocity, the intake valve can be kept open for an appreciable time after the piston has started to ascend on the upstroke of compression, without a reversal of the gasflow. Similarly, at the end of the exhaust stroke, the overlap period, with both valves open, can be extended over a long period of crankshaft movement; the induction of the descending piston on the intake stroke is augmented to a marked degree by the extractor action of the exhaust gas column. This kind of tuning applies to any OHV engine, especially any that has inclined valves. Particularly on powerplants such as the Ford/Foyt Indy V8, with the intake tracts entering ports that are between the camshafts, the fullest advantage can be taken of the phenomenon.

Although not so important today as in the past because of the present highly developed turbocharging, fuel injection, and carburetion systems, the relationship between the two valves in a "hemi" head provides a further aid to reliability at high speeds. The intake, which is directly opposite the exhaust, can be expected to direct some of its flow of relatively cool gas onto the hot exhaust valve. Besides helping to keep down the exhaust valve temperature, this characteristic helps to vaporize by "frying" any wet globules of fuel in the intake mixture.

Fewer hemispherical combustion chamber designs are used on production cars today than were twenty years ago. Even in racing engines, it is rare to encounter a combustion chamber that approaches a true hemisphere. This trend to other combustion chamber shapes will probably continue, and so it is worth a brief discussion here.

First, the absence of any "squish" features that control turbulence has undoubtedly prejudiced the design in the eyes of designers to whom smoothness and good torque at low speeds are matters of major importance. Nevertheless, the "hemi" possesses remarkable features of natural turbulence; it makes for general all-round excellence in combustion and expansion, which, allied to its good volumetric efficiency characteristics, render the design desirable from almost every point of view. Later designs have, in fact, combined a squish feature in the layout by introducing a circumferential "flat" around the combustion chamber edge, which corresponds with a similar shape of piston crown.

Even this degree of turbulence, which is adequate for high-compression operation on high octane fuel, is not sufficient to ensure complete and uniform burning of the lean mixtures that are now demanded for exhaust emission control. Considering the many highly inefficient design features that have been forced on designers by emission considerations, there is little incentive for adopting the complexity of the "hemi" because most of its merits would be lost. Therefore few new "hemi" designs are likely to be seen in production cars.

Another limitation concerns compression ratios. The considerable volume of the hemispherical head makes careful design mandatory for achieving a high compression ratio. Obviously a flat-topped piston is desirable, and if the stroke of the engine is sufficiently long, such a shape may give the desired ratio. However, apart from any question of piston speed, too long a stroke may reduce the bore diameter unduly, making adequate valve sizes impossible to accommodate. In practice, it is nearly always necessary to adopt a very convex shape of piston crown to take up some of the head space. Most such pistons are also provided with valve cutouts to enable the valve edges, at their lowest points, to clear the piston fully at tdc.

While high compression ratios are needed for any normally aspirated competition engine, supercharging makes a lower ratio desirable. This is not simply for antiknock reasons; it provides a greater combustion chamber space for the supercharger to compress mixture into. Because longer strokes and lower compression ratios favor low exhaust emissions, one might speculate that the "hemi" would be a desirable feature to adopt in new designs. There is still the need for turbulence, of course, but at this point the future of the hemispherical combustion chamber in production cars is totally unpredictable.

If the "squish" is absent in a hemispherical combustion chamber, a remarkable degree of turbulence can be obtained by carefully planning the intake port shape and direction. It has already been shown that suitable design makes it possible to induce the gas to flow in to take the maximum advantage of valve overlap without wastage through the exhaust. Similarly, the charge can be directed slightly to one side of the chamber instead of aiming straight at the exhaust valve or straight down into the cylinder. By giving the gas a bias, a definite swirl can be set up. Then by placing the spark plug in correct relationship to this, extremely good torque figures are obtainable at medium rpm, without losing any of the good points of the design that enhance maximum speeds.

Torque Requirements

The essential difference between torque and horsepower being that the former is quite independent of engine rpm, it follows that horsepower is dependent on the number of turns that the crankshaft makes in a given time by virtue of the torque applied to it. The piston pressure is applied for about 120° per two revolutions by each cylinder. If the engine is a single cylinder, it must have a sufficiently heavy flywheel to store up energy during 120° period and give it back to the crankshaft during the remaining 600° of rotation. The turning effort in a "single" is impulsive, but it is necessary for our purpose to take the average torque over one revolution without considering whether it is produced by the expansion of gas in the cylinder or by the flywheel's rotating mass. As cylinders are added, the degree of impulsiveness becomes less. In the case of a six-cylinder engine, for example, a power stroke starts at each 120° of crankshaft rotation, so that—in theory—no flywheel effect is necessary to tide over the completely "idle" periods. (See Fig. 6-13.)

Since torque is produced by the power stroke, its magnitude is absolutely related to the pressure developed on that stroke, the total force applied being the bmep multiplied by the total piston area. This force is applied throughout the length of stroke to turn the crankshaft. Since both piston area and stroke are involved (dimensions that make up the swept volume, or piston displacement, of the engine), it is possible to calculate the torque for any engine by using a constant that converts the volume in cubic centimeters to give torque in the conventional pounds/feet. This is described in the appendix.

From the formula makeup, it will be evident that the value

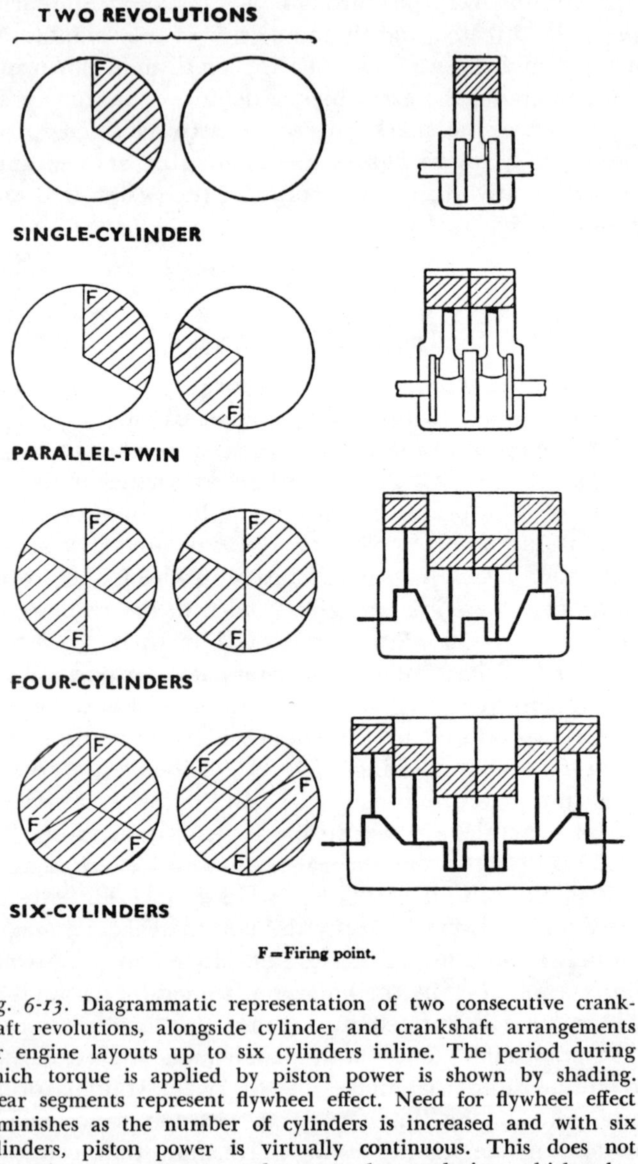

Fig. 6-13. Diagrammatic representation of two consecutive crankshaft revolutions, alongside cylinder and crankshaft arrangements for engine layouts up to six cylinders inline. The period during which torque is applied by piston power is shown by shading. Clear segments represent flywheel effect. Need for flywheel effect diminishes as the number of cylinders is increased and with six cylinders, piston power is virtually continuous. This does not affect the average torque value over the revolution, which takes no account of whether it is produced by one or many power strokes.

of torque is unaltered by the number of cylinders. However, if a ceiling of, say, 6500 rpm is fixed, and given perfectly adequate charging and a rational piston speed, the best torque can be obtained with only four cylinders. This does not mean that the "four" has inherently a better torque range than a "six", "eight", or "twelve". What it means is that an oversize "four" such as the "Offy" that has been made strong enough to handle the big cylinders and the rpm can often show a considerable torque advantage over an engine with more cylinders—particularly in vital situations such as entering the straight at Indianapolis.

The torque curve of a Wankel rotary engine is generally less "peaky" than that of a comparable reciprocating engine. Also, the Wankel tends to develop its best torque at a slightly lower percentage of its usable range of rpm. In a reciprocating piston engine, the inertia of the reciprocating masses causes and influences torque fluctuations between cylinder firings. In the Wankel rotary, the centrifugal force of the masses rotates about the eccentric shaft center. Balance weights on the eccentric shaft are not included, of course, because they are a part of the shaft and are rotating with it. The effect of centrifugal loadings is so slight on the Wankel engine's torque fluctuations that they need not be considered at all so long as the rotor's center of gravity coincides with its geometric center.

A gas turbine engine is capable of generating its maximum torque at zero rpm. This has been demonstrated not only in the Rover/BRM car that raced at Le Mans a number of years ago and by the drag racing cars that have used turbines but also in the gas turbine tractors that are used in tractor pulling competitions in the American Midwest. The procedure is to rev up the compressor while the brakes are locked. Then, when maximum gas generation is achieved, the brakes are released and the vehicle departs—with a great smoking of tires as the gases from the combustion chambers apply their maximum force to the turbine.

Today, of course, the matter of supercharging versus turbocharging also enters the picture concerning power output. Mechanically driven superchargers like those used in drag racing always run at a speed proportional to engine rpm. Thus, if engine rpm builds up slowly, it may be some time before the supercharger reaches its maximum output. This is no problem, of course, with a blown dragster; the revs build up immediately because of the spinning rear wheels at the start, and there is no delay waiting for the blower to catch up with the engine as there would be with a turbocharger.

For most other applications, however, the turbocharger has the advantage. The time lag that occurs before the turbocharger has reached its maximum rpm is of little importance in forms of racing that last for more than a few seconds and where the engine speeds are always kept well up in the rpm range. During wide-open throttle acceleration in the upper speed ranges (as in leaving a turn on an oval track) or when full throttle is being applied on an uphill portion of a road course, the turbocharger quickly reaches its maximum output because of the large quantity of exhaust gases being generated. Conversely, a mechanically driven supercharger would not be able to supply full boost under these conditions because the engine had not yet reached its maximum rpm.

The gas turbine and the turbocharger obviously have something in common: their ability to increase power output at a point that is farther down on the output rpm range. A turbocharger corresponds almost directly to the compressor stage of a gas turbine. Perhaps when it is realized that a reciprocating engine is neither the most efficient gas generator nor the most efficient medium for turning pressure into rotary motion, there will be a resurgence of gas turbine competition cars—the kind that almost took over at Indianapolis in the mid-1960s.

7 / Efficient Combustion

Improving Performance

In chapter 6, we reviewed in general terms the main characteristics that distinguish various kinds of combustion chambers. Now we will consider how to improve the performance of these layouts. Efficient combustion has been much in the minds of engineers lately, not merely for reasons of improved performance but also as the primary means of gaining better fuel economy and lower exhaust emissions. Both new and old ideas are being evaluated—for example, multiple spark plugs have reappeared for the first time in three decades, this time in the Mazda rotary engine.

Today's engines, which are not supercharged, are turning out more power than were blown units in the 1930s. Horsepower is still on the increase even though the basic design of engines has not changed much. Obviously these increases are coming from somewhere. The basic fact is that to obtain power we burn a fuel/air vapor mixture; the more of this we burn in a given time, the more power there is.

Breathing

As the mixture burns, it must pass through the engine cylinders, and it thus performs the operating cycle in so doing. If it is passed through and burned indifferently at astronomical revolutions, it will produce a certain amount of power. If it is burned more efficiently at lower rpm, quite probably it will produce the same amount of power. Best of all, it can be burned efficiently and produce an adequate range of rpm.

In considering how best to obtain the maximum power for each charge expanded in the cylinder, it is necessary to start with the atmosphere. (We are considering only unsupercharged engines here; blowing will be dealt with in chapter 14.) The atmosphere is being used for charging; its pressure is available for nothing and provides unlimited quantities of air. If the pressure can be allowed to push the air into the cylinders in the proper manner, we shall go a long way toward getting full value from the subsequent expansion, since the speed at which the air enters and its direction have considerable bearing on turbulence, an essential of good combustion.

The conditions inside the combustion chamber and the actual shape of the cylinder head certainly influence the efficiency of the expansion process. But modern engines do not vary much in this respect—with one exception. The stratified charge engine (Fig. 7-1), which first reached the public as the Honda CVCC, has made possible the efficient burning of what a few years ago would have been an impossibly lean mixture. Exactly what influence that the stratified charge principle may have on future racing engine designs is uncertain. However, from the performance delivered by Honda cars in Showroom Stock Sedan class racing, it is obvious that this kind of engine is certainly no handicap.

The excellent power outputs obtained from such types as the inclined overhead valve hemispherical head result far more from the excellent breathing afforded by this layout than from the shape of the combustion chamber itself. In designing the prototypes for the Rolls-Royce V8 engine, engineers tried both hemispherical and squish-type combustion chambers, in both cases with pushrod valve gear. The latter type was finally adopted; it enabled the engine to be made narrower, and there were no disadvantages to the required power characteristics. Specific power output was not inferior, the engine was not

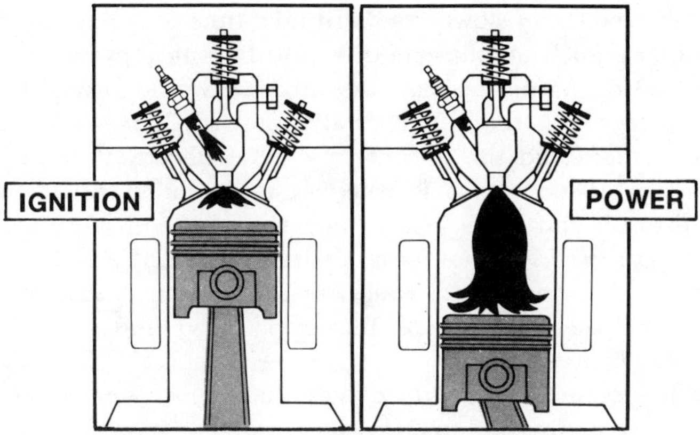

Fig. 7-1. Stratified charge engine. Rich mixture in small chamber is ignited by spark plug. Rich mixture ignites lean mixture in main combustion chamber and cylinder. Mixture in cylinder is too lean for plug to ignite.

quite so sensitive to carbon deposits, and considerably less ignition advance was necessary in comparison with the hemispherical head version.

It was shown in chapter 2 that claims of high thermal efficiencies from certain head designs, based on the use of a so-called compact form of combustion chamber presenting the minimum surface area to the gases, had little basis in fact and that even if all the heat lost could be trapped, the net gain would be under 3 percent on the thermal efficiency figure. Any modern form of head, regardless of the valve arrangement, permits only the minimum jacket loss, and no one form shows outstanding advantages in this respect. It is in the port arrangement, and not in their area of metal, that well-designed heads reap the benefit of high thermal efficiency figures.

The charging process starts at the air intake. From this point until the moment of ignition, the mixture should ideally be in a state of agitation or turbulence. The rate and completeness of combustion depends on it. If we imagine an engine inhaling a charge that is completely devoid of movement after passing into the cylinder and being compressed, it is not difficult to visualize that the area of mixture ignited by the spark plug will burn relatively slowly and will take time to spread throughout the gas. Such a contingency would be unacceptable even in a slow-speed engine, because the mixture would probably still be burning when the exhaust valve opened. The whole speed of combustion, and the pressure rise generated by it, is dependent on turbulence. This is true not only under conditions of wide throttle and high speed when the gas speed through the intake port will promote the desired disturbances. On light throttle and at low speed, complete combustion is still needed; it is under these conditions that good head and port design show to advantage.

The design of the intake port and valve presents several problems. For high volumetric efficiency (the inhalation of the maximum weight of charge in a given time), a large valve is called for. This provision also reduces the pumping work necessary. On the other hand, a high gas velocity is required for turbulence, which means a reduced size of valve. Thus, a balance has to be arrived at in determining the size of the intake valve, since it is of no use to obtain a high volumetric efficiency and reduction of pumping losses if the combustion and expansion processes are going to suffer because of the insufficient agitation of the mixture. In practice, it is not difficult to arrive at a figure for gas velocity through the valve that promotes good turbulence without causing losses in the other departments.

Intake Valve Size

Increases of intake valve size should not be undertaken without full consideration of everything involved. It must be obvious that a small increase in valve diameter can have no effect on the initial cost of a passenger car, and some readers may well have wondered why the makers had not thought of it. The point is that an increase in valve size may show an increase in bhp at the highest rpm on wide throttle openings. In a car used almost exclusively under those conditions, a modification may well be worthwhile. Lower down in the speed range, however, the reduction in turbulence may lower the available torque so that acceleration from very low rpm can quite easily be inferior to that of the unmodified engine.

When car manufacturers increase the power output of a basically unchanged engine to meet current demands, the valve sizes are sometimes increased, along with such items as cubic capacity and compression ratio. Often the result is to elevate the rpm at which maximum torque occurs; power is thus increased over a wide speed range because of the greater volume of mixture. The highest volumetric efficiency is obtained when the intake valve opens straight into the combustion chamber. This feature is used on all OHV types, but there is more likelihood for the valve to be a little cramped on the popular all-valves-in-a-row head than on engines where there are two rows of valves. Thus, uncramping the valves can often do more to improve engine output than merely increasing their size (Fig. 7-2).

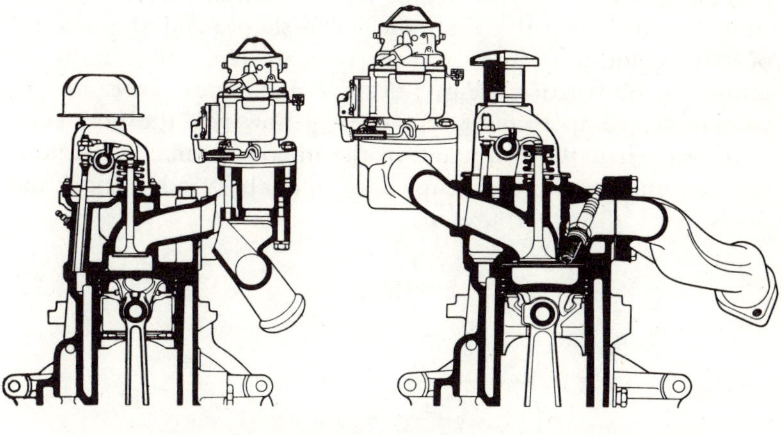

Fig. 7-2. Early Ford Cortina with single port face head (left) has bathtub combustion chambers that shroud valves. 1968-1970 version of this engine (right) has crossflow head with most of combustion chamber in piston, thus greatly increasing flow space all around valve heads.

Intake Port Characteristics

The shape and the direction of the intake port have a considerable influence on the degree of turbulence obtainable. Although existing data can give a reliable guide to the results expected from any particular formation, much of the work in the laboratories consists of painstaking trials with various dimensions and shapes of port, noting their influence on the airflow through the setup.

The Jaguar DOHC six-cylinder engine is an example of what can be achieved. This engine, which is well known for its excellent power output, also obtains good fuel economy. The intake passage configuration reveals the amount of thought that has gone into the design, the result of close collaboration between the Jaguar concern and the Weslake Laboratories.

In plan, the intake ports are slightly tapered toward the entry to increase the gas velocity. Additionally, the port (again looking down on the head in plan) enters the cylinder head not across its diameter but slightly to one side. The port then follows a curved path to the valve opening. This curvature has a great influence on the degree of swirl imparted to the mixture. Finally, the valve seat, the valve shape, and the amount of valve guide projection into the port are laid out to form the minimum obstruction (Fig. 7-3). For maximum power on the six-cylinder competition E-type engine, however, the port curvature was virtually eliminated in the interest of maximum flow; the straight ports were coupled to specially graded induction ducts from three carburetors.

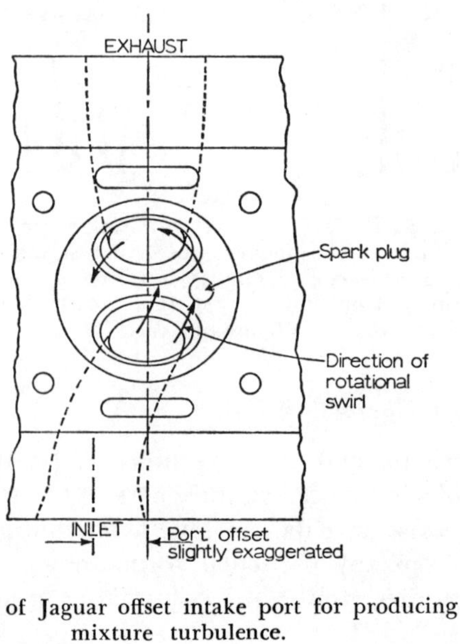

Fig. 7-3. Design of Jaguar offset intake port for producing intake mixture turbulence.

This engine is just one example of the care being taken in designing an important item. Today, most designers devote quite a lot of time to such matters; some results are better than others, however, and in general there is room for further improvement. Placing the intake ports between the camshafts of racing engines has in many cases made a more direct form of intake tract possible. Work is being done on experiments with both side ports and peripheral ports on Wankel rotary engines. In nearly all kinds of unsupercharged engines, it has been found that bigger is not always better when intake ports are being refined.

Combustion Chamber Requirements

Apart from the turbulence set up at the time of entry of the charge, the agitation should continue right up to the moment of ignition for maximum thermal efficiency. To this end, the free entry of gas is most important. The combustion chamber should be devoid of nooks and crannies where stagnant gas can accumulate or become trapped. And the agitation should be sufficient to keep that layer of gas next to the cylinder walls moving; otherwise it will chill off and possibly escape complete burning. Any improvement in this area will not only bring about more power from the fuel being burned but will also lower exhaust emissions.

The use of squish is a quite effective way of ensuring continuance of turbulence and is easily arranged on the bathtub combustion chamber with inline valves. Probably, however, for consistency of turbulence (in regard to its persisting right through the intake and succeeding phases up to the moment of ignition), the hemispherical shape is unsurpassed; its freedom from corners and odd shapes brings about its success. Still, extremely high outputs are obtainable from the simple old bathtub layout. This type has performed well in the past on relatively poor qualities of fuel, indicating its very good combustion characteristics.

Exhaust Pressure

There is always a limit to the pressure that is available to push the mixture into the cylinder (assuming the engine is not supercharged). We are much better off when looking at the exhaust arrangements. Although we are accustomed to thinking of the last upstroke of the operating cycle as the exhaust stroke, it will be obvious that, at the time of exhaust valve opening, about 55° or so before bdc, the cylinder will contain gas at a pressure of at least 67 or 70 psi with the engine under load. This pressure will ensure speedy evacuation of a large proportion of the exhaust gases before the commencement of the exhaust stroke proper. Further, with a cam profile that gives the required quick opening characteristic, the sudden release of the pressure will initiate the rapid gasflow down the pipe that is required for good scavenging. An unduly large exhaust valve and port could possibly prejudice the scavenge effect without any compensating advantage that might be thought to arise from giving the outgoing gases an easier flow. The existing cylinder pressure is well able to look after the eviction process, even when the proportions of valve and port are such that the gas velocity through them is 300 to 350 feet per second or more. This figure ignores the speed of pressure waves propagated in the gas, which is very much greater.

By the time that the piston has passed bdc, a large proportion of the exhaust gases will be gone. Even if the valve size is such that a back pressure of a few psi persists throughout the exhaust stroke, this would represent a loss calculated on the bmep of only the same amount. The scavenge at the tdc overlap period is much more important.

A serious disadvantage of any exhaust valve is its large uncooled area. This used to be such a major snag that many designers went to considerable lengths to do away with it altogether, substituting some other means of controlling the breathing—rotary or sleeve valves, for example. Cooling jacketing today receives proper attention from designers; spark plugs, which represent another uncooled area, tend to be smaller, but the exhaust valve head still takes up considerable room. Obviously, the smaller it can be made, the better. Not only will it pick up less heat, but it will more quickly dispose of the heat that it does absorb.

The power required to open the valve against the cylinder

pressure can be considerable. This becomes greater with a large valve, which throws additional stresses on the operating mechanism. It is evident, therefore, that the fullest advantage is to be had by accepting as normal a gas velocity through the exhaust valve of up to 50 percent greater than that permissible through the intake valve. This can reduce the size of the valve head to proportions that will keep its nuisance value as a hot spot within bounds and ensure that too much power is not absorbed in operating it.

If supertuning operations designed to push up the top end of the power curve are undertaken, an increase in intake valve size may be called for even at the expense of low-speed torque. In this event, the exhaust valve size might be advantageously increased also. It is the opinion of many successful speed tuners that such experiments are best conducted in two stages; first, enlargement of the intake valve only; second, both valves. Furthermore, on some engines, such as the VW air-cooled 1600 powerplant, a valve size increase does no good at all unless there is also an increase in cubic capacity.

Undue emphasis on sheer bhp at peak rpm can be most misleading, even in oval track work. For almost every other purpose, a sensible torque curve, starting not too low down but reasonably so, is preferable to a few rpm gained but seldom used, at the top end, which may prejudice reliability.

Spark Plug Position

It has been suggested that good engine design should start with the spark plug location. Since this component is where the power stroke originates, the principle has merit. Such factors as detonation and compression ratio are influenced to a large extent by plug position, the former being, of course, tied up with the amount of ignition advance that is permissible.

Since the spark plug must necessarily be inserted either into the top or the side of the combustion chamber, a position should be aimed at that will give as nearly as possible an approximately equal distance from the combustion chamber wall in all directions. (We are at present not taking into consideration the Honda CVCC or the Wankel rotary engines.) For example, with an engine having hemispherical heads and inclined overhead valves, it might be feasible to have the plug vertically in the center of the head. This position would be ideal, and most designers using hemispherical combustion chambers get as near to this location as they can. It is not, of course, possible with two large valves, but it is with four. The situation has been improved over the years by a reduction in spark plug thread diameters which, in turn, has been made possible by better oil control and fuel metering that help to reduce spark plug fouling.

The spark plugs have an extremely hard life in modern competition engines, and it may well be that the popular 14-mm size represents a sensible compromise between physical size, heat dissipation, retention of sparking efficiency, and effective insulation. Nevertheless, 10-mm spark plugs are being widely used in Grand Prix racing engines. Dimensional reductions of this kind are most helpful in the plotting of combustion chamber layouts. With some modification to the valves and the use of the smallest possible plug, the central position for the latter could be achieved in a two-valve hemispherical head. A major concern, of course, must be that cracks do not develop between the valve seats and the spark plug boss.

It is also possible to modify the position of the spark gap without altering its location. When projected nose spark plugs were developed in the 1950s, the design purpose was to extend the spark plugs' heat ranges. However, the consequent placement of the spark gap deeper in the combustion chamber resulted in improved power—particularly in wedge and bathtub combustion chambers. Subsequently, similar advantages were found in certain "hemi" head engines. The benefits in many cases have been great enough that it has been worth machining a spark plug clearance in the piston to take advantage of the deeper gap position.

A change in the spark plug's gas volume can also help overcome a poor plug location. Gas volume is the space between the plug's threaded shell and the insulator nose. Larger gas volume tends to reduce fouling tendencies and to help ignition if the spark plug is located in a rather stagnant turbulence area. Thus, testing spark plugs with different electrode configurations and different gas volumes on supertuned or blueprinted production engines will often turn up a few horses that nobody knew were there.

An adequate mixture turbulence near the plug at the moment of ignition should be an important consideration in combustion chamber design work. It is possible that under conditions of varying throttle opening, with consequent variations of intake mixture velocity, a plug position slightly to one side of the center will be more likely to ensure a proper degree of swirl past the plug electrodes. Nevertheless, the necessity for the shortest possible flame travel should receive first consideration. This problem is obviously more serious in large-bore engines (with relatively small bores, the dimensions are such that the plug can hardly ever be put in an impossibly poor position). Here again, positioning the spark plug deeper in the combustion chamber, as by a projected nose spark plug, takes advantage of the fact that turbulence is always somewhat greater at a point a bit distant from the chamber walls. Since 1972, projected nose spark plugs have been recommended in every American production car engine.

In general, the shorter the flame travel, the less sensitive the engine will be to throttle opening and load in the matter of ignition timing control. The tendency toward detonation with inferior fuels will also be reduced. With careful design, the makers of hemispherical combustion chamber engines can put the plug where they choose. In the case of the more popular bathtub and wedge chambers, any suggestion of a central plug position is out of the question. The usual head of these types manages to accommodate two valves of adequate size in a most creditable manner, but there is certainly no room for anything else. A position for the plug must therefore be chosen at the side of the head. To counteract the long flame travel from this position across the piston crown, we can fortunately arrange a squish formation almost directly opposite; in effect, the mixture comes out to meet the spark plug just at the moment of ignition. Because the squish areas are themselves almost in contact at tdc, the combustion chamber dimensions as far as flame travel is concerned are further reduced. Bore/stroke ratio also has some influence; the smaller the cross-dimension of the head, the better.

In any form of combustion chamber, the use of a flat-topped piston is desirable for efficient combustion. However, it is hardly ever possible to achieve the necessary compression ratio for competition purposes without compromises. In any case, piston domes and humps must be considered not only from the standpoints of valve clearance and compression ratio but also with thought to their influence on the spark plug (Fig. 7-4). Obviously a piston crown that separates the spark plug from a large quantity of mixture at tdc will not aid efficient combustion.

Fig. 7-4. Racing piston for Chevrolet big block V8 must have dome for adequate compression ratio. Notice that a spark plug relief has been milled in center of dome for better combustion.

Ignition Current

Efficient combustion cannot begin until the mixture has been ignited by the spark plug. Before this can happen, sufficient voltage must be available to ionize the air in the spark plug gap. Without ionization to make the air molecules conductive, the spark current cannot pass from one spark plug electrode to the other and thus create a spark. Ionization becomes more difficult at high cylinder pressures, so the first requirement of a competition ignition system is that it produce sufficient secondary voltage to fire the plugs under wide-open-throttle, high-rpm racing conditions.

To help the available voltage do the job, various other things are done to assist ionization of the spark gap. First, a narrower spark plug gap will ionize at lower voltages than will a wide gap. Therefore, the racing tuner may find it necessary to gap the spark plugs to a smaller dimension for racing than would be practical for highway driving. The tendency for narrow gaps to misfire at low speed and during idle is, of course, of no importance in racing. Second, spark plug electrodes with square, sharp-edged corners promote ionization. Therefore, new racing-type spark plugs are installed frequently—and always right before the race itself, in the case of high-rpm racing engines.

Conventional Ignitions

Four kinds of ignition systems are currently in competition use. One of these, magneto ignition, was once almost a necessity for racing engines, but it is now fading in importance because of the development of electronic ignition systems. The virtue of the magneto (Fig. 7-5) is that its voltage output increases as engine rpm rises. Thus, at high speeds where high secondary voltage is most needed for reliable spark gap ionization, the magneto is able to deliver all that is required.

Coil and battery ignition, which was until recently the only kind of ignition found in passenger cars, is not able to maintain its voltage output at high rpm. This is because the coil core is magnetized by primary (battery) current only during the time intervals when the ignition distributor's breaker points are closed. As engine rpm increases, the breaker points are closed for an ever-shorter period of time between firings. The magnetic field of the coil becomes progressively weaker (or the coil becomes less fully saturated by magnetism).

Fig. 7-5. Scintilla Vertex magneto installed in place of normal distributor on six-cylinder drag racing powerplant.

Some high-rpm improvement can be obtained by adjusting the breaker points to a narrower gap. This change, of course, causes the engine to misfire at lower speeds and to idle irregularly. However, these are not serious problems during a motor race, and the increased periods of time that the points are closed improves coil saturation.

In addition to the obvious scheme of building a better coil for high-speed service—the sports and high performance coils that are commonly sold in speed shops—other improvements have been made to coil and battery systems for competition use. One approach is to have two sets of breaker points in the ignition distributor (Fig. 7-6). The two point sets are wired so that if either set is closed, battery current will reach the coil's primary windings. The second set of points is almost closed when the first set opens to fire the spark. A very brief time after the coil has discharged to fire a spark, the second set of points closes and begins charging the coil for the next spark. This system increases coil saturation time considerably over that possible with a single set of breaker points.

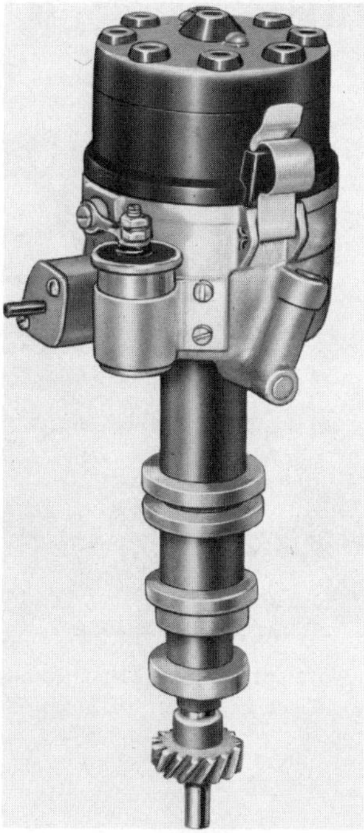

Fig. 7-6. Mallory Double-Life distributor. This favorite hot rod and drag racing unit has dual breaker points. Models are available for almost all American engines.

Another kind of dual-point system employs two separate coils. In effect, it is two separate ignition systems built into a single distributor. These systems are usually seen only on engines that have eight or more cylinders. Each set of points, with its own coil, fires half the cylinders of the engine (Fig.

7-7). There have even been variations on this theme in which there are four sets of breaker points operating two coils. The principle is that of two independent ignition systems, each with overlapping dual points. A related system is to have two separate distributors, each with its own coil.

Fig. 7-7. Breaker plate of Grant Flamethrower compound distributor. Each point set fires four of engine's eight cylinders:

Electronic Ignitions

A basic form of electronic ignition is the transistor-switched coil and battery system. With this setup, the primary current to the coil is switched on and off by a transistorized device. The switching transistor is signaled to interrupt the primary circuit, thus firing a spark, by the opening of the breaker points. However, the points themselves carry only very low current and voltage; the heavy switching work is handled by the transistor. The kind of system normally has no advantage over an ordinary coil and battery system. Its principal merit is reduced wear and tear on the breaker points, which last longer and require less attention at tune-up time.

For competition purposes, however, there is a potential for some gain. In production cars that use full battery voltage to the coil only during starting and then switch back to a reduced voltage in the interests of longer breaker point and spark plug life, it is possible with transistor switching to use full battery voltage at all times. The transistor will prevent this from taking a toll on the breaker point contacts.

A somewhat better form of electronic ignition is essentially the same as the system just described, but with the breaker points eliminated. In this system, the transistor receives its signal from an electronic signal generator rather than from the mechanical breaker point mechanism. The electronic signal generator that replaces the points has taken a number of interesting forms. One system employs a cup-shaped device in place of the normal distributor cam. Slits are cut in the periphery of the cup, and the intervening sections of the cup rim interrupt a beam of light. When a slit uncovers the light source, a photoelectric sensor signals the switching transistor to fire a spark.

A more common practice is to use the principle of magnetic reluctance as a signal generator. The normal distributor cam is replaced by a timing wheel (reluctor) that has spokes but no rim; it is a kind of gear-like fixture with as many teeth as there are cylinders. At one side of the distributor housing is a small electrical coil with an iron core that functions as a weak electromagnet. When one of the timing wheel teeth aligns with the coil core, magnetic reluctance changes the field strength of the electromagnet and signals the transistor device that a spark should be fired.

Similar sysems that are in wide use have a timing wheel that consists of a ceramic drum in which ferrite rods are embedded (Fig. 7-8). Another approach is to use rotating magnets that pass alongside a coil to generate a signal to the switching transistor. Thus, the principle of magnetic permeance can be employed, as well as that of magnetic reluctance.

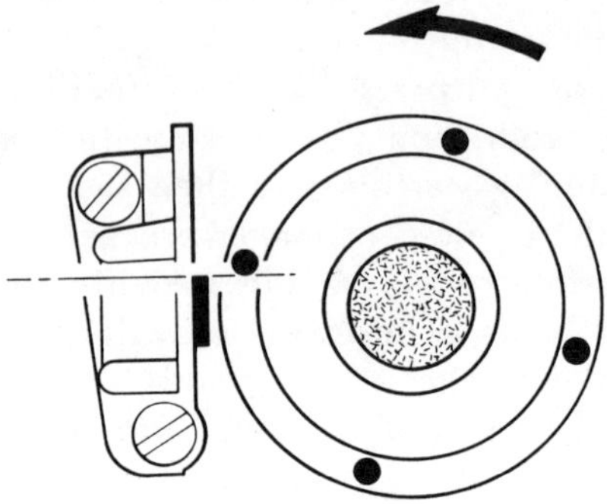

Fig. 7-8. Timing wheel for Lucas distributor used in Triumph TR7. As wheel turns (arrow) ferrite rods align with pickup core (line at left), thus producing the signal that triggers spark.

These breakerless systems have several advantages from a racing point of view. First, there is no breaker point contact erosion or mechanical limitation. More important, the transistorized switching device can begin recharging the coil almost instantaneously after a spark has been fired, greatly increasing the coil saturation time and making possible reliable operation at higher rpm.

The form of electronic ignition that has made the most solid place for itself in competition is the capacitor discharge (CD) system. A CD system is essentially similar to a transistor-switched coil and battery system. Some CD setups use breaker points and others use breakerless signaling; but there is one very important difference between the CD and ordinary transistorized systems. Instead of having the coil receive battery voltage as its primary current, the coil receives its primary current at elevated voltage from a capacitor.

A capacitor, or condenser, has the unique ability to store electricity. In a CD system, a transistorized charging device is used to step the battery's nominal twelve volts up to 250 volts or more. This higher voltage is stored in the capacitor. So when the coil's primary winding is charged, it is to about 250 volts instead of to the 12 volts of other systems. A special coil is necessary, of course.

Because the CD spark kicks off from a much higher voltage, a very high secondary voltage can be made available to the spark plugs. In addition—and perhaps more important—the spark voltage rises to its peak at the plug electrodes almost instantaneously. This will fire spark plugs that are badly fouled. The high voltage is reached so suddenly that it ionizes the spark gap and jumps across before the fouling deposits have had a chance to drain the current off to ground. Fig. 7-9 shows a comparison with a normal coil and battery system.

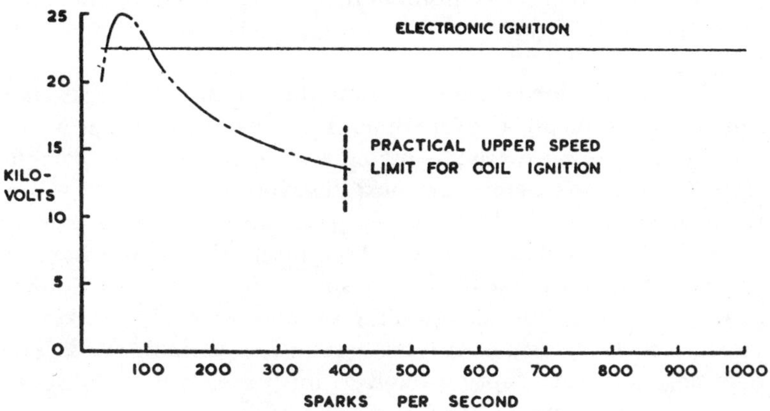

Fig. 7-9. Voltage drop on coil ignition system, compared with steady output of electronic systems.

Timing

Many racing engines have fixed spark timing. They operate at all times within a narrow band of high rpm and so never need to have the spark timing retarded from the degree of advance that is necessary for maximum output. There is, however, usually some provision for temporarily retarding the spark for starting purposes. In engines of this kind, it is far more important that the timing remain accurate than whether the engine can be made to run slowly. Obviously a careless driver who allows the rpm to fall off and then opens the throttle wide can easily induce severe preignition that will destroy the engine almost instantly. Few people realize how dangerous it can be mechanically to drive slowly in a Grand Prix or Indy type racing car.

The engine's operating cycle becomes compressed into exceedingly short time intervals at higher speeds, and very minute imperfections in the distributor drive can become important. At high rpm, a spark theoretically timed at 40° before tdc might arrive at the plug when the piston is by no means at this position because of a combination of drive backlash, torsional distortion, and bending in the various shafts and components involved.

Mechanical breaker points limit the number of sparks that can be produced in a given time. This is because when a certain rpm is reached, the point spring does not have time to close the contacts before the next distributor cam lobe pushes them open. This is called breaker point float. If a stiffer spring is used, there is danger of point bounce; that is, the breaker points are thrown together with such force that they bounce apart again. On most Grand Prix racing engines, a breakerless ignition such as the Lucas Opus system is invariably used. The elimination of cams, springs, and so forth means that the speed of operation can be vastly increased.

The first stage of the operation of the Lucas system is to time the number of sparks required per revolution. Because of the necessity for accuracy of timing, which was one of the reasons for developing the system, the flywheel end of the crankshaft is used for preference since it is least subject to torsional oscillations. The flywheel itself can be used for timing and carries a number of pole-pieces equal to the number of sparks per revolution (four in the case of a V8 engine). Mounted alongside the flywheel so that the pole-pieces run past it with only air-gap clearance is an electromagnetic pickup. If there is no room at the flywheel end of the engine for this assembly, the pole-pieces can be mounted on a pole-wheel at the front end, but the general principle is the same. These components are shown in Fig. 7-10.

Fig. 7-10. Polewheel and pickup and spark generator of Lucas electronic ignition system for racing.

When the crankshaft rotates, a voltage impulse (or trigger impulse) is produced at the electromagnetic pickup each time that one of the pole-pieces runs past it. The pickup is connected electrically to the trigger amplifier (Fig. 7-11), which is the transistorized switch that is normally closed, thus sending primary current to the spark generator (capacitor and coil). At each trigger impulse, the primary is interrupted and a spark is fired. A conventional sort of distributor rotor and cap sends the spark to the appropriate spark plug.

Before moving on to the matter of spark plugs, which have a great deal to do with efficient combustion, it is worth saying a few words about variable ignition timing. In the past, the spark timing could be controlled by the driver via a dashboard- or steering-column-mounted lever (remember the old Model A Ford?). Because of the higher and much wider range of rpm typical of modern engines, the driver would have little time to do anything else except fiddle with the spark control. Consequently automatic spark advance is universally used today.

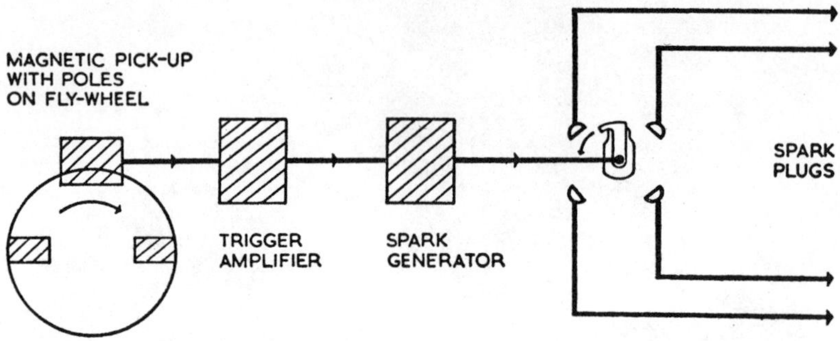

Fig. 7-11. Schematic diagram of Lucas electronic ignition system for four-cylinder engine with poles on flywheel.

There are three kinds of automatic spark advance. First, there is the centrifugal advance, which advances and retards the spark timing in direct proportion to engine rpm. Second, there are vacuum advance systems that provide additional advance when the engine is under light load. The vacuum advance is wholly a device for fuel economy and has little purpose or function in competition. The vacuum retard systems found on emission-controlled passenger cars are used for the reduction of exhaust emissions and have a negative influence on economy. Finally, there are a few systems that vary ignition timing electronically. At present, these are being used as economy/emission control devices on some passenger cars, but an electronic timing system has been used on some Maserati racing engines. For the purposes of this book, we will limit our consideration to the centrifugal advance only.

The advance curve of a competition engine should be arrived at by means of dynamometer testing. The curve must provide the degree of spark advance that makes combustion most efficient at the rpm or range of rpm over which the engine will be operated in competition.

Generally, it will be necessary in preparing a production car distributor for racing to replace the springs in the centrifugal advance and possibly to replace the weights. In dyno-tuned production cars the primary spring is often very light so that it will keep the spark retarded only for starting purposes. Once the engine is running, nearly full primary advance should be available. The secondary advance will receive the most attention during testing in order to establish the best curve for the engine. If the range of advance is insufficient for maximum output, then the weights must be exchanged for those that permit greater advance. Spring and weight kits are widely available in speed shops for all popular production car distributors. In addition, there are many proprietary distributors on the market that have been designed for easy modification and tuning when used on competition engines. Heavier springs are usually needed to convert a stock distributor for use in competition engines that are raced in a narrow band of high rpm.

Spark Plugs

Five factors are of utmost importance in selecting spark plugs for competition use. First, the spark plugs must be designed to fire with the lowest possible voltage requirement. Second, the design should place the spark gap at the most advantageous point possible in the combustion chamber. Third, the design must have good resistance to fouling. This is somewhat tied in with the fourth requirement, which is that the plug must have the coldest possible heat range. Last, the plug must be designed so that it will not interfere with the movements of the piston or the valves.

To reduce the voltage requirement, racing plugs have electrodes that are designed to place sharp edges in opposition to one another. For example, on projected nose racing plugs, which are used in most production-based competition engines, the side electrode extends only halfway over the center electrode (instead of overlapping the center electrode completely as on a "street" plug). This places the sharp edge of the side electrode's tip in proximity with the two opposite sharp edges of the center electrode.

In the projected nose racing plug, the electrodes are of

heavier section than those of a conventional projected nose plug. This is partially to improve heat dissipation and thus preclude preignition. However, it also helps the electrodes to last longer. Cutback gaps suffer off-center electrical erosion, causing the gap to grow rapidly. For this reason, cutback side electrodes are seldom used in passenger cars because the plug life is greatly shortened. In racing, rapid erosion is inconsequential since new plugs are installed before the race, and the plug gaps will not show noticeable erosion before 500 to 1000 miles.

Some engines, for reasons of clearance between the spark plug and moving parts inside the combustion chamber, cannot use a projected nose spark plug. There are two kinds of clearance gaps available in racing designs for them. One resembles an ordinary projected nose racing plug, but the side electrode is opposed to the side of the center electrode instead of lapping over the end. The other is a standard gap (nonprojected nose) racing plug that has the cutback style side electrode.

The projected nose spark plug is highly desirable for its extended heat range. At low speeds, the projecting insulator nose heats up rapidly to burn off fouling deposits. This feature can be of vital importance if several laps have to be run under the caution flag at reduced speeds—particularly on large-displacement, high-output machines such as NASCAR stockers. At high speeds, the rush of cold fuel/air mixture being drawn in through the intake valve passes directly over the projected nose, cooling it off and preventing preignition. Thus, these spark plugs have the miraculous ability to run "hot" at low speeds and "cold" at high speeds.

In an engine where clearance limitations prevent the use of any kind of projected nose plug, but where the extended heat range feature is desirable, a fine wire electrode racing plug is often the best choice. These plugs were once known as platinum gap plugs because the fine wire electrodes were made of platinum. Today most of the fine wire electrode plugs used in America have electrodes made from gold palladium and similar precious metal alloys. In any case, the metal must be highly corrosion resistant and must not distort when heated to incandescence.

The principle of extending the heat range is similar with both fine wire electrode and projected nose spark plugs. At low speeds, the fine wire electrodes heat up, burning off fouling deposits. At high speeds, the electrodes are cooled by the blast of fuel/air mixture coming in through the intake valve.

Racing engines need very cold spark plugs. The hot plugs used in production cars to prevent fouling in city traffic would become so hot under racing conditions that the electrodes would burn away, and the engine might be destroyed by preignition/detonation damage. Racing plugs are made only in the colder heat ranges. When an engine is tested for the first time, the wise tuner always starts with the coldest plug that will fit. It is much cheaper to throw away spark plugs that have fouled because they are too cold than to throw away an engine that has been ruined by preignition/detonation damage caused by spark plugs that are too hot. From the coldest available plug, you can always work back one step at a time until you find a plug that remains clean under racing conditions.

Grand Prix racing engines, oval track racing engines, highly supercharged engines of all kinds and "fuel-burning" drag racing engines naturally need the coldest plugs of all. In fact, it is a difficult–if not impossible–job to design plugs that are cold enough in any form other than with the retracted gap configuration (Fig. 7-12).

Fig. 7-12. Firing tip of retracted gap racing plug. These plugs can be made in extremely cold heat ranges for high-output racing engines.

Retracted gap or push-wire racing plugs have an insulator nose that does not reach even to the end of the threaded portion of the spark plug shell. The side electrode consists of a piece of heavy-gauge wire pushed through the side of the threaded part of the shell. Thus, the spark gap is contained wholly within the plug shell, retracted from the combustion chamber. These spark plugs cause no clearance problems and can be made in very cold heat ranges. However, because of their narrow heat ranges, they will quickly foul and misfire if they are operated for even a few minutes at less than full engine output.

The information now available from spark plug manufacturers concerning selection of a suitable plug to meet any operational conditions is extremely comprehensive and of high technical quality. These services should be fully used in any case where selection proves difficult.

8 / *Mixture Production*

Mixture Supply

Although carburetors and carburetor induction systems have become very efficient, there are several inherent snags associated with this method of introducing the fuel into the intake air and distributing the resultant mixture to the cylinders. Leaving aside for the moment any possible limitations of the carburetor itself, it will be evident that the large-bore passages that form the induction system, although the only available means of transferring the mixture to the cylinders, constitute a serious handicap when dealing with a medium compounded of gas (the air) and a liquid (the gasoline).

It is quite wrong to assume that the fuel, having met the air in the carburetor, emerges in the form of a gas. What happens is that the air emerges holding minute drops of liquid fuel in suspension, and these droplets can separate out easily. In passenger car engines, the intake manifold is water jacketed or heated by the exhaust gases to encourage the formation of a more homogeneous mixture. Desirable as this is from the viewpoint of smooth running and rapid warm-up, it has the unfortunate effect of reducing the engine's power output.

With a fuel injection system, these problems are largely eliminated. The large-bore passages of the induction system handle air only; there is no need to heat the intake manifold (or intake air distributor, as it is generally called). Unlike a carburetor, fuel injection does not depend on the velocity of the incoming air to draw fuel into the engine. Instead, fuel is injected into the airstream under pressure. Thus, the restriction of the carburetor's velocity-increasing venturi is eliminated, and the induction tracts can be designed so that air pressure is high in the intake air distributor and airflow most rapid at the intake valve itself. The fuel is generally sprayed into the airstream just ahead of the intake valve so there is no difficulty in maintaining a thoroughly homogenous mixture.

Volumetric Efficiency

If the mixture is heated before it reaches the cylinders, its volume, but not its weight, will increase; in other words, the amount of air contained in the manifold will be less when it is warm than if it is cold. Less weight will thus be drawn into the cylinders, with a corresponding reduction in power. It is, of course, true that the mixture will become considerably heated as soon as it enters the cylinder ports because of the proximity of the water jackets and combustion chambers, but this is inescapable. The heat thus absorbed by the mixture can be kept to a sensible minimum and need not necessarily be augmented by that derived from a deliberately heated manifold.

Tuners try to keep the mixture cool in production-based cars that are prepared for competition. Many oval track stock cars have elaborate cold air systems. Sometimes air is brought in from a removed headlight location or through the ventilation slots ahead of the windshield. In any case, underhood air, which is already heated by passing through the radiator and by its proximity to the engine, is totally avoided. The cool outside air is ducted by one means or another to the carburetor intake. In drag racing, hood scoops, generally in favor, make possible a direct path for outside air entering the carburetor(s).

Of course, when we talk about the weight of the mixture, we are really talking about the number of fuel or air molecules contained in a given volume of the mixture. At higher temperatures, the molecules expand so that fewer can be contained within the limits of the induction system. Therefore, most drag racers do not stop with obtaining cold air for their engines, They also chill the fuel to increase the number of gasoline molecules in a given volume and to help keep the entire mixture cool right up to the point where it enters the combustion chambers. Normally the fuel is chilled by passing a spiral section of the fuel line, just ahead of the carburetor, through a canister packed with dry ice.

Designers of passenger cars, even very fast cars, know that they must also produce a car that will run smoothly at low speeds around town. Thus, they are faced with a thorny problem. Should they design a cold induction system in the interest of power or heat the manifold for smoothness and flexibility? Usually they choose the latter. And because of the current legal limits on exhaust emissions, the designer actually has no choice left at all. It would not do, after all, to have inefficient combustion during the kind of driving where air pollution is the greatest concern—in the city. Just as obviously, heated manifolds, exhaust gas recirculation, and intake air preheating systems must be eliminated from production cars if they are to be successful on the race track.

Another problem with carburetor systems concerns the shape of the intake manifold. Numerous bends are required in the manifolds of production cars to conduct the mixture from the carburetor to the cylinders. Each of these bends represents a place where, because of the velocity of the mixture (which will also vary at the inside and outside radii of the bend), inertia tends to throw out the fuel globules, thus separating the gasoline from the air. When detached in this manner, the fuel naturally falls onto the pipe wall. If the wall it hot, it may "fry off" the separated fuel and enrich the air flowing at that particular instant, causing a patchy mixture. If the manifold is cold, the fuel may just lie in liquid form until a sudden throttle opening and surge of air causes it to be picked up and carried to one or more particular cylinders. In either case, an uneven mixture (and rough running) results.

These difficulties can be avoided by having a separate carburetor choke for each cylinder with a more or less straight and direct pipe right into the intake port. This is the system universally used on all competition engines where the rules permit it. But in nearly all forms of stock car and production car racing, the rules limit the tuner to either the kind of carburetor installed on the car at the factory or to another carburetor that has the same number of venturis. In this case, a compromise must be arrived at, and this is usually to use an intake manifold that helps to keep the mixture mixed.

With careful manifold design and attention to all aspects of installation to the best advantage from the volumetric efficiency point of view, the fuel separation snag can be largely overcome. All bends are made to the greatest possible radius, and sharp turns of all kinds are totally avoided. The factory-installed intake manifolds of many inline four- and six-cylinder engines are excellent from this point of view, especially on small-displacement imports. V8s tend to have the poorest intake manifold designs because some sharp bends are necessary to fit the manifold and carburetor beneath the car's hood. High-rise manifolds that project upward through the hood, with trade names such as "Tunnel Ram" and "Tarantula", are therefore seen on a great many V8s used in drag racing.

Though carburetor systems can be made to work with commendable efficiency, it should be obvious that a simpler way to avoid mixture troubles is to have the manifold carry air only. It is the presence of liquid fuel in the air that introduces all the headaches. So if a liquid fuel such as gasoline or alcohol is to remain our combustible agent, it appears that injecting it near the intake valves will solve a lot of problems.

Fuel Injection or Carburetor?

The elimination of the carburetor and the use of fuel injection in no way alters the engine's operating cycle. There is no connection with diesel principles, as some laymen assume, and spark ignition is retained. In fact, there is no difficulty in converting engines designed for use with carburetors to fuel injection operation.

At wide-open throttle, a fuel injection system offers only a slight advantage over a good carburetor system, and this small advantage is a result of having fewer restrictions to airflow. Carburetors must have a venturi and, for the most part, a butterfly-type throttle valve. An injection system needs no venturi, and on racing engines, slide-type throttles are used that present absolutely no restriction to airflow at wide-open throttle.

A fuel injection setup has an insurmountable advantage over a carburetor at lower rpm and on intermediate throttle openings. This is because an injection system does not depend on air velocity and the consequent depression to draw fuel into the intake air. Thus, with a carburetor, low-speed throttle response can be very poor. When the throttle is suddenly opened, there is insufficient airflow to draw in the fuel that is required for immediate acceleration. With a fuel injection system, the fuel is sprayed in under pump pressure; there is no lag while the engine tries to pull itself up by its own bootstraps.

In addition to these performance considerations, fuel economy will always be better with a fuel injection system. Moreover, the economy is greatest in the speed ranges most often encountered in highway driving. The engine receives only the fuel it needs because an injection system can be made responsive to many engine factors, for example, temperatures at various points on the engine, the air temperature, the airflow quantity, the manifold vacuum, the rpm, the temperature of the air inside the intake air distributor (manifold), and almost any other operating factor that can be imagined. On the other hand, a carburetor is sensitive mainly to the driver's foot, and all communication between the carburetor and the engine is via a tenuous and fickle column of air.

Fuel injection is being used on more and more passenger cars because of these advantages. Another reason for its increasing popularity is its ability to reduce exhaust emissions without reducing engine output to the degree that would be necessary with a carburetor. All things considered, today's designer would, if given a free hand, probably employ fuel injection on all engines regardless of the application, be it racing or stop-and-go traffic in the city. Nevertheless, for both these purposes, the carburetor will undoubtedly remain with us for some time.

Constant Depression Carburetors

Constant depression carburetors have also been known as variable choke carburetors and constant vacuum carburetors. Though renewed interest in the principle has been stimulated by emission control concerns, we will limit our discussion to matters that have a bearing on competition engines.

Fig. 8-1 is a simplified drawing of a constant depression carburetor. The variable choke (variable venturi) feature of the design consists of a vacuum chamber (L) that contains a

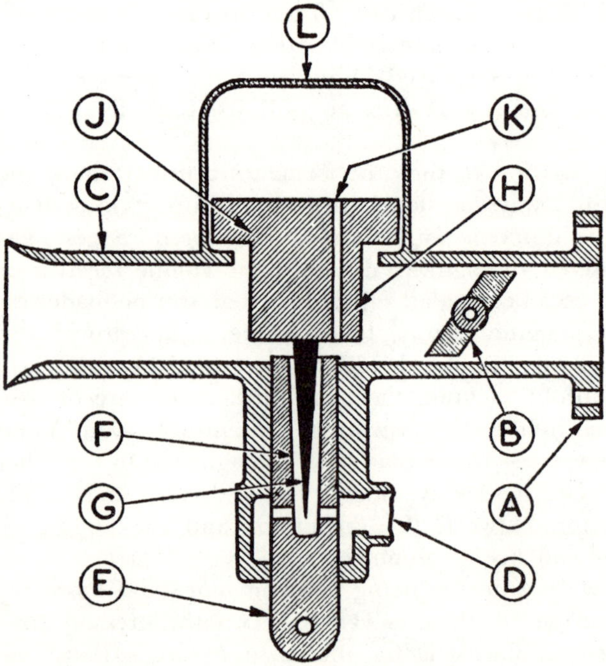

A. Mounting flange
B. Throttle valve
C. Air intake
D. Fuel inlet
E. Jet head
F. Jet
G. Needle
H. Air slide
J. Piston
K. Air passage
L. Vacuum chamber

Fig 8-1. Simplified diagram of a constant depression carburetor, based on the British SU carburetor design.

piston (J). The lower end of the piston is in the form of an air slide (H). The vacuum chamber is connected by an air passage (K) to the area between the air slide and the throttle valve (B). Any increase in vacuum in this area will raise the piston together with the needle (G), and any decrease in vacuum will lower the piston and the needle. The needle is tapered so that as the piston rises and admits more air, more fuel becomes available from the jet (F); the mixture remains of approximately uniform strength, though the needle is designed to produce a considerably richer mixture when the piston is fully up. The jet head (E) can be lowered mechanically for increased richness during starting and warm-up.

It is a mistake to assume that manifold vacuum is the sole factor in determining the rise or fall of the piston. If this were so, the piston would not rise fully at wide-open throttle when manifold vacuum is least, thus restricting the airflow and the engine's performance. The word *venturi,* after all, is actually the name of an Italian scientist who discovered that when air velocity increases, its pressure is reduced. Thus, as the air passes through the restricted venturi—in this case between the face of the air slide and the face of the jet—it must speed up and hence lose pressure. This pressure loss, or vacuum, draws the fuel from the jet (as it does in all carburetors) and raises the piston. Therefore, air velocity through a constant depression carburetor has a more important influence on piston movement than does manifold vacuum.

Fig. 8-2 is a cross-section through an actual SU carburetor of the kind that has been used in competition on countless MG and Austin Healey Sprite sports cars. This carburetor is somewhat more complex than its simple principle might lead one to suppose (compare to Fig. 8-1). The area around the jet is elevated, producing a more pronounced venturi, and the needle has a rather subtle taper. The jet has a restricted orifice, and there is a coil spring atop the piston that tends to force it down.

In tuning a constant depression carburetor for competition purposes, it is the practice to use springs of different strength, needles with different tapers, and jets with different orifice diameters to modify the mixture to meet the demands of racing. The oil well is similar to a miniature hydraulic shock absorber and prevents the piston's movements from being sudden or violent. Another kind of constant depression carburetor is the Stromberg CD shown in Fig. 8-3. This unit differs from the SU; in place of the piston of the SU unit, there is a sealed diaphragm that operates the air slide and needle. These carburetors are used in competition on many Triumph and Jaguar sports cars. In addition, a great many Hitachi constant depression carburetors have seen competition in Japanese sports cars.

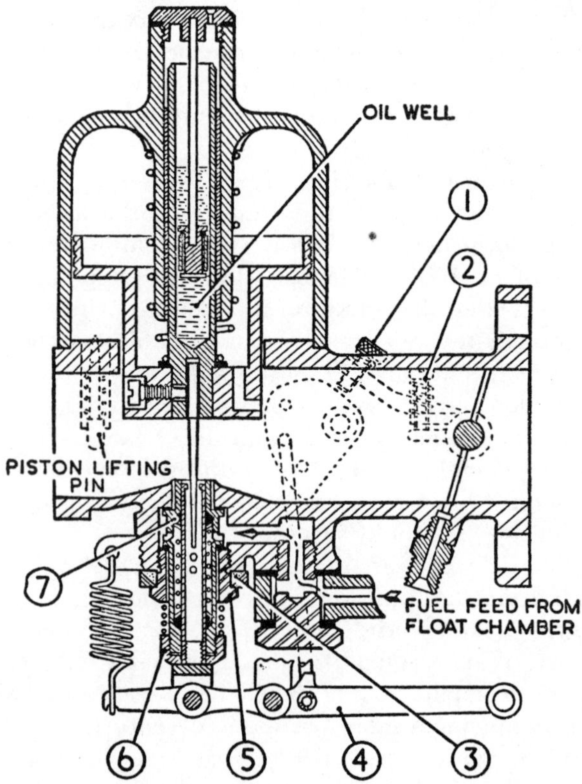

1. Fast idling adjusting screw
2. Idle speed adjusting screw
3. Washer
4. Jet control lever
5. Nut
6. Jet head
7. Top guide

Fig. 8-2. Section through typical SU carburetor.

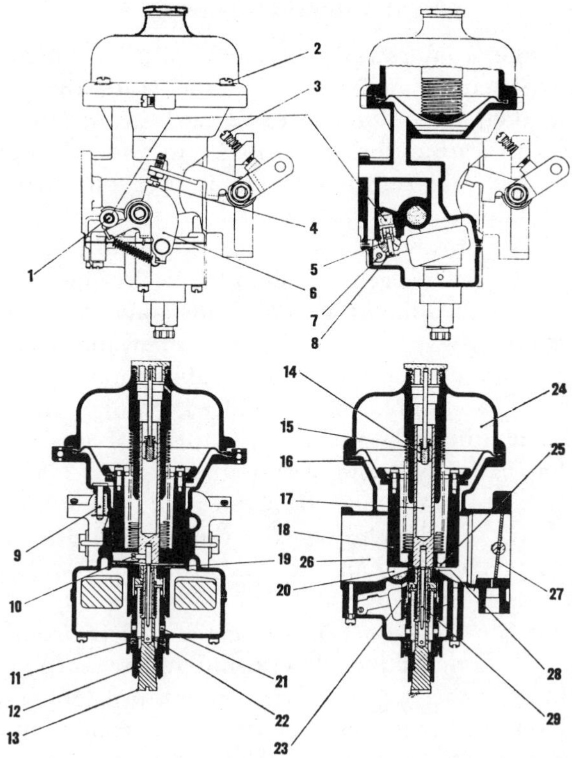

1. Fuel inlet to float chamber
2. Top cover (suction chamber) screws
3. Throttle-stop idling screw
4. Fast-idling adjusting screw
5. Needle valve seating
6. Lever and cam for fast-idle
7. Float lever pivot
8. Needle valve
9. Lifting pin for air-valve piston
10. Jet needle retaining screw
11. Jet base O-ring
12. Jet holder
13. Jet screw for adjustment
14. Air valve damper piston
15. Air valve return spring
16. Diaphragm
17. Air-valve guide rod
18. Air-valve piston
19. Jet orifice
20. Starter bar
21. Fuel inlet to jet base
22. Fuel inlet to jet bore
23. Jet orifice bush
24. Suction chamber
25. Air passage to suction chamber
26. Main air intake
27. Throttle valve butterfly
28. Bridge
29. Jet needle

Fig. 8-3. Arrangement of components in Stromberg CD carburetor.

Fixed Venturi Carburetors

For all the troubles that variable venturi carburetors have caused inexperienced mechanics over the years, the principles involved in their operation are more easily seen than those of a fixed venturi carburetor. We will begin with the extremely simple Solex 28 PCI carburetor that is used on Formula Vee race cars.

Fig. 8-4 is a cross-section of the 28 PCI, which has not been used on a new Volkswagen for almost two decades but is used in Formula Vee because it is more amenable to tuning than later Solex units, which do not have a removable venturi and have less accessible air correction jets and pump jets. Because every jet, emulsion tube, and air bleed can be interchanged with others of different size, the versatility of this unit is as great as that of many racing carburetors.

It will be obvious that the depression (vacuum) created by the high air velocity through the venturi will draw fuel out of the discharge openings in the spraying well. These openings are at the height of minimum venturi diameter. However, if it were not for the air correction principle used in fixed venturi carburetors, the rate of fuel flow would increase rapidly as air velocity through the venturi increased, resulting in a proportion of fuel to air that was ever greater as rpm went higher. Since air velocity through the venturi increases at a faster rate than does air volume, an overly rich mixture at high speeds would result.

The air correction principle is to "dilute" the fuel in the spraying well with air via the emulsion tube. This emulsion tube has a number of holes in it that are submerged in fuel (the fuel level, of course, being determined by the float and its valve). The amount of air that can enter the emulsion tube is controlled by the size of the air correction jet. Thus, a richer mixture can be obtained—particularly at higher speeds—through a decrease in air correction jet size with no change in the size of the main jet. The main jet, however, determines the maximum quantity of fuel available to the engine under all conditions above idle speed.

The main circuit of the carburetor will function reason-

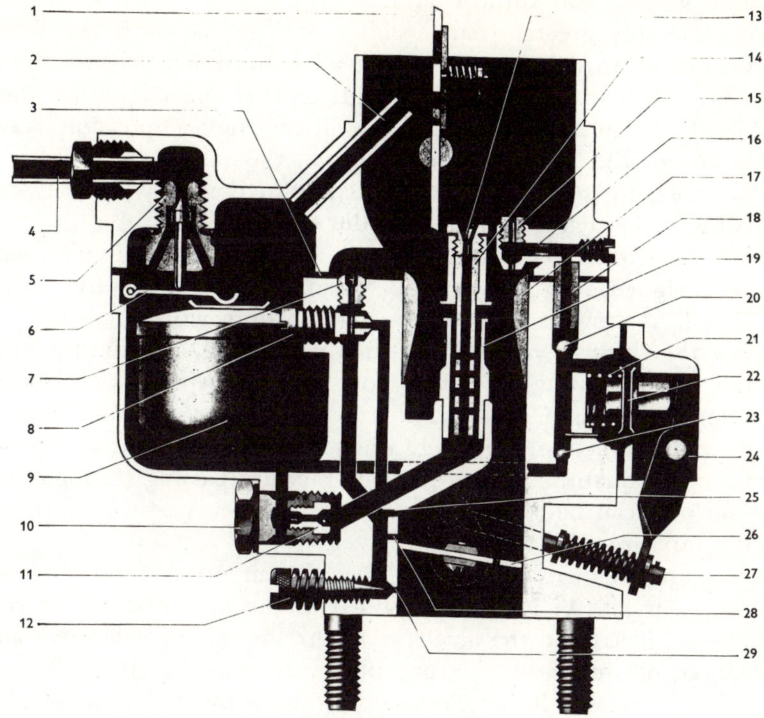

1. Choke valve
2. Float bowl vent tube
3. Gasket
4. Fuel line
5. Float needle valve
6. Float toggle
7. Pilot jet air bleed
8. Pilot jet
9. Float
10. Main jet carrier
11. Main jet
12. Volume control screw
13. Air correction jet
14. Emulsion tube
15. Pump air correction jet
16. Pump jet
17. Venturi
18. Fitting tube
19. Spraying well
20. Pump ball check valve, upper
21. Pump diaphragm spring
22. Pump diaphragm
23. Pump ball check valve, lower
24. Pump connector link
25. Idle air bleeder passage
26. Throttle valve
27. Throttle connector rod and spring
28. Accelerating port
29. Idle port

Fig. 8-4. Cross section of Solex 28 PCI carburetor, as used in Formula Vee racing cars.

ably well at full throttle and at part throttle when airflow is considerably greater than at idle. At idle, however, the fixed venturi setup will not operate at all because the venturi diameter, which must be great enough to feed the engine at high speeds, is too large to be effective in causing a depression (vacuum) at idle. By way of comparison, the variable venturi carburetor can make its venturi smaller so that it is entirely capable of efficient operation at idle speeds.

In a fixed venturi carburetor, a separate idle circuit must be designed into the overall system. With the throttle valve closed, as shown in Fig. 8-4, a high vacuum will be created below the throttle valve. This will draw a fuel/air mixture into the engine through the idle port (29). At speeds well above idle, the vacuum is not able to do this, and the idle circuit ceases to function; however, when the idle circuit is in operation, the quantity of the fuel/air mixture passing through the idle port can be adjusted to suit the engine's needs by turning the mixture adjusting screw.

The quantity of fuel in the idle circuit is controlled by the pilot jet. At idle, air is added to the idle mixture via the idle air bleed and the accelerator port. As the throttle valve advances off idle, the mixture obviously must be increased in volume. When the accelerator port is below the edge of the throttle valve, it too begins discharging mixture into the carburetor throat—in considerably greater quantity than was possible through the idle port alone. With increased throttle opening, mixture also begins to enter through the idle air bleed passage.

At this stage, the venturi should begin to function. However, the richness of the mixture available through the air bleed passage and the accelerator port determines how smooth the transition will be from the idle circuit to the main circuit. A "flat spot" at this point will result if the idle mixture is too lean, and a fuel-wasting surge will occur if the mixture is too rich. The smoothness of the transition can be adjusted by means of replacing the pilot jet air bleed, the pilot jet, or both with similar jets of different sizes.

Finally, there is the accelerator pump circuit. Instantaneous opening of the throttle valve tends to make the mixture lean because the quantity of the fuel flow does not increase as quickly as the air quantity. To augment briefly the fuel available at the venturi, an accelerator pump is built into the carburetor. It discharges additional fuel into the venturi whenever the throttle valve moves to a larger opening.

The amount of fuel discharged is controlled by the adjustment of the pump's stroke and by the diameter of the pump jet, which is very simple in the Solex 28 PCI. Because the venturi will be delivering more fuel at high speeds, the amount of additional fuel from the pump must be less. Therefore, the pump circuit has its own air correction jet (15), which serves to dilute somewhat the pump's discharge with air.

The choke valve (1) enriches the mixture for starting and warm-up by restricting the amount of air that can enter the carburetor, thus making all mixtures proportionally richer. Fig. 8-5 shows a cross-section of another kind of Solex carburetor, which has a "bi-starter" in place of the choke valve. The bi-starter is simply a miniature starting carburetor built into the side of the main carburetor to provide an additional rich mixture for starting and warm-up. Choke valves hinder airflow, so few carburetors designed for high performance have them. The Solex and Weber carburetors used on fast GT cars have instead the kind of starter device shown in the illustration. Aside from the starting device, however, it will be informative to compare the fuel circuits of the larger Solex carburetor shown in Fig. 8-5 to those of the Solex 28 PCI shown in Fig. 8-4.

The Weber Carburetor

For some years, virtually all carburetor-equipped road racing cars, and a considerable number of drag racing cars and oval track cars, have employed the carburetors manufactured by Soc. p. Az. Edoardo Weber, Bologna, Italy. While these carburetors are undoubtedly a high-precision masterpiece, their operating principles are straightforward. In addition to the simple single-choke variety, however, the racing world also has available Weber carburetors developed as a twin-choke assembly of very compact design so that they take up the minimum of installation room. Units with three barrels for the Porsche flat-six also rank among the marvels produced by the Weber craftsmen. These multiple-choke units have an obvious advantage when compared with separate, single-choke carburetors in controlling layout, fuel piping, and so on and especially in their simplicity of throttle synchronization.

The Weber carburetor is now being installed in increasing numbers on European production cars as standard equipment. Because of this, the wide use in competition, and the availability of Weber carburetor conversion setups for nearly all cars, a full description is merited here. We will begin with an examination of a simple single-barrel unit, the Type 32 shown in Fig. 8-6. This carburetor has been used on several popular continental vehicles, notably Fiat, Lancia, Renault, and Simca.

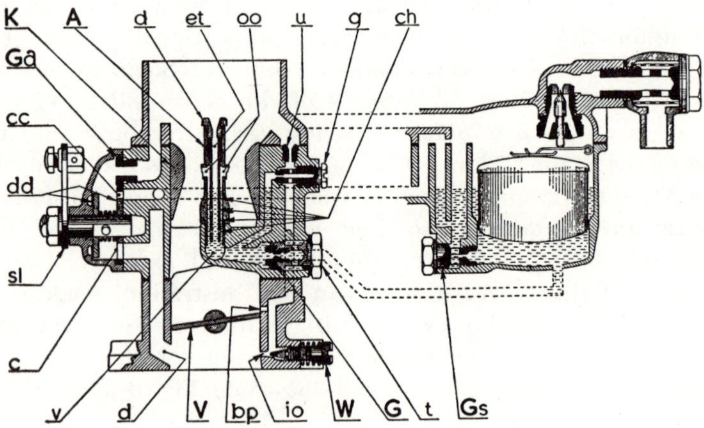

BI-STARTER

Ga. Starter air jet
cc. Starter valve duct
dd. Spring-loaded disc valves
c. Starter mixture exit duct
sl. Starter lever
Gs. Starter fuel jet
d. Starter mixture delivery duct

MAIN CARBURETOR

A. Spraying well
a. Air correction jet
et. Emulsion tube
oo. Spraying orifices
u. Pilot jet air bleed
g. Pilot jet
ch. Emulsion holes
G. Main jet
t. Main jet holder
K. Choke tube
bp. By-pass
W. Volume control screw
io. Idling mixture orifice
V. Throttle butterfly
v. Reserve well

Fig. 8-5. Section through typical Solex carburetor.

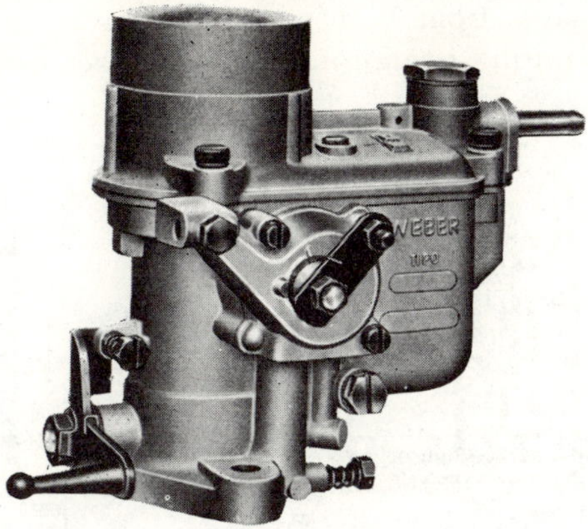

Fig. 8-6. Weber downdraft single-choke carburetor.

As with the Solex 28 PCI of Formula Vee fame, we will make reference to a cross-section of the Weber Type 32 (Fig. 8-7). Assuming intake air to be flowing downward through the auxiliary venturi (26), this air mixes with fuel drawn from the nozzle (27) using principles exactly the same as those previously described. The auxiliary venturi is a significant feature; its purpose is to increase the velocity, hence the vacuum, in the immediate vicinity of the nozzle and to direct the fuel/air mixture into the center of the choke tube.

The remainder of the main circuit is quite similar to the Solex unit. Its emulsion tube (22) has air admitted to it via an air correction jet (air adjusting jet, 1) and a main jet (16) that controls the volume of fuel available to the main circuit. The auxiliary venturi principle is also widely used on American high-performance carburetors.

What is unique in comparison to most American carburetors is that the auxiliary venturi and the choke tube (23) are replaceable with other choke tubes and venturis of different size. This makes the Weber carburetor highly tunable and the painstaking precision with which the units are made ensures that the tuning will be noticeable during dynamometer testing. Also available as standard Weber tools are metric drill sets and precision plug gauges so that jet sizes can be checked and altered accurately. Several technical books, tool lists, and parts lists are available from Weber representatives—in English—that describe the tuning operations for all Weber carburetors.

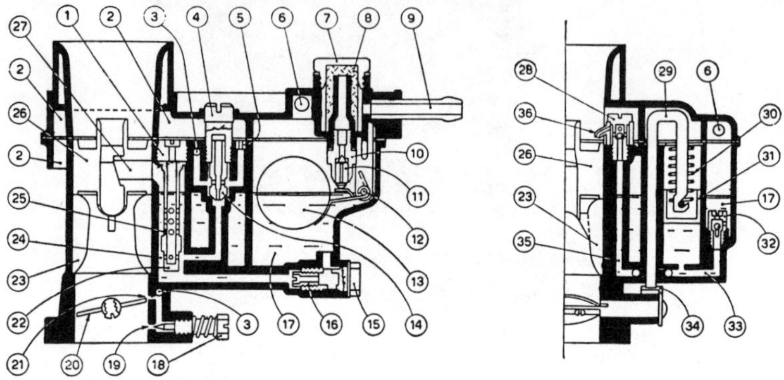

1. Air adjusting jet
2. Air intake
3. Idling mixture duct
4. Idling jet holder
5. Idling air intake bushing
6. Air intake
7. Strainer inspection plug
8. Strainer gauze
9. Fuel inlet connection
10. Needle valve
11. Needle
12. Pivot
13. Float
14. Idling jet
15. Main jet holder
16. Main jet
17. Bowl
18. Idling mixture adjusting screw
19. Idling hole to intake pipe
20. Throttle
21. Progression orifice
22. Emulsioning tube
23. Choke
24. Emulsioning tube well
25. Emulsion orifices
26. Auxiliary venturi
27. Nozzle
28. Pump delivery valve
29. Pump control shaft
30. Pump spring
31. Pump plunger
32. Pump intake valve
33. Pump exhaust duct
34. Pump control lever
35. Pump delivery duct
36. Pump jet

Fig. 8-7. Section through Weber Type 32 IMPE downdraft carburetor.

When the engine is idling, fuel is carried through a passage from the bottom of the emulsion tube well (24) to the idling jet (14), where it mixes with air from the idling air intake bushing (5). The idle mixture then passes through the idling mixture duct (3) into a system of ports near the throttle valve that are not unlike those of the Solex unit. The idling jet holder can be removed for a change of idling jets without carburetor disassembly or fuel loss. This practice is applied fully to the more complex Weber carburetors in which all jets are accessible from the top. On the Type 40 DCOE 2, shown in Fig. 8-8, a small round cover held by a wingnut can be seen on top. Ths is the jets inspector cover, which allows the tuner to check the sizes and condition of the jets. When the entire top cover, which is held by five fillister head screws, is removed, every jet is available for quick replacement.

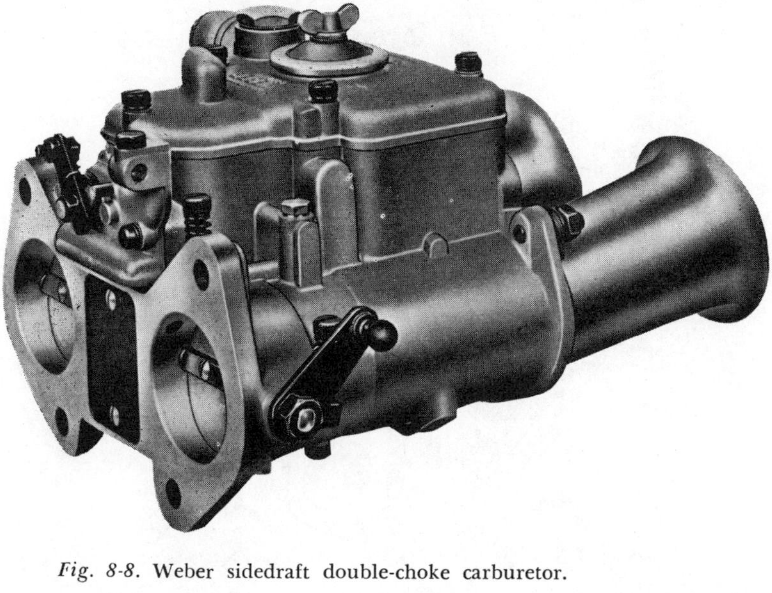

Fig. 8-8. Weber sidedraft double-choke carburetor.

Weber carburetors use metal piston-type accelerator pumps; the principle is different in no real way from that of the Solex 28 PCI. It is the easy precision interchange of tuning components that accounts for the Weber carburetor's wide acceptance among competition engine tuners and designers. On most American carburetors, many of the jets and air bleeds are simply (and cheaply) drilled into the carburetor's body material, but virtually every jet and drilling is a replaceable precision component on the Weber carburetor. As Fig. 8-9 shows, this results in a multiplicity of small parts—and the units are expensive. However, there are nice touches, such as ball bearings for the throttle shaft. The replaceability of all components means, in addition to tuning ease, that the carburetor will never need to be discarded because of wear.

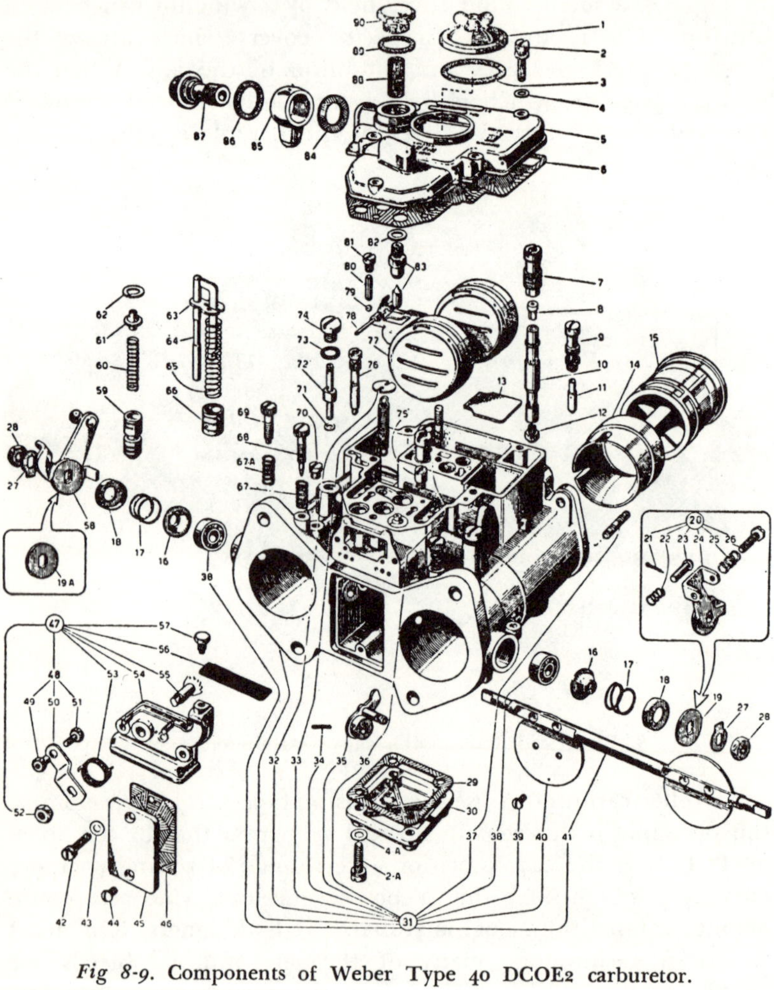

Fig 8-9. Components of Weber Type 40 DCOE2 carburetor.

1. Jets inspection cover
2. Cover fixing screw (5)
2A. Fixing screw (4)
3. Gasket
4. Washer for screw (5)
4A. Washer for screw (4)
5. Carburetor cover
6. Gasket for carburetor cover
7. Emulsioning tube holder (2)

8. Air corrector jet (2)
9. Idling jet holder (2)
10. Emulsioning tube (2)
11. Idling jet (2)
12. Main jet (2)
13. Plate
14. Choke (2)
15. Auxiliary venturi (2)
16. Dust cover (2)
17. Spring (2)
18. Small lid for spring retainer (2)
19. Distance washer (rear carburetor)
19A. Distance washer (front carburetor)
20. Throttle control lever, complete (front carburetor), including:
21. Split pin
22. Spring
23. Pin
24. Throttle control lever
25. Spring
26. Screw
27. Lock washer (2)
28. Fixing nut (2)
29. Gasket
30. Bowl bottom small lid
31. Carburetor body, including:
32. Plate for spring
33. Shaft return spring
34. Spring pin
35. Pump control lever
36. Stud bolt
37. Stud bolt (2)
39. Throttle fixing screw (4)
40. Throttle (2)
41. Throttle shaft
42. Screw securing support (2)
43. Washer for screw (2)
44. Fixing screw (2)
45. Plate
46. Gasket
47. Starting control, including:
48. Starting control lever, complete with:
49. Nut for screw
50. Starting control lever
51. Screw securing wire
52. Lever fixing nut
53. Lever return spring
54. Sheath support cover
55. Starting shaft
56. Strainer
57. Screw securing sheath
58. Throttle control lever (rear carburetor)
59. Starting valve (2)
60. Spring for valve (2)
61. Spring guide and retainer (2)
62. Circlip (2)
63. Spring retainer plate
64. Pump control rod
65. Spring for plunger
66. Pump plunger
67. Spring for idling mixture adjusting screw (2)
67A. Spring for throttle adjustment screw (rear carburetor)
68. Idling mixture adjustment screw (2)
69. Throttle adjustment screw (rear carburetor)
70. Progression holes inspection screw (2)
71. Pump jet gasket (2)
72. Pump jet (2)
73. Gasket (2)
74. Screw plug (2)
75. Inlet valve
76. Starting jet (2)
77. Float
78. Pivot
79. Valve ball (2)
80. Stuffing ball (2)
81. Screw for stuffing ball (2)
82. Gasket for needle valve seat
83. Needle valve seat
84. Gasket for fuel filter casing
85. Fuel filter casing
86. Gasket for fuel filter casing
87. Fuel filter bolt
88. Strainer
89. Gasket for strainer inspection plug
90. Strainer inspection plug

The double-choke Weber carburetor shown in Fig. 8-10 with its top removed is standard equipment on the Ford Cortina engines that are used in Formula Ford racing. All of the jets and venturis are accessible for cleaning, inspection, and replacement. This unit is, in fact, very similar to the single-barrel Type 32 shown previously, but now it is a "Siamese twin". On production Ford Cortinas, the second barrel is opened mechanically, but only at accelerator positions that would correspond to a "straight-and-level" speed above 85 mph. When converted for use in Formula Ford racing, the linkage is sometimes modified so that both throttle valves open simultaneously.

On many Showroom Stock Sedan class cars, twin-choke,

Fig. 8-10. Venturi arrangement of Weber downdraft double-choke carburetor. At present, most Formula Ford tuners are returning to differential throttle opening, instead of simultaneous opening.

progressive-opening carburetors are used in their original configurations. The Ford Pinto is typical of these. Its carburetor is a Weber design, but built in the United States by the Holley Company, and a derivative of the "Cortina" Weber shown in Fig. 8-10. Similar designs are used on Toyota, Opel, Datsun, Fiat, and Volkswagen sedans. However, in some cases—the 1976 VW Scirocco, for example—the second barrel's throttle valve is opened by a vacuum unit.

Fig. 8-11 shows cross-sections of a Weber carburetor that has a progressive-opening differential-throttle arrangement. The toothed sector (47) is mounted loosely on the primary throttle shaft (32). The sector has a slot (46) in which slides a lug (43) on the stop sector (44) that is fixed to the primary throttle shaft. This sector on the primary throttle valve shaft engages a toothed sector (48) that is fixed to the secondary throttle valve shaft (36).

With this kind of differential-throttle arrangement, the primary stage of the carburetor is used for most normal highway driving. Because the unit is essentially a single-barrel unit of modest proportions, the throttle response is good at low speeds and fuel economy is greatly benefited. Only when the primary throttle valve has moved sufficiently for the lug to begin moving the meshed sectors does the secondary barrel come into play, providing ample breathing for high-speed acceleration and power.

Fig. 8-12 shows the Weber Type 58 DCOE, which, in its many versions, is much favored for racing and sports-racing engines. Mickey Thompson and Moon, among American speed equipment manufacturers, have available cross-ram intake manifolds for mounting this type of Weber carburetor on the small block Chevrolet V8 engine. Originally developed for the 2.5-liter Coventry Climax Grand Prix engine of the 1950s, the Type 58 has remained in constant demand—particularly in America—because it is the largest made by Weber. (It might be mentioned here that the *DC* in such designations as DCO or DCOE stands for *doppio corpo,* "double body". Type numbers, such as 40 or 58, represent the bore diameter at the throttle valve in millimeters.)

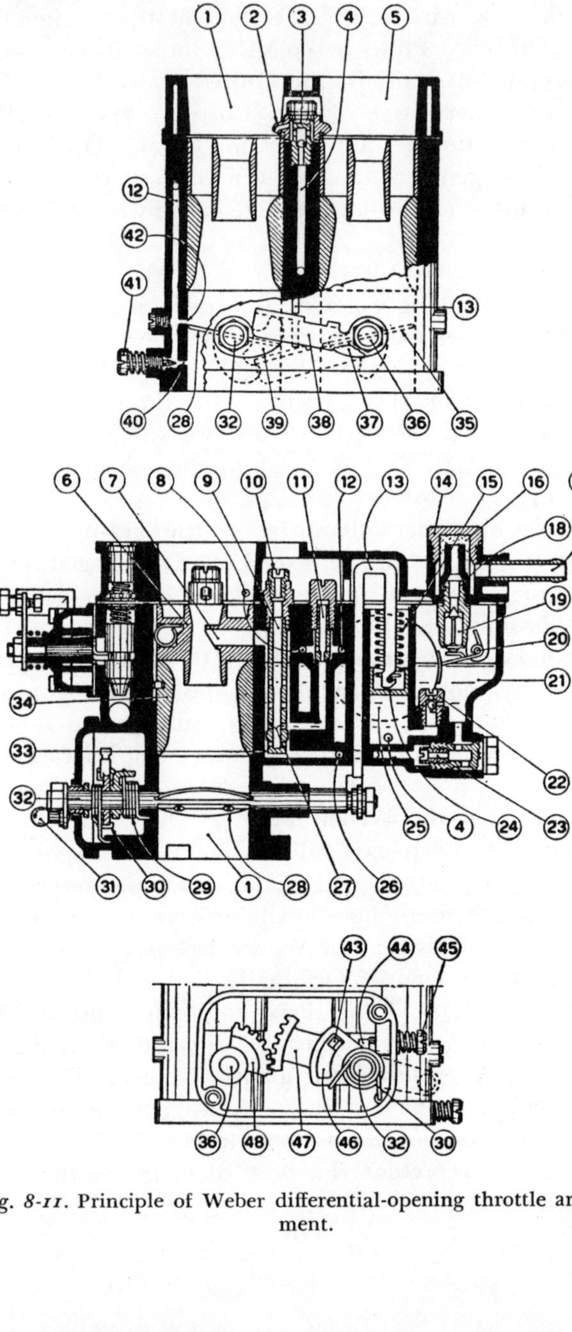

Fig. 8-11. Principle of Weber differential-opening throttle arrangement.

1. Primary intake pipe
2. Pump jet
3. Pump delivery valve
4. Pump delivery duct
5. Secondary intake pipe
6. Auxiliary venturis
7. Nozzles
8. Idling air orifice
9. Emulsioning tubes
10. Air adjusting jet
11. Idling jet
12. Idling mixture duct
13. Pump control rod
14. Strainer inspection plug
15. Pump spring
16. Strainer
17. Fuel inlet connection
18. Needle valve seat
19. Needle valve
20. Float pivot
21. Bowl
22. Pump inlet valve with discharge orifice
23. Main jets
24. Pump plunger
25. Float
26. Jets—emulsioning tube ducts
27. Emulsioning holes
28. Primary throttle
29. Primary throttle return spring
30. Secondary throttle return spring
31. Throttle main control lever
32. Primary shaft
33. Emulsioning tube wells
34. Chokes
35. Secondary throttle
36. Secondary shaft
37. Secondary pump control lever
38. Pump control neutral lever
39. Primary pump control lever
40. Idling hole to intake pipe
41. Idling mixture adjusting screw
42. Progression hole
43. Lug
44. Stop sector
45. Idling adjusting screw
46. Primary sector slot
47. Primary toothed sector
48. Secondary toothed sector

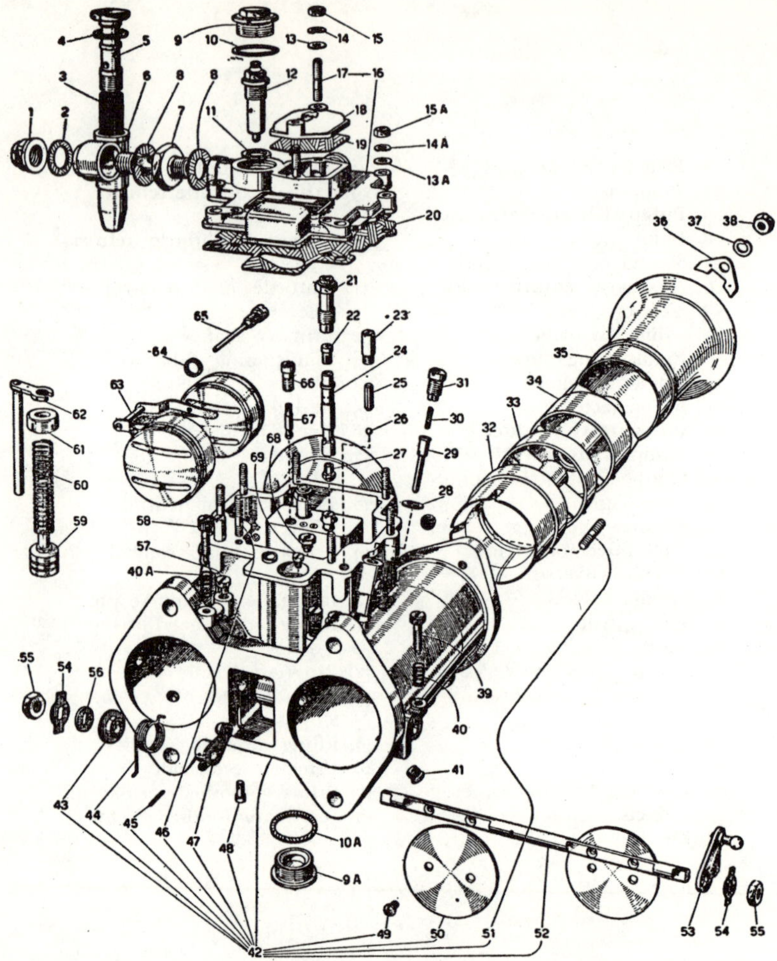

Fig. 8-12. Components of Weber Type 58 DCOE2 carburetor.

1. Box nut for fuel filter casing
2. Gasket for fuel filter casing
3. Strainer
4. Gasket for tap
5. Strainer plug
6. Fuel filter casing
7. Connection for fuel filter casing
8. Gasket for connection (2)
9. Needle valve seat inspection tap

9A. Float bowl exhaust screw plug
10. Gasket for needle valve seat inspection tap
10A. Gasket for bowl exhaust screw plug
11. Gasket for needle valve seat
12. Needle valve seat
13. Washer for carburetor small lid fixing nut (2)
13A. Washer for carburetor cover fixing nut (7)
14. Spring washer for carburetor small lid fixing nut (2)
14A. Spring washer for carburetor cover fixing nut (7)
15. Carburetor small lid fixing nut (2)
15A. Carburetor cover fixing nut (7)
16. Carburetor cover
17. Stud bolt securing small lid (2)
18. Small lid for jets inspection
19. Gasket for small lid
20. Gasket for carburetor cover
21. Emulsioning tube holder (2)
22. Emulsioning tube air corrector jet (2)
23. Screw for stuffing ball (2)
24. Emulsioning tube (2)
25. Stuffing ball for high speed valve (2)
26. Ball for high speed valve (2)
27. Main jet (2)
28. Gasket for pump jet (2)
29. Pump jet (2)
30. Spring for pump jet (2)
31. Pump jet holder (2)
32. Choke (2)
33. Auxiliary venturi (2)
34. Auxiliary venturi extension pipe (2)
35. Additional air horn (2)
36. Plate securing air horn (4)
37. Spring washer for carburetor air horn fixing nut (4)
38. Nut securing air horn (4)
39. Idling adjusting screw
40. Spring for idling adjustment screw
40A. Spring for idle mixture adjusting screw (2)
41. Tapping screw for inspection holes (2)
42. Carburetor body, including:
43. Washer for spring
44. Throttle return spring
45. Spring pin
46. Stud bolt securing carburetor cover (7)
47. Pump control lever
48. Spring retainer
49. Throttle fixing screw
50. Throttle valve (2)
51. Stud bolt securing air horn (4)
52. Throttle shaft
53. Throttle control lever
54. Lock washer with double tab (2)
55. Fixing nut (2)
56. Distance washer
57. Transition holes inspection screw (2)
58. Idle mixture adjusting screw (2)
59. Pump piston
60. Pump piston spring
61. Retainer for pump piston spring
62. Pump control shaft
63. Float
64. Gasket for float fulcrum screw
65. Float fulcrum screw
66. Idling jet holder (2)
67. Idling jet (2)
68. Pump intake valve
69. Pump discharge screw

Downdraft models, such as the 48 IDA shown in Fig. 8-13, are also very widely used. Manifolds are available for grafting these carburetors onto almost any engine, including the air-cooled Volkswagen. Similar Weber carburetors on the "big" VW air-cooled have helped to turn that unlikely powerplant into one of the dominant engines—perhaps *the* dominant engine—in USAC Midget racing.

Fig. 8-13. Weber 48 IDA downdraft double-choke carburetor installed on left cylinder head of VW air-cooled engine used in off-road racing.

Holley Carburetors

The foremost carburetors made in America, insofar as speed tuners are concerned, are those manufactured by Holley. As with the Weber, this acceptance by people who prepare competition engines is based largely on the tunability of the units. A big Holley four-barrel, such as the model 4160 shown in Fig. 8-14, seems to be made of rubber. The gaskets are rubber, countless rubber O-rings surround virtually every fuel and air passage where components join, and the acceleration pump, the power valve, and the secondary vacuum unit all have rubber diaphragms. The fame of these carburetors, however, rests mainly with the metering bodies, one of which is shown in the illustration. All of the critical jets are contained in this one device.

Fig. 8-14. Partially disassembled Holley four-barrel carburetor. Hand-held component is metering body with power valve screwed into it.

Replacement metering bodies, with different drilling sizes, are readily available in speed shops for the alteration of carburetor tuning during dynamometer or race track testing. Some secondary metering bodies have the secondary main jets drilled directly into the metering body metal. To change secondary jet sizes, you must change the entire metering body. Approximately twenty-five different sizes are available. Other Holley models have replaceable screw-in main jets (Fig. 8-15) for both the primary and the secondary sides of the carburetor. These jets are quickly accessible by removing the float bowls. Different power valve assemblies (one of which is shown screwed into the metering body in Fig. 8-14) are readily available for competition applications, and nearly every speed shop in America stocks a vast array of other tuning components for the various Holley carburetors.

Fig. 8-15. Main jets for Holley carburetor.

Holley carburetors can be obtained in a great many types and sizes. In addition to the four-barrel variety, there are three-barrel, two-barrel, and single-barrel versions available. A triple-carburetor setup using Holley two-barrel carburetors is shown on a Chevrolet 427 V8 engine in Fig. 8-16.

Fig. 8-16. Holley carburetors installed on Corvette engine. Notice extension tubes added to bowl vents; these keep fuel from being drawn out of float bowls during hard acceleration or cornering.

Fuel Injection Systems

Fuel injection systems have reached a mature state of development. Regrettably, however, the American automobile industry has taken no significant role in this development and thus many of the key patents are held abroad. Several promising starts were made in the late 1950s and the early 1960s along lines that later proved to be fruitful overseas. From our present viewpoint, it would appear that the termination of these projects was exremely short sighted.

High cost has been the main stumbling block for Detroit, since the typical American engine for the past twenty years has been a large V8. In Europe, the cost has been greatly reduced by the predominance of four-cylinder engines and by production in quantity. The various Bosch fuel injection systems, for example, have components that have been designed to be interchangeable from almost any car model to another.

A second impediment has been the allegation that mechanical dependability of an injection system is inferior to that of a multiple-carburetor installation. There is, however, very little evidence to support this contention. The main problem may be that most mechanics and corporation engineers are loath to become acquainted with anything new. Carburetors are a comfortable old shoe; the fuel injection takes some getting used to.

Fuel injection systems are much easier to troubleshoot and repair than are carburetors. This is because different functions in fuel injection are likely to be handled by separate components. Thus, temperature sensors, cold-start valves, injector nozzles, air regulators, pressure regulators, and similar devices can be quickly tested individually and faulty units replaced without disturbing other parts of the system.

By way of contrast, carburetors have all of their various functions concentrated in a relatively small space, with countless minute drillings, air bleeds, and orifices—all controlled by jets and valves that are impossible to test separately. Therefore, carburetor troubleshooting usually means removing, disassembling, cleaning, and rebuilding.

The rejetting complexities that are involved in tuning carburetor mixtures are in many cases reduced to the mere turning of a screw on a fuel injection system. Also, because each cylinder is supplied with fuel individually, there is less opportunity for a small particle of dirt to knock out the entire engine. Now that fuel injection is being used on a great many production cars (it is, for example, used on every Volkswagen model sold in America), it may not be long before people begin to wonder why we ever bothered with carburetors. There is, after all, nothing in the Bosch K Jetronic system of the VW Rabbit that could not have been duplicated easily with the technology that existed fifty years ago. The arguments still occasionally made against fuel injection are the same that would probably be heard against carburetors—if injection systems had preceded carburetors.

Many complicated features of early fuel injection systems

were unnecessary and were less effective in producing performance than the simpler means in use today. For instance, injection directly into the engine cylinders was initially considered an advantage because the fuel was fed straight to where it would do its work. The most prominent sponsor of the method for many years was Mercedes Benz, and the success of direct cylinder injection on the DOHC straight-eight Grand Prix cars of 1954 and 1955 is well remembered. Direct cylinder injection was also applied to the 300 SL and 300 SLR sports-racing cars. A schematic view of the 300 SL layout is given in Fig. 8-17.

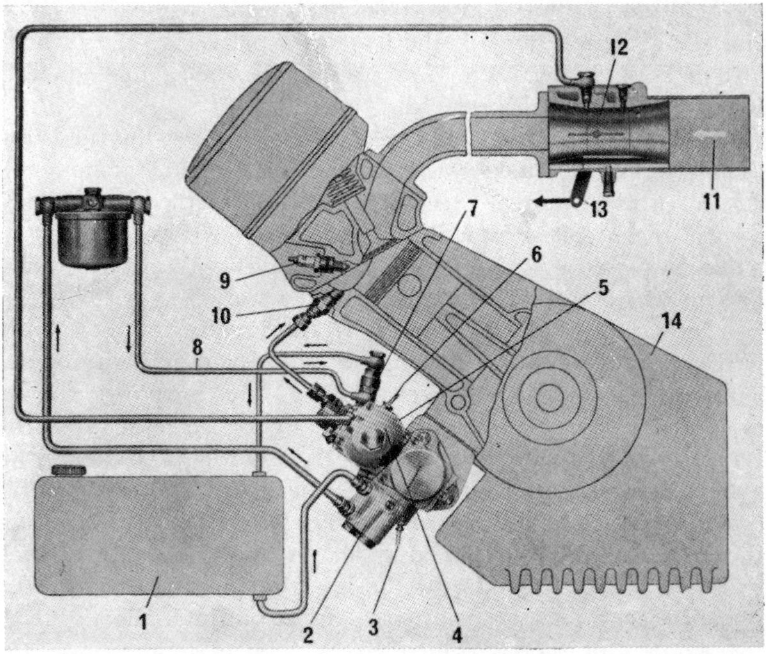

1. Fuel tank
2. Feed pump
3. Oil union
4. Diaphragm unit
5. Injection pump
6. Non-return valve
7. Overflow valve
8. Feed pipe
9. Spark plug
10. Injector
11. Air intake
12. Venturi control unit
13. Throttle pedal connection
14. Engine

Fig. 8-17. Layout of Bosch system for direct cylinder fuel injection.

Having both a spark plug and an injector nozzle to contend with in each combustion chamber is not the easiest condition for a designer to live with. Furthermore, the fuel sometimes sprays onto the piston or the cylinder instead of remaining a combustible vapor. By the time that Mercedes Benz started to attract international notice with the 300 SL, the once universally used Riley carburetor had already made an unlamented disappearance from the American oval-track scene. For a number of years, American racing engines had been using a continuous-flow port injection system that was developed by Stuart Hilborn and that provided excellent vaporization of the mixture. Hilborn, using his homemade system, powered himself to hot-rod fame in the late 1940s by seeing off 150 mph with a Ford flathead V8. The system was almost immediately adapted to the ever-present dirt-track Offies.

Injecting the fuel into the intake air at a point just upstream from the intake valve is the only method being used now in competition engines. Direct cylinder injection has returned to that place from whence it came, the diesel. The next question to settle is whether it is necessary to time the injection pulses to the opening of the intake valves.

Insofar as fuel vaporization is concerned, timing the injection pulses offers no measurable advantage at high speeds. The intervals of time between intake valve openings are so brief that some test engines were found to run with undiminished power and rpm when the pulses were timed to occur just after the intake valve had closed. The main difference in the timed systems and the continuous-flow systems used on production cars today is that timed systems vary the amount of fuel by changing the duration of the injection pulse; the flow rate remains consant for the most part. In a continuous-flow system, the amount of fuel is controlled by changes in the flow rate and pressure.

What was once considered a firm rule—that the injection spray had to be directed into the intake port—has likewise proven to be more theoretically than practically axiomatic. The Crower injection system used on many drag-racing machines has nozzles that direct the fuel spray straight across the injection tract—against the opposite wall. At wide-open throttle, the flow is deflected downward by the high-velocity airstream, which seems to aid the vaporization of the mixture.

It should not be assumed that any crude system that will dump fuel into the intake ports is equal to a sophisticated and costly high-pressure, mechanically timed system. The point is that one kind of injection system should not be automatically considered superior to another *for all engines and for all applications.* The type of fuel injection that proves to be best for a four-cylinder economy sedan, which must cope with both highway speeds and city driving, may not be the best type of injection system for a drag racing powerplant that is operated only at wide-open throttle.

The best system for achieving power on the drag strip is also not the best for oval track or road racing. It cannot even be assumed that two different makes of engines in the same racing class will both develop their best power with the same kind of injection system. Furthermore, a complex timed system may be necessary to obtain the best performance from a normally aspirated engine, whereas a simple, one-jet continuous-flow setup might be best if the same engine is turbocharged.

Lucas Injection

Earlier we explained the principles involved in carburetor operation by examining the simple Solex 28 PCI. The principles involved in fuel injection can also be learned by examining a single system. Of course, the means of executing these principles may be vastly different with different systems—for example, the metering that is accomplished by shuttle valves in the Lucas system is accomplished by an electronic computer on the Bosch L Jetronic system. Nevertheless, both systems apply the principle of fuel metering—just as the SU constant depression carburetor and the Weber fixed-venturi carburetor both achieve variable fuel metering but by entirely different means.

The Lucas system is widely used in formula car racing both in America and abroad. Fuel injection systems used on passenger cars sold in the United States are designed mainly for fuel economy and low exhaust emissions, but the Lucas system is designed for performance. A number of British cars that must be shipped to America with very mild carburetor setups in order to meet the emission laws are sold in England and Europe with the Lucas system. For example, the 1972 Triumph TR6 sold in the United States delivered only 111 bhp, but the same car was available in other countries with a Lucas injected engine that produced 150 bhp.

The Lucas system also achieves most of the design features for which high-performance fuel injection systems are noted: elimination of venturis, which are necessary in a carburetor but which restrict gasflow and reduce volumetric efficiency; manifolding that is designed to take full advantage of pressure-wave effects and the air inertia or ram principle (Fig. 8-18); and elimination of the float bowls that are generally needed on carburetors, thus avoiding the erratic fuel levels that can occur in a carburetor during hard cornering.

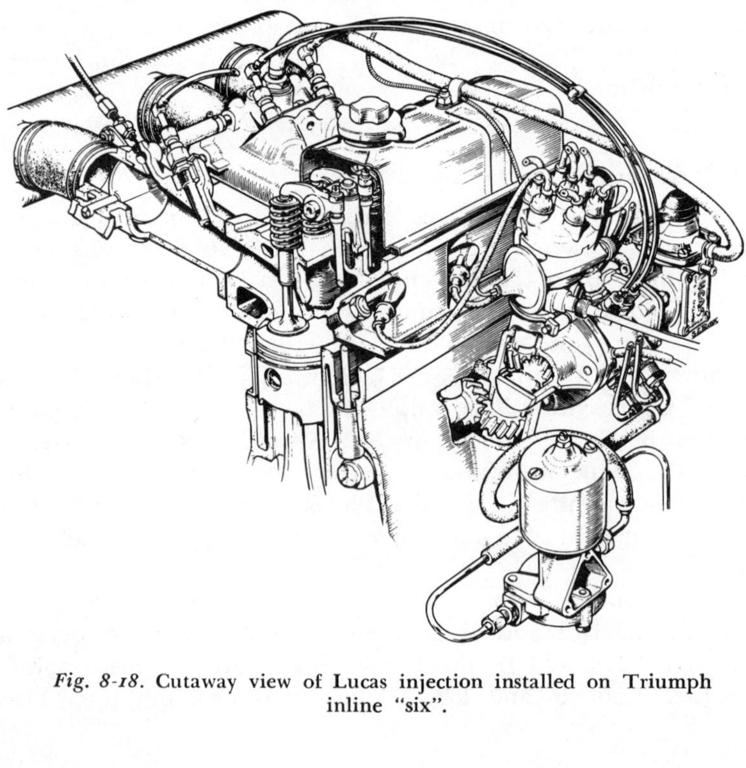

Fig. 8-18. Cutaway view of Lucas injection installed on Triumph inline "six".

From a competition point of view, fuel injection also offers the potential for eliminating the butterfly throttle valves, using instead a slide-type throttle that leaves the throttle body bore totally unrestricted at wide-open throttle. In addition, this kind of fuel injection also permits quicker throttle response. The better vaporization offered by fuel injection may also make higher compression ratios practical for some engines, and, especially with timed systems, greater valve overlap periods may become workable.

In the Lucas system (Fig. 8-19), a self-priming, electrically driven fuel pump draws the fuel from the tank via a filter. The pump output is for the most part directed to the metering and distribution unit. But a pressure relief valve allows some surplus to return to the fuel tank, thus maintaining uniform fuel pressure. The fuel is routed to the individual injectors, in the correct quantities and at the correct times, by the metering and distribution unit.

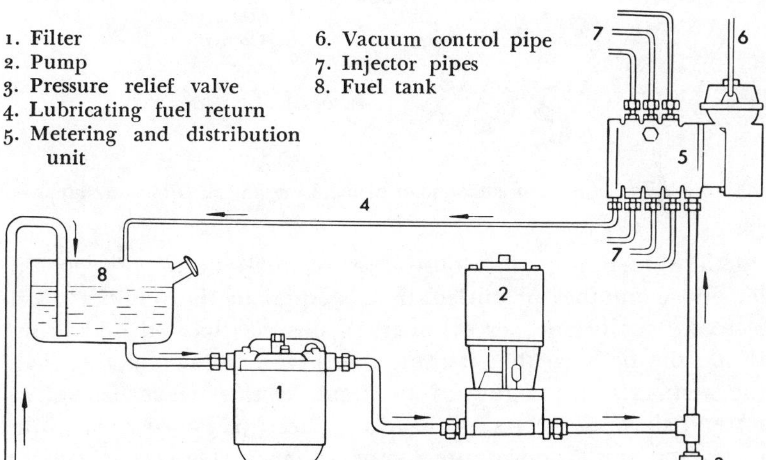

Fig. 8-19. Schematic view of Lucas injection system.

The metering and distribution unit is engine driven and is similar in principle to the more elaborate pump used with direct cylinder injection, but it has a built-in manifold vacuum-controlled compensation unit. This part of the metering and distribution unit is known as the *control unit.* The part that does the actual distribution and metering work is called the *metering unit* (Fig. 8-20).

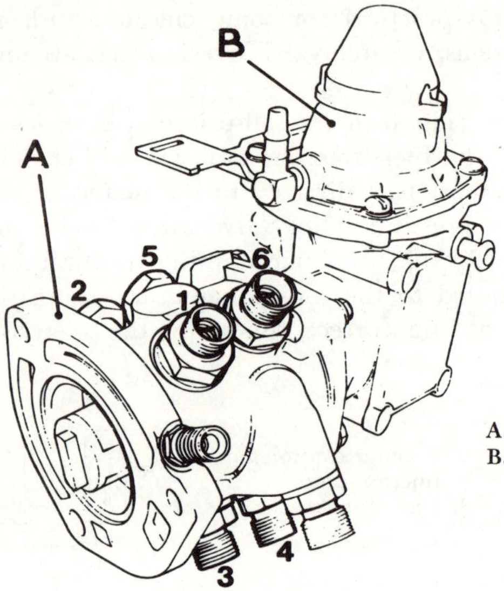

A. Metering unit
B. Control unit

Fig. 8-20. Two main components of metering and distribution unit.

The metering unit consists of an outer body with one inlet and a number of outlets that is equal to the number of injectors—usually one per cylinder. A sleeve is located and seated inside the body so that it cannot revolve or move axially. Taking a six-cylinder engine as an example, this sleeve has six inlet and six outlet ports arranged in spaced pairs 60° apart, inlet ports and outlet ports alternating. A space between the body and the sleeve forms a reservoir for pressurized fuel from the pump. See Fig. 8-21.

The six outlet ports are coincident with the outlet ports in the body, and sealed unions with integral nonreturn valves connect the sleeve ports and the body ports to the injector delivery pipes. A rotor, which has two radial ports to a central bore and is driven by the engine at one-half crankshaft speed, revolves within the sleeve. The central bore of this rotor contains a shuttle with a fixed stop at one end and a variable stop at the other end; variations are effected by the control unit.

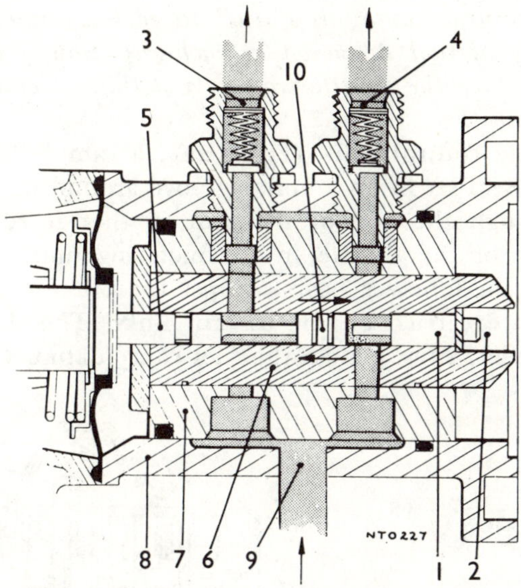

1. Fixed stop
2. Rotor drive
3. Outlet to No. 1 injector
4. Outlet to No. 2 injector
5. Variable control stop
6. Rotor
7. Sleeve
8. Body
9. Fuel inlet from pump
10. Shuttle

Fig. 8-21. Cross section of Lucas metering unit.

When the engine is started and the rotor turns within the sleeve, the rotor port at the variable stop end becomes coincident with the port in the sleeve leading to the fuel reservoir in the body. Fuel at high pressure enters the rotor bore and drives the shuttle to the fixed stop end of the rotor. This movement of the shuttle displaces fuel in the rotor bore through the ports in the rotor and the sleeve and out through the nonreturn valve in the union that serves the injector for No. 1 cylinder.

A further 120° of rotor rotation causes the rotor ports at the fixed stop end to align with the sleeve port leading to the pressurized fuel reservoir. Fuel now enters at the fixed stop end of the rotor and drives the shuttle back toward the variable stop end. The displaced fuel from the rotor bore ports passes to No. 5 cylinder via the sleeve port and the union with its nonreturn valve.

The shuttle continually moves between the two stops, displacing an accurate amount of fuel to each cylinder in turn. The quantity of fuel delivered at each injection is dependent upon the distance the shuttle travels, and this is determined by the control unit.

Inside the control unit (Fig. 8-22), a cam follower, with a diaphragm seal set in an annular groove around its periphery, projects through the end of the unit so that it contacts the variable stop of the shuttle in the metering unit. The other end of the cam follower bears against the outer two of three rollers that are carried on the control links. The third roller, which is smaller, runs against the fuel cam (datum track).

1. Cam follower
2. Rollers
3. Control links
4. Fuel cam (datum track)
5. Calibration springs
6. Calibration screws
7. Fuel cam carrier
8. Full load setting screw
9. Point X
10. Excess fuel lever
11. Balance spring
12. Depression chamber
13. To manifold
14. To atmosphere

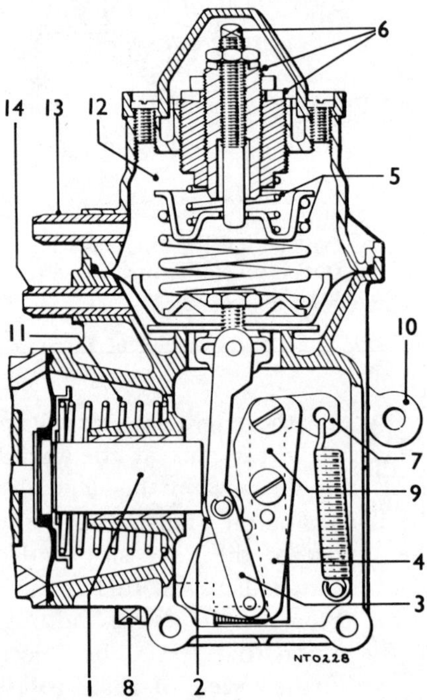

Fig. 8-22. Components inside control unit.

The control links are pivoted at the top, where they are attached to the center of a spring-loaded diaphragm. The lower ends of the control links are free. Two springs are positioned between the diaphragm and the three concentric calibration screws. These springs are in the depression chamber, which is connected by a pipe to the intake manifold.

The fuel cam (datum track) is secured by two screws to a carrier that is in contact with an external adjustment, the full-load setting screw. The fuel cam carrier is pivoted at point X, and the pivot extends through the rear face of the unit. The excess fuel lever is pivoted at the rear face of the control unit and has a cam face at the lower end, which contacts the fuel cam carrier pivot.

The engine's fuel demands, which are proportional to throttle openings and load, are reflected in changes in intake manifold vacuum. Each change is sensed by the spring-loaded diaphragm, which takes up a position balancing the loading for the springs against the vacuum in the depression chamber. The central links are thus raised or lowered along the cam track, allowing the cam follower to move in or out, thus altering the position of the variable stop for the metering unit shuttle.

To prevent the full hydraulic force of the variable stop from impinging on the control linkage, a balancing spring is installed on the cam follower, which results in only light pressure between the follower and the rollers. Movement of the excess fuel lever for cold starting is effected by pulling the choke control knob on the dashboard. This alters the position of the excess fuel lever, and the fuel cam carrier is drawn away from the cam follower, thus causing the shuttle to travel farther. When the control knob is pushed back, the carrier is returned to the normal operating position by the action of a tension spring.

Adjustments to the mixture are made by altering the positions of the calibration screws and the full load setting screw. Also, it is possible to install fuel cams (datum tracks) that have different profiles. However, these alterations require experience and can be made correctly only with the engine on a dynamometer and with an accurate exhaust gas analyzer attached to the engine.

Although the construction of other kinds of fuel injection systems is vastly different from that of the Lucas system, we might make one comparison to show how various approaches can be made using similar principles. With the Bosch K Jetronic system (Fig. 8-23), used on the VW Scirocco for example, the flow of fuel is continuous to the injectors instead of timed, but there is a plunger that performs somewhat the function of the Lucas shuttle. The movement of this plunger is controlled by an airflow sensor—a gasflow meter that measures the quantity of air being drawn into the engine—instead of by a manifold vacuum. Comparisons between the Lucas system and various other mechanical fuel injection systems will reveal other equally interesting and entertaining parallels.

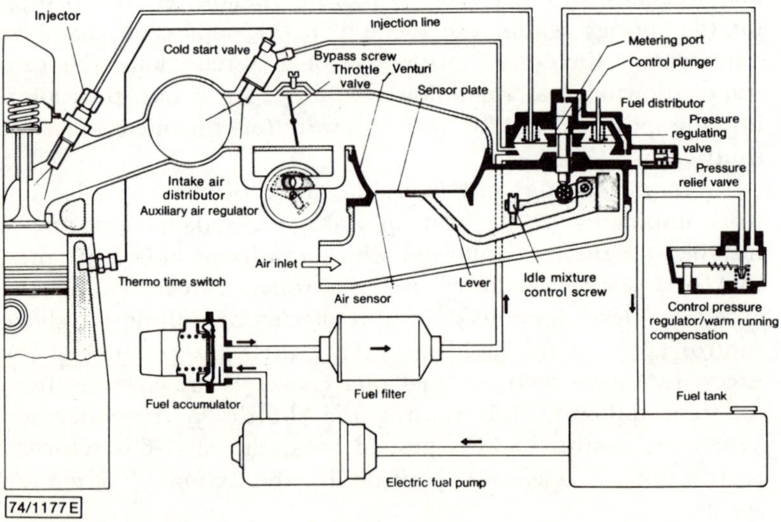

Fig. 8-23. Schematic view of Bosch K Jetronic fuel injection system.

High-speed Engine Application

For a six-cylinder, high-speed sports-racing engine, an injection period of some 240 crankshaft degrees is required. To obtain this, dual rotors can be driven at one-quarter crankshaft speed and two deliveries obtained by having two ports in the rotor to serve each of the shuttle spaces. A section view of this type of metering and distribution unit is given in Fig. 8-24. The driveshaft, turning at half engine speed, is geared 2:1 to the dual rotors.

The kind of injector used with the Lucas system is shown in Fig. 8-25, and its position on the engine is clearly indicated in Fig. 8-26. Fig. 8-26 also shows the clean throughway that can be obtained by employing a slide-type throttle in conjunction with six separate intakes. This layout gives great scope for ramming the charge since it is simple to change the length of the intake trumpets to suit the characteristics of a racing circuit *vis-à-vis* the rpm where maximum torque is needed.

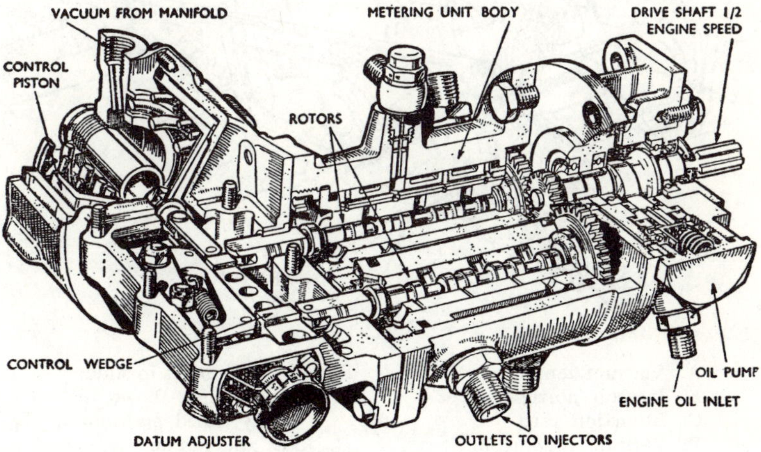

Fig. 8-24. Lucas shuttle-metering unit with control mechanism for six-cylinder engine. This unit employs dual rotors.

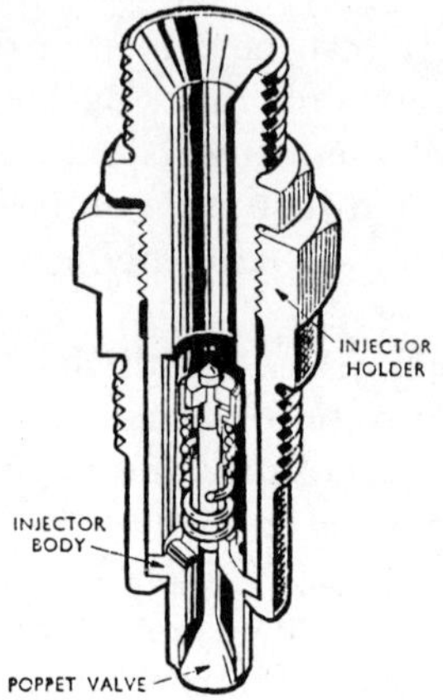

Fig. 8-25. Lucas injector for fitting in intake air manifold.

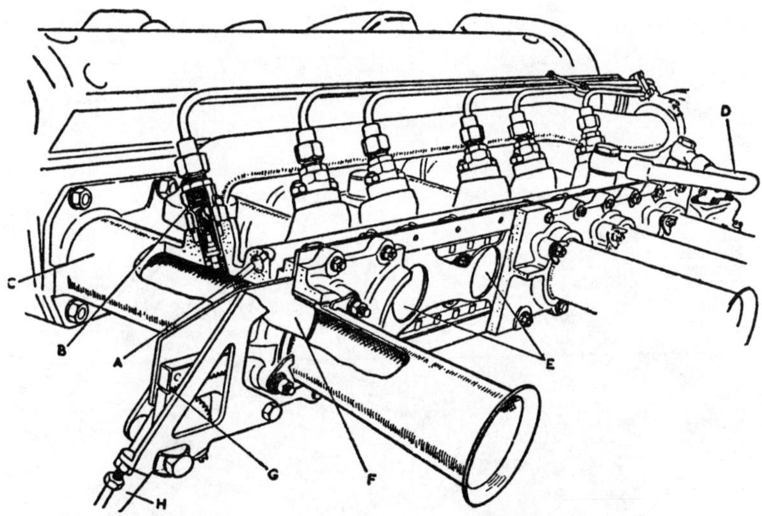

A. Vacuum tapping
B. Injector nozzle
C. Induction tract
D. Vacuum tapping to mixture control
E. Air intake holes in slide
F. Throttle slide shown in fully closed position
G. Rack and pinion
H. Connected to pedal

Fig. 8-26. Layout of Lucas system on Le Mans Jaguar engine.

The principle of using intake trumpets of various lengths for tuning purposes can have a great influence on cylinder charging, as shown in Fig. 8-27. This ram tuning system is widely applied not only to road racing engines, which encounter more course variations from one race to the next, but also to oval track cars and drag racing machines. However, in drag racing and oval track racing, the tuned length of the ram pipes tends to be selected more or less permanently, rather than being changed to suit the course as is the case with road racing formula cars.

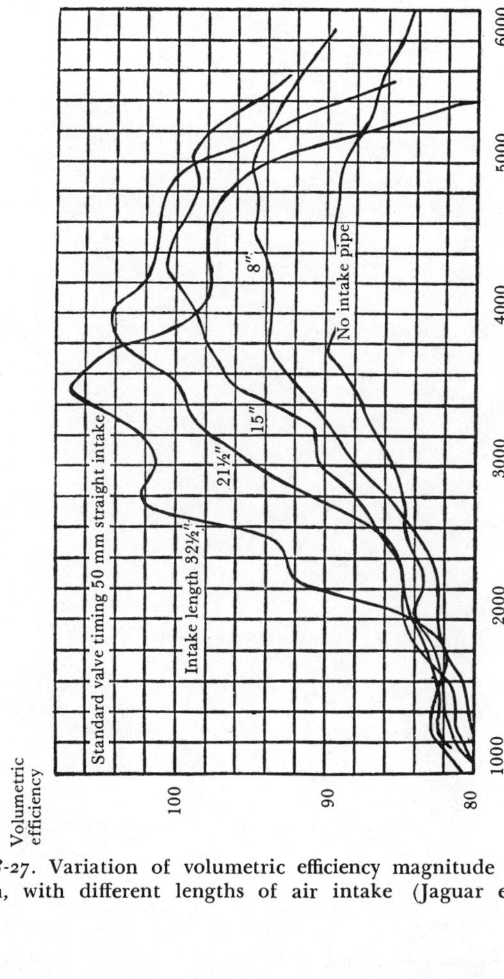

Fig. 8-27. Variation of volumetric efficiency magnitude and position, with different lengths of air intake (Jaguar engine).

PART II

Practice

9 / The Cylinder Head

General Condition

Because of the number of books and service manuals available that treat in detail the precision examination of engine components, we will not go into much detail on straightforward items. We assume that the owner or builder of a competition engine who is considering the various implications of looking for more power will be conversant with the normal methods of checking for wear of the cylinder head and fitness for the job. It will be obvious that any examination of an engine component intended for use in competition must be ruthless, particularly if the part has had a good deal of service.

Admittedly, it is not very helpful to suggest that components showing undue signs of wear should be replaced, since what might be allowable in ordinary highway driving might spell trouble at high speed and load. The aim should always be to have all the clearances and dimensions correct. Valve seats are, of course, the main point. In production car cylinder heads with integral valve seats, there must be enough metal left for a flawless competition valve grind without recessing the valves into the head, which is the equivalent of reducing their lift, or running the periphery of the intake valve seat into that of the exhaust.

The cylinder head casting must be faultless. Cracks, casting flaws, valve seats with hairline fissures, and localized heat distortion are very frequently overlooked because the cylinder head was not thoroughly clean when it was inspected. Many of these flaws will never cause trouble on the race track because they will be discovered during later stages of cylinder head preparation. Thus, the main problem is the wasted time, effort, and money that has been put into the faulty head before the defect is uncovered.

Before any work begins, the cylinder head should be cleaned in a hot tank and then bead-blasted. The hot tank boil-out, done in a corrosive solution, will remove scale from the water passages and oil sludge from the oil passages, valve spring gallery, and so on. Bead-blasting is a process by which tiny glass beads are blown against the metal with great force by compressed air. The glass beads remove rust, stains, corrosion, and remaining hard deposits. Because the beads are smooth, they will not cause the abrasive action that sandblasting does.

The cylinder head should be checked for warping before it is cleaned. Go over the head gasket surface with a piece of fine emery cloth wrapped around an absolutely flat and rigid piece of steel (Fig. 9-1). Brightened areas or depressed areas surrounded by bright areas indicate irregularities in the gasket surface that may be too small to detect with a normal straightedge check for warping. On a competition engine, the gasket seal must be perfect, and the head should be milled or ground to an absolutely true surface by a competent machinist if any irregularities are discovered. (In many racing classes, milling is always a part of cylinder head preparation.)

Improving Volumetric Efficiency

The highest possible breathing efficiency must be aimed at. We have seen how the designer tackles this job. In its passage through the factory, however, the product may become subject to minor imperfections in machining and so on, which must be corrected. This, of course, is blueprinting; in some classes

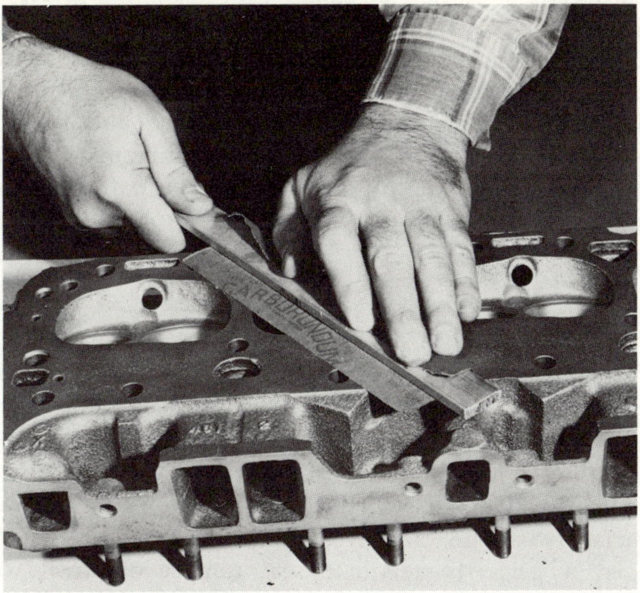

Fig. 9-1. Gasket surface being lightly polished with emery cloth.

it may be all that is permitted, but it is also an essential preliminary to any subsequent supertuning operations. The cylinder head and the valve gear will provide a useful start to these operations.

It is practically certain that an engine of the OHV type will be dealt with since flatheads have almost totally disappeared. Thus, the valves will be arranged either in a straight row in the head itself or in two inclined rows at an angle to each other. Most of the modifications apply with equal force to either layout.

In the design of the intake tract, there are three factors to consider: valve size, port and manifold shape, and carburetor or injection system throttle bore size. These three factors are closely interrelated, and any modifications must be designed with regard to the effect on the whole combination. For example, the installation of a larger-bore carburetor without any other modifications would have no advantage (unless the original carburetor was far too small); in fact, the performance would probably be impaired because of the reduced gasflow velocity. Increasing the valve and port diameters, of benefit at very high speeds, would likely affect the low-speed torque. However, larger carburetors plus an increase in valve and intake tract size will undoubtedly provide an appreciable horsepower increase through about the top one-third of the power curve; this is therefore a modification of interest to those concerned with production-car racing.

Intake Valves

For those who desire a performance rather better than standard for normal road use, an inexpensive modification can be made to the existing valves that does not impair the low-speed pulling power and gives a little more pep at high rpm. The existing valve seats will probably be found to have a contact surface with the valve facing that is about 3/32 in. wide. However, a perfectly adequate seal can be obtained with a seat having a width as narrow as 1/16 in.—or even less. The object aimed at is best explained by the diagrams given in Fig. 9-2.

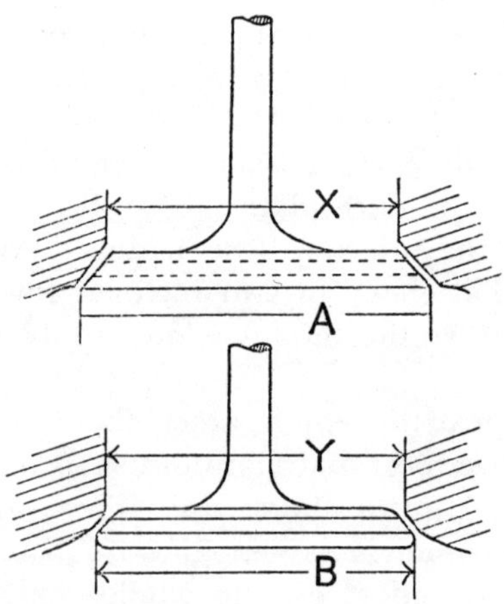

Fig. 9-2. Modification to intake valve. Port diameter increased from *X* to *Y*. Valve head diameter reduced from *A* to *B*.

The top part of Fig. 9-2 shows the valve and seat as standard; the dotted lines indicate the proposed narrow seat. The lower part of the illustration shows how the outside diameter of the valve head can be reduced from *A* to *B*, while at the same time the port size is increased from *X* to *Y*. The increased diameter of the port and the reduction in valve diameter will give slightly easier gasflow at higher engine rpm.

The work involved is not difficult to carry out, but it must be carefully done. The reduction in valve diameter can be accomplished by mounting the valve in a drill chuck or, better still, a lathe or a valve grinding machine (Fig. 9-3). The surplus metal can be removed with a file or grinder, finishing it off with fine emery cloth. The outer diameter should be nicely radiused, but the valve should not be reduced in thickness since the radius is used to merge the narrowed seat into the valve body.

Fig. 9-3. Reducing valve diameter or grinding additional inner bevel. This work can be done with greatest precision on a valve grinding machine, as shown.

A cutter is the ideal method to enlarge the port. If this cannot be managed, files and scrapers provide a satisfactory substitute; professional tuners use a high-speed power-driven rotary file (shown in Fig. 9-4). Care is necessary to ensure that there is plenty of room past the valve around its circumference when it is in the open position. There should be at least 3 mm between the valve edge and the nearest combustion chamber wall. It will be obvious that engines that have had the valve diameters increased by the manufacturers in the course of development will little benefit from this kind of work.

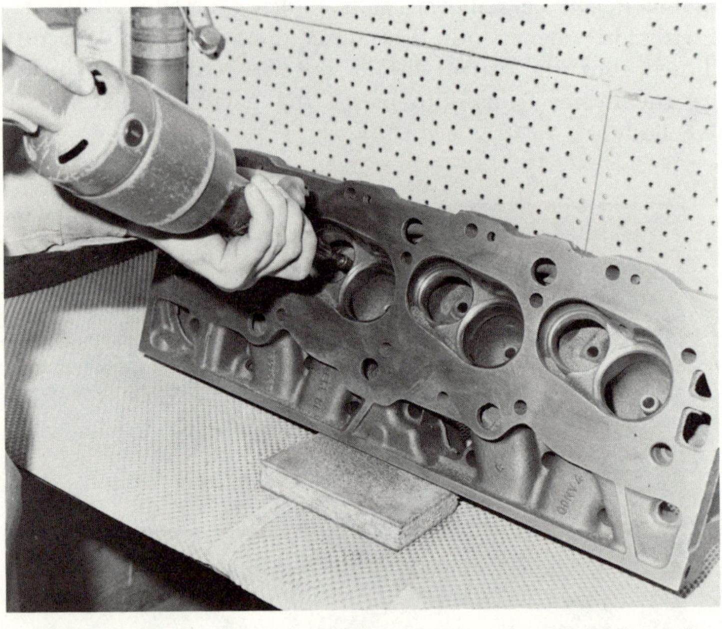

Fig. 9-4. Ports being enlarged with rotary file and high-speed porting tool.

The protruding valve guides form what might be called a major obstruction, and some tuners make a habit of cutting the guide flush (Fig. 9-5) with the inside of the port—especially in drag racing, where a few seconds of terrific power are worth far more than long component life. If the removal of the projecting guide leaves a reasonable length *in situ,* there is probably no harm done, though the work is best restricted to the intake valve guides only. At least equal, and usually better, results will be obtained if the protruding part of the guide in the port is streamlined in the direction of gasflow so as to present a knife-edge on the approach and exit sides. At its narrowest part (at right-angles to the gas stream) the guide thickness should leave about 1/16 in. of metal between the bore of the guide and its outside wall. The obstruction to gasflow presented by a streamlined guide of this type is much less than that of the cylindrical valve stem.

Fig. 9-5. Fully-ground-off valve guide in VW head modified for drag racing. In aluminum heads, this should always be done during porting, and a new guide installed later. Having guide installed prevents tool from cutting a pocket around valve guide bore.

The Intake Tract

The intake ports and the intake manifold should have a smooth internal finish. A high polish is not necessary, but it does set the seal of craftsmanship on the job. The port shape should not be altered unless the valve openings have been enlarged; in the latter case, the endeavor should be to obtain a consistent bore size right through from the valve seat to the manifold flange.

Cleaning up the ports and the manifold in this manner is a tedious job. The work is easier if small rotary files and polishers are used in conjunction with a high-speed portable motor similar to the one shown in Fig. 9-4. Care must be taken in using this equipment not to remove too much metal from the wrong places.

The intake manifold flanges form a frequent source of obstruction to smooth gasflow because it is comparatively rare for these either to meet perfectly or to be of identical size of mating aperture (Fig. 9-6). Any error can be seen by clamping a sheet of white paper between the flanges and torquing the bolts (when the flanges should be smeared lightly with graphite). When the paper is removed, the dimensional differences between the flanges are reproduced on either side of the paper; correction is carried out by filing or by using the mechanical equipment mentioned above. The same procedure can be carried out on the flanges between the manifold and the carburetors. Another method is to use the gaskets as guides (described later in a discussion of exhaust valves and ports).

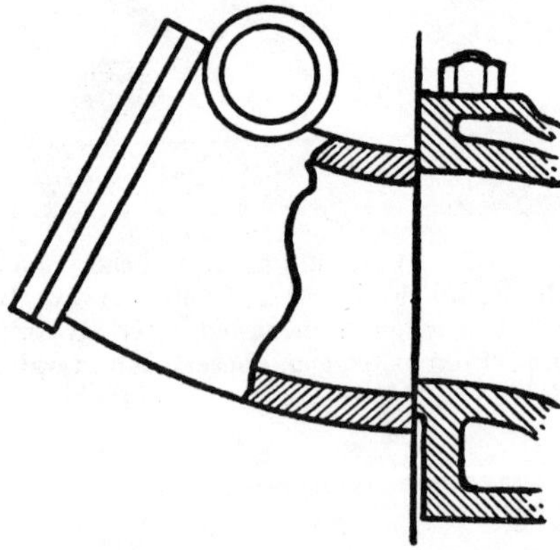

Fig. 9-6. Misalignment of manifold flange/port in head. This must always be corrected.

Some engine manufacturers meet this point by making the outlet ports from the manifold slightly smaller than the intake ports into the head so that any step will be in the nonobstructive direction (the same principle being followed in reverse for the exhaust manifold). This certainly meets all requirements except one, which assumes greater importance as engine and gas velocities begin to rise: the obstruction caused to high-speed pressure waves in the gas column, which will be affected by any step in whichever direction it faces. Perhaps the best production-line technique is the one on the Ford Cortina GT engines used in Formula Ford racing. On these engines there is a steel ring pressed into each intake port to align it with the manifold. Any correction of flaws involves matching the intake port or the manifold outlets to the inside diameter of the rings.

Oversize Valves

This attention to standard fittings will result in considerable improvement to the induction side and will show throughout the power range. However, it may be desired to increase peak performance, even at the expense of less power. For this purpose, larger-size intake valves can be usefully employed. Bear in mind, though, that this is a fairly far-reaching modification, and the characteristics of the power flow in normal road usage may not be so pleasant as on the unmodified engine.

Some makers market oversize valves since there is sufficient metal in the head casting to allow these to be fitted into suitable enlarged ports. Instructions for fitting are provided. Because of the large amount of metal to be removed, the only practical way of doing the job is by means of a reamer of correct size, using the valve guide as a pilot hole.

When modifying an engine in this manner, the amount of metal between the valve seats is usually the factor controlling the size. The main point is not to be too ambitious; a millimeter or two on valve diameter means quite a lot in terms of opening area. Also, the exhaust valve can usually quite well be left alone and the enlargement confined to the intake with little disadvantage.

When the engine makers are unwilling or unable to supply oversize valves, these can be obtained from specialist manufacturers; in some cases valves from a larger engine can be adapted. The starting point for your search should be the specification catalog of TRW, Eaton, Teves, or some other major valve manufacturer. Ensure that the material is adequate for the job and be careful in choosing the shape. The tulip shape shown in Fig. 9-7 is effective in directing the gasflow but is not necessarily better than a flat shape when gas velocity is very high. Attention to the ports required for larger valves does not differ materially from the operations already described when modifying standard valves.

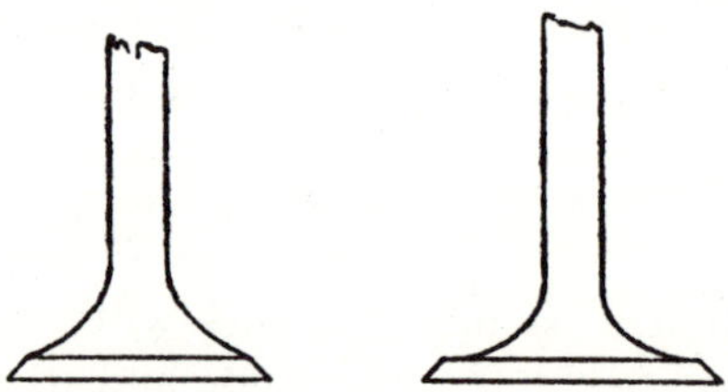

Fig. 9-7. Tulip-shaped intake valve (left) and flat-type (right).

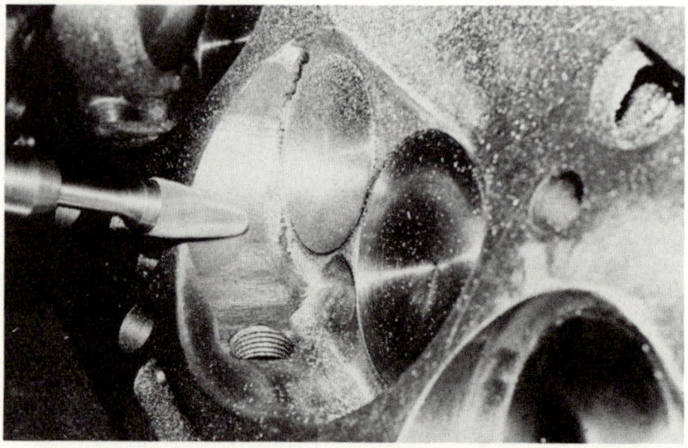

Fig. 9-8. High-speed rotary file being used to enlarge combustion chamber wall adjacent to valves.

There is less point in making the valve of larger diameter if there is still an obstruction caused by the proximity of the combustion chamber wall when the valve is fully open. When removing metal to correct this (Fig. 9-8), remember also that the compression ratio will be lowered. Where subsequent modification to the compression ratio is to be made, therefore, the actual head volume must be obtained by measurement after all work on the latter is complete. (This work is described in chapter 12.)

Exhaust Valves and Ports

The problems associated with the exhaust valves and ports are quite different from those applicable to the intake side. The residue of the heat produced has to be ejected via the exhaust valves and ports, and for this reason care must be taken in removing metal. There is plenty of pressure available to clear the exhaust, and the time factor is adequate. As a consequence, obstructions in the port that would be a grave disadvantage in the intake tract are less important as far as the exhaust is concerned. Really large increases in exhaust valve and port diameters are necessary mainly in highly supercharged engines.

The exhaust valves have to stand a lot of heat, the bulk of which is dissipated from the valve itself by conduction through its contact with the valve guide and the valve seat. The guide should therefore not be shortened, except perhaps in "expendable" drag engines, and the seating must remain of ample width. If there is plenty of metal in the guide, it can be streamlined in the direction of gasflow in the same way as was described for intake guides, but the wall thickness should be left on the generous side. The top edge of the valve can be radiused off in the same manner as for the intake valve, removing only the minimum amount of metal.

Generally an exhaust valve with a slightly convex head is preferable to a flat one, even though it may be a shade heavier. In view of the heat, it is policy to play safe, particularly when high compression ratios or supercharging is contemplated. For very high outputs, valves of very special alloy steel are obtainable, and both the exhaust and the intake valves may have convex heads (Fig. 9-9). Sodium-filled or similar types with augmented cooling are available for some applications, but they are an unnecessary luxury in most cases where drag racing or sports car engines are concerned. Drag racing engines are run for brief periods while sports car engines are prepared for flexibility and durability in long events at varying speeds. Sodium-filled valves generally have larger-diameter stems and must therefore have large-bore guides fitted to accommodate them.

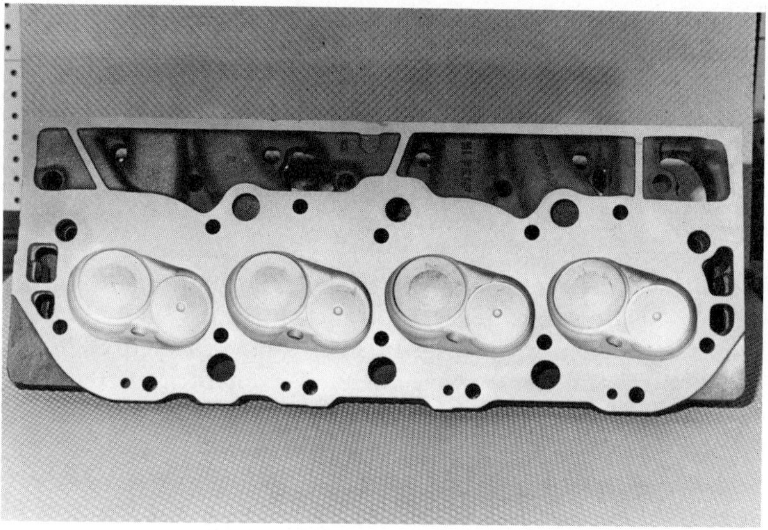

Fig. 9-9. High-strength valve heads. These extra-large valves have a combination of concave and convex contours that help to prevent warping, a problem that increases in seriousness with increased valve diameter.

There is no benefit in modifying standard exhaust valves in the manner described for the intakes, but it is sometimes possible to fit larger ones. If larger exhaust valves are installed, the ports should be merged into the larger diameter seats; if there is a sufficient thickness of metal available, the ports can be enlarged right through to the outlet flanges.

Exhaust Port Matching

Many times, backyard tuners bolt on a set of exhaust headers with no thought of whether the openings in the header flange match the exhaust ports of the cylinder head. This error can destroy most of the benefit that the headers might have provided in terms of better output. Even if the racing class rules forbid headers, the openings in the head and the manifold should be carefully matched (if this is not in violation of the rules).

Perhaps the best procedure is to fit the gasket (s) to the manifold or header flange, using bolts and nuts to keep them in place. With a cast manifold, the next step is to remove metal until the openings in the manifold exactly match those of the gasket; the gasket material should also be removed if it overlaps an opening. With headers, there is usually need to remove only a little metal here and there from the flange in order to match them to the gasket because quality speed equipment is made with considerably more precision than production components.

The gasket is then fitted to the exhaust port face of the cylinder head. If headers are being installed, the gasket openings are likely to be considerably larger than the exhaust outlets in the head; if a stock manifold is being used, the condition you will encounter is unpredictable. The gasket openings are carefully marked on the head's port face with Prussian blue or a waterproof felt-tipped marker. Then the ports are beveled to match the marks (Fig. 9-10). It is frequently necessary to match square ports of production cylinder heads to the round pipes used in exhaust headers.

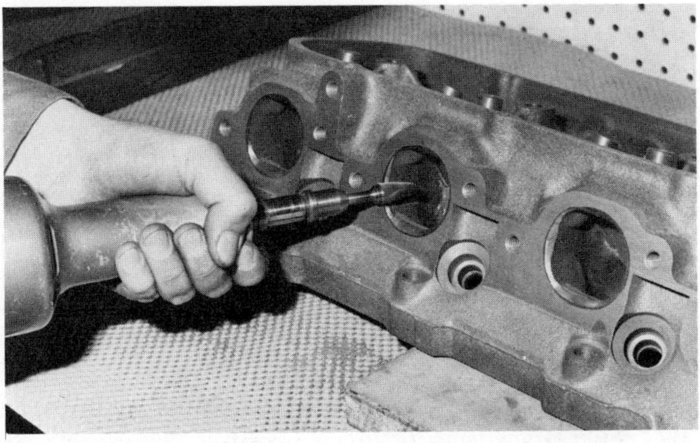

Fig. 9-10. Ports of High-Perf big block Chevy head being beveled to line marked using gasket for headers as guide. Square ports must be matched to round primary pipes.

Once the shape of the port has been beveled to match the gasket openings (this may be all that the rules permit in some cases), the exhaust ports can then be enlarged further into the head. If larger exhaust valves have been installed, the ports will be opened up uniformly all the way back to the valve seat. Fig. 9-11 shows the beveled ports being enlarged in this manner.

Fig. 9-11. Head shown in Fig. 9-10 having ports reshaped to round section all the way back to valve area.

Combustion Chamber Polishing

Polishing of the combustion chambers to a smooth finish assists turbulence and delays the formation of carbon; when applied to ports, polishing reduces the mixture flow friction. In the case of the combustion chambers, it is by no means unusual for the volumes to differ in relation to each other—another source of varying combustion pressure. The volume is again affected by variations in piston height at tdc because of manufacturing tolerances in crankshaft stroke, connecting rod centers, bearing dimensions, and so on.

Cylinder-to-cylinder combustion chamber volume differences are generally insignificant concerning variations on performance, though uniformity does seem to help an engine to rev freely to higher rpm. The main factor involved is compression ratio. After polishing the combustion chambers—and possibly enlarging them also—mill the head or select pistons that will produce the compression ratio that has been decided upon. In the case of stock classes, some modification to the combustion chamber may be permitted so long as the compression ratio is not increased. In others no machining is permitted though compression ratios can be juggled somewhat by variations in valve seating depth. (These will be discussed in more detail in chapter 12.)

It is important to obtain uniform combustion chamber volumes during polishing. (Its importance will become apparent later when we direct our attention to compression ratios.) The usual way of checking the volumes between sessions with the polishing tool is to mount the cylinder head upside down on the workbench so that the head gasket surface is level. The valves must be in position, complete with springs, and the spark plug holes sealed by old spark plugs that have been solidly filled with solder or epoxy cement. These plugs must be identical and should completely fill the spark plug holes so that there will be no capacity contained in them.

A mixture of kerosene, with a little ATF in it to give color, is run into each combustion chamber, either from a glass graduate or from a burette; the latter is preferable because of its greater accuracy. The volume of liquid run in is read off from the appropriate scale as the combustion chamber becomes completely filled flush with its machined gasket face. Most tuners seal the combustion chamber with a piece of Plexiglas stuck to the gasket face with vaseline and having a small hole for the fluid to be introduced. If this is not used, it is advisable to check the result once or twice. With all the volumes established, the next operation is to equalize them, using the largest one as the target.

Reworking is done by removal of metal. Because reshaping of the combustion chambers will already have been done, a fine tool should be used to leave a very high polish (Fig. 9-12). The entire job is quite a long trial-and-error process. Capacity checks using the kerosene/ATF mixture and burette must be alternated with polishing sessions with the power tool. Caution is important; if too much metal is removed from one combustion chamber, all the others must be modified to suit. Impossible standards of accuracy should not, however, be attempted; ±0.02³ cm is the professional tuner's tolerance range.

Fig. 9-12. Polished combustion chambers in modified Chevy small block head. Second chamber is being touched up in order to increase its volume slightly, thus matching it to the others in the engine.

Though some combustion chamber profiles are quite smooth and should remain so during reshaping and polishing, others are irregular. Thus, it may not be difficult to decide where to take off the surplus metal. For example, in one popular sports car head design, there is a definite tongue of metal between the valve ports; this may project unequally and can be ground back to compensate (as indicated in Fig. 9-13). This work should not be done to excess because the projection is instrumental in preventing heat transfer between the incoming and outgoing gases. If irregularities in outline are not apparent, the rule is to remove the minimum thickness of metal from the maximum area of surface. With volumes uniform, a final polish should be applied with fine emery cloth.

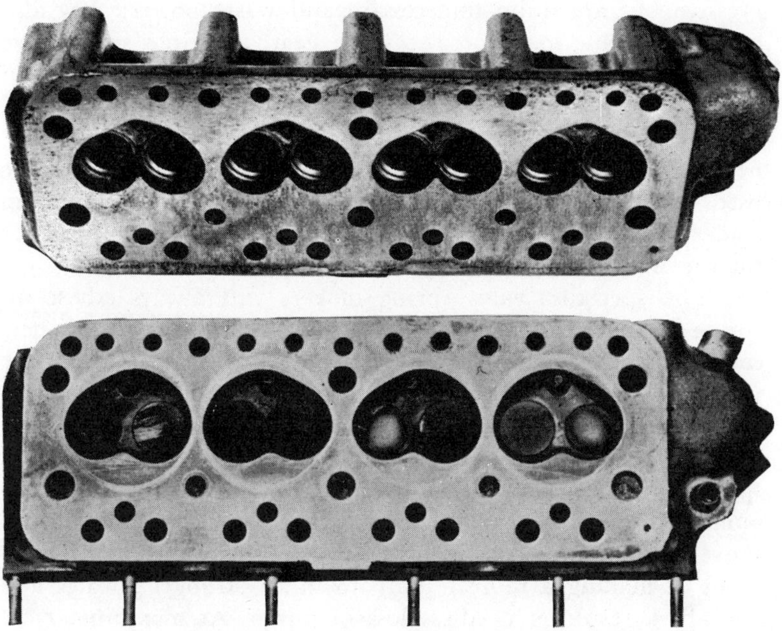

Fig. 9-13. MG cylinder heads. Top head shows shape of stock combustion chambers. Bottom head has had metal removed from vicinity of valve edges for easier gasflow.

Valves, Springs, and Retainers

Modified standard valve guides will meet most requirements. A very small frictional loss will be saved by using phosphor bronze intake valve guides, but this metal is unsuitable for exhaust valve guides in most engines. An alloy such as Barronia metal will serve for both. With this kind of guide, ample lubrication is required, but the standard system should provide this without modification, unless it incorporates tight-fitting valve stem seals (often the case). Guides can be selected from a manufacturer's catalog.

Coil springs are normally standard on engines of the kind being considered and are perfectly satisfactory for rpm in the range to be expected. Enormous spring pressures may feel impressive but are quite unnecessary and waste power; they also put undesirable loads on the valve gear. Multiple springs are generally used on competition engines because they keep down the cross-section of the spring wire, thus preventing coil-to-coil interference. The spring pressure chosen should be only that necessary to prevent valve float at the maximum rpm to be used. At some point, there must be a limit to the maximum rpm, and it is preferable for valve float to set this limit rather than a broken crankshaft.

The specialist valve spring makers will always advise on spring strength, and it is a good idea to study some of their catalogs (such as the *Crane Cams Winner's Handbook* or the various books by Ed Iskenderian that are commonly available at speed shops). It is not easy to give any firm rule concerning spring strength. The best method is to start with standard springs unless the camshaft manufacturer makes a specific recommendation for springs and valve gear. If standard springs prove inadequate, select springs with slightly greater tension instead of making a radical jump to much stronger springs and perhaps a resulting needless loss of power. At maximum rpm, the inertia of the valve gear amounts to a considerable load, which is influenced by such items as valve lift and cam contour. Thus, only after experimentation will you be able to ascertain correct spring strength.

Dual or triple springs will probably be standardized. If these are a push fit within each other, some damping will be provided, which prevents coil surging at high rpm. This surging, which is a sympathetic vibration of the coils running up and down the length of the spring, can lead to fracture if it reaches excessive proportions.

Anything that you can do to lighten the load on the springs and the valve operating mechanism is advantageous, since it not only reduces inertia loss and stresses but increases reliability. Thus, weight reduction of reciprocating parts should be given careful attention. The valve spring retainers will probably have a diameter fully equal to the outside diameter of the outer spring coil. It is quite feasible to turn down the outer diameter of the retainer until it is equal to a little over the diameter of the sparing coil measured at its mean diameter. See Fig. 9-14.

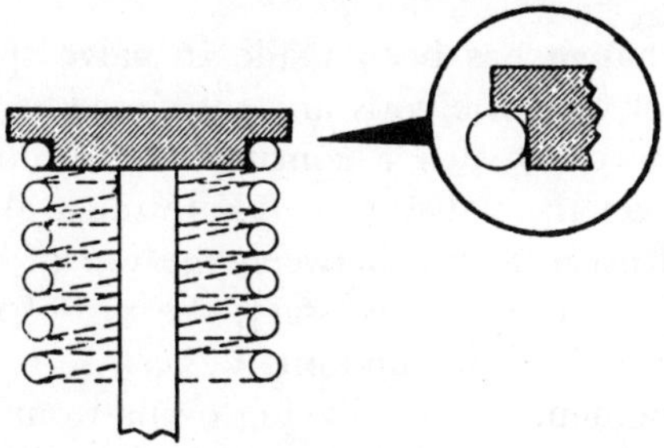

Fig. 9-14. Weight of valve spring retainer may be reduced by turning down its outside diameter as shown.

Some valves are equipped with a shroud designed to prevent excessive oil from reaching the valve guide; this shroud is located beneath the retainer inside the spring coils. With valves and guides in really good order, it is sometimes possible to remove these without excessive oil consumption or plug fouling resulting. In most cases about half the shroud can be cut off without impairing its designed function, thus giving a further saving in reciprocating weight. See Fig. 9-15.

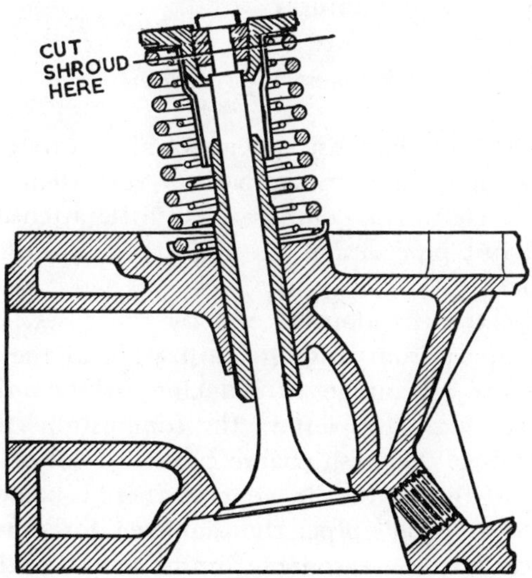

Fig. 9-15. Certain types of oil shroud on valve stems can be lightened without fear of over-oiling, assuming clearance in guides is correct.

When a change has been made in valve springs, it is important to check that the coils are not closed right up at maximum lift of the cam; such a contingency would throw an impossible strain on the valve gear. There must also be adequate clearance (at least 1/8 in.) between the top of the valve guide and the spring retainer. If necessary, the top of the guide must be ground to provide this amount. Otherwise contact may take place should momentary overrevving occur from any cause, and this could be serious if it led to valve stem breakage.

When the valves are assembled in the head, they should be lapped in, even when new, finishing off with metal polish. Valves and seats refaced in routine manner may not be adequate, because a mere retention of compression pressure is insufficient. If you do not do the work yourself, valve grinding should be entrusted only to a skilled speed tuner/machinist, never to an ordinary repair shop. Valve stems should be polished between head and guide, rubbing along the stem and not around it with extrafine emery cloth to erase any turning ridges. Split keepers can be checked for correct fitting by a light tap or two on the retainer itself after they are assembled, along with the springs and retainers.

Exhaust Pipe Layout

A glance at the exhaust systems of some cars competing in speed events shows that, in comparison with their counterparts in the motorcycle world, comparatively little attention has been given to exhaust pipe design, an important adjunct to the scavenging system.

It is true that freedom of exit for the gases is a desirable feature, but apart from this, the utilization of the exhaust system to greatest advantage in reducing to the minimum the amount of residual gases left in the combustion chambers and helping to induce the fresh charge calls for a lot of design work on this part of the engine. In general, there is some latitude in the design of a "good" pipe; the diameter, for example, is not highly critical within reasonable limits. On the other hand, it is easy to stray into the "bad" pipe category by committing the common fault of making the pipe unduly large in diameter. An outsize pipe is quite likely to have inferior power output compared to that of a well-designed standard pipe and muffler system.

With a given camshaft, variations in pipe length and diameter will result in more power over a certain range of engine rpm but no more, and possibly less, power outside that range. The point to decide, therefore, is where the useful rpm range lies and to concentrate on designing a pipe that will give the greatest power increase within this range. If the engine is of a type that with modifications peaks at, say, 6000 rpm, it is probable that below 3500 rpm there is only moderate power available and that in driving, the rpm would be kept within the 3500 to 6000 range. These figures are just an example, but there is no difficulty in arriving at those for any particular engine.

Exhaust Design

An entire book is available from the present publisher on this subject—*The Scientific Design of Exhaust & Intake Systems*. Therefore, this section will only make recommendations.

As far as straight-through, unmuffled racing systems are concerned, the possible alternatives are to use a completely separate pipe for each cylinder (an "independent" system) or a subdivided system working on the "interference" principle. In this, cylinders firing at equal time intervals have their pipes merged so that the idling branches act as fluid buffer columns to augment and extend the low-pressure period in the working branches.

In the case of the independent system, the length of the individual port pipes can be arrived at from the following formula:

$$P = \frac{ASD^2}{1400\ d^2}$$

Where P = pipe length in feet

A = exhaust valve opening in degrees of crankshaft rotation

S = engine stroke in inches

D = engine bore in inches

d = exhaust valve port diameter in inches.

The pipe diameter internally is based on the dimensions of the valve throughway.

This formula will do no more than ensure that the pipe is of adequate length to give a good pressure-wave action, particularly at higher speeds. There is no formula that enables a system to be designed completely on paper; experiment is essential for the final result.

The use of megaphone ends on the single pipes can be beneficial in prolonging the low pressure in the pipe, and it may be feasible to combine a megaphone effect in one component, as shown in Fig. 9-16; an independent system with megaphones is shown in Fig. 9-17.

Systems designed for interference working are shown for four- and six-cylinder engines in Fig. 9-18. Variation of the

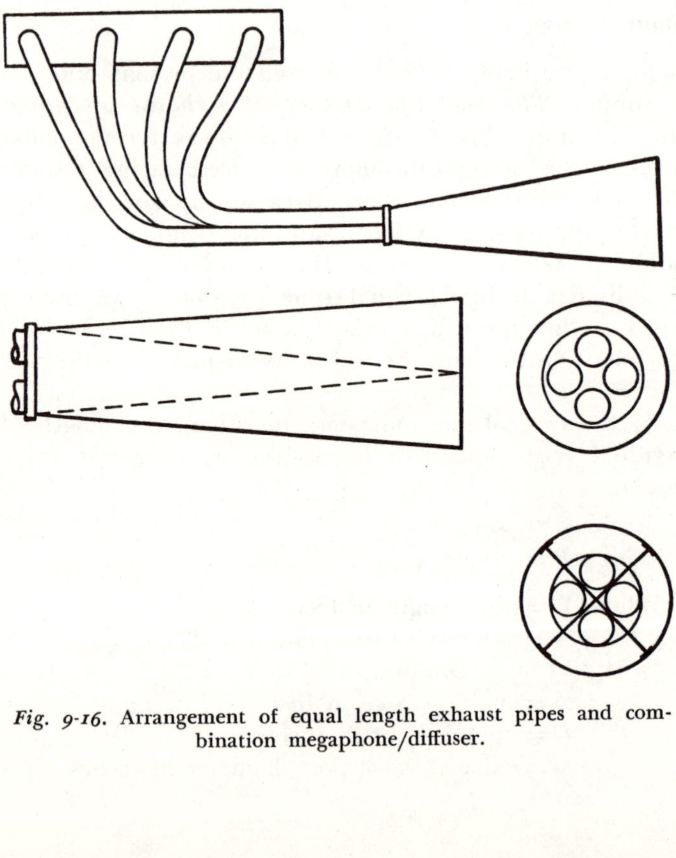

Fig. 9-16. Arrangement of equal length exhaust pipes and combination megaphone/diffuser.

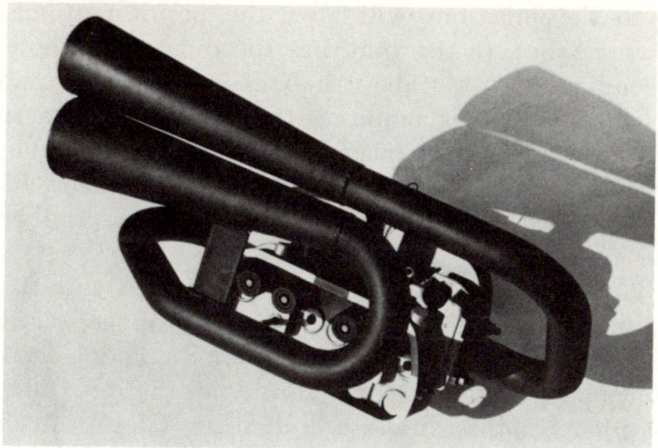

Fig. 9-17. Megaphone pipes for VW engine used in off-road racing.

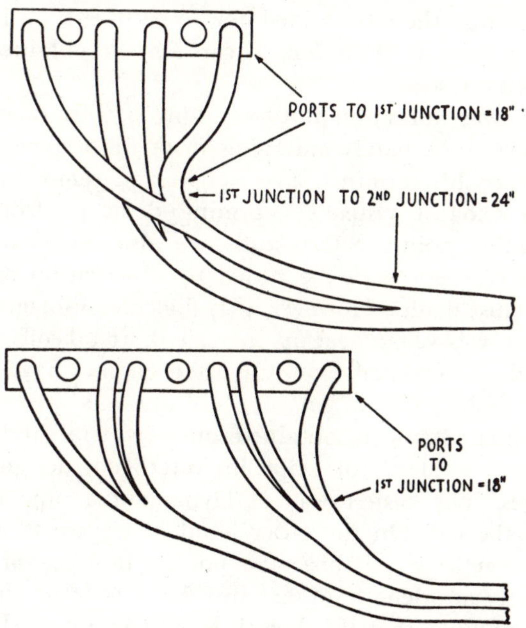

Fig. 9-18. Arrangement of subdivided exhaust systems with dimensions giving good average performance on most engines.

length to the junctions will alter the position of maximum torque in relation to the rpm, but the dimensions shown have given good results from about half maximum rpm upward.

When a muffler is required, this should be of a straight-through absorption type (glass-packed types are popular), preceded in the case of independent pipes by an exhaust box into which the pipes are led. The box, or collector, can be built into the muffler.

A large-diameter main pipe (whether used with a muffler or not) may look impressive, but apart from unnecessary weight, may actually be disadvantageous because an undue increase in diameter soon after the port will slow down the initial discharge. On the other hand, an increase in diameter later on in the pipe has been beneficial in some cases because it prevents a pressure buildup without affecting the energy of discharge.

The length should be as short as will allow the gases to be clear of the cockpit but not carried to the extreme rear. Extension pieces can then be added for experiment until the best length is arrived at. This length can also be obtained through dynamometer testing.

There was a fashion not very long ago in racing for the use of short stubs barely emerging from the engine cover. Fittings such as this certainly save weight and remove all anxiety about the exhaust system's becoming detached from the car, both desirable points. Short stubs are also useful in enabling individual carburetors to be tuned by observation of the individual exhaust flames. However, it is difficult to imagine any useful pressure wave being set up in such short takeoffs, though in fact it could be claimed that stub-fitted engines have performed well from time to time.

The final result is obtained only by trial and error, although there is plenty of scope for intelligent designing in the early stages. The easiest way of laying out a pipe is to make one to fit the car. On the other hand, if we are to accept that there is something in this pipe energy business and thereby choose a certain length of exhaust branch from the cylinder port, it is obvious that this length should be as nearly the same as possible for all ports. Reconciliation of this fact with a workable design may be difficult on an inline engine, but it is by no means impossible in most cases.

The Induction System

The effect of an excessively lengthy pipe between the carburetor venturi and the atmosphere is unpredictable. Certainly the air column contained in the pipe will assist in damping blowback, but the effect on the action of the carburetor itself has to be taken into account. We do not want excessively fluctuating pressure across the choke. Too great an air intake length will restrict the airflow unless the intake design is well done. On the other hand, the air should be as cool as possible without affecting carburetion.

In conjunction with a suitable exhaust system, it is possible to obtain a further torque increase at certain engine speeds by using a specific length of air intake to augment the pressure waves in the air column, but this effect is significant only on highly tuned racing engines.

The idea that an ordinary sort of engine can be made to push its air into its own intake has gained credence in many quarters. There is, of course, reason for this idea. The ramjet aircraft engine obtains its pressure air supply by virtue of having its air intake pushed at very high speed through the atmosphere so that there is a perfectly good mechanical reason for the occurrence. To obtain a similar effect on a wheeled vehicle would necessitate a forward-facing air intake; to build up any appreciable pressure, the intake mouth would have to be quite large, with a consequent increase in air resistance to the passage of the vehicle. Further, it will be evident that the low speed of the vehicle (in comparison with that of a ramjet aircraft) and the problem of variations in wind direction relative to the direction of travel may affect the performance unpredictably.

The success of systems that seemingly make use of this kind of pressurization often comes from the cool air obtained, not from its pressure. More important than pressurized air is tranquil air. Thus, in cases where some pressurization gain is found, it is usually where the ducted air is led into a plenum chamber (Fig. 9-19) rather than directly into the carburetor or fuel injection intakes. The high-mounted scoops seen above present-day Formula One cars are intended to obtain air from above the turbulence created by the cars themselves, thus helping to maintain a calm at the air intakes.

Fig. 9-19. Carburetor air collecting chamber with forward intake, fitted to an Austin-Healey Sprite.

Manifold Pressure

The matching up of the mixture intake to operate in sympathy with the exhaust pulsations is a line that can be followed with a prospect of good results. What we have to do is to obtain an increased positive pressure wave in the intake pipe (in which there is the same form of pulsating pressure as in the exhaust) at the time when the inlet valve is open. This is particularly valuable when the cylinder is being "supercharged" by exhaust pipe backflow. It is, of course, the critical period; a combination of this supercharge plus the piston rising on the compression stroke will eventually reverse the flow of fresh mixture in the intake pipe. The longer this reversal can be delayed, the more mixture will get into the cylinder.

We then have to consider occurrences when the intake valve shuts rapidly in the face of a high-speed column of intake gas. Obviously, if a negative pulsation in the pipe can be arranged at just this moment, the shock will be lessened. Further, the amount of gas in the pipe will naturally influence the effect of the bounce. In fact, it might be hoped that, using a sufficiently long pipe, the column might pile up and compress itself against the closed valve because of its own momentum, ready for leaping into the cylinder as soon as the valve opens again. This is an oversimplification of the problem. It is just as likely that an adverse wave motion will induct the column in the reverse direction—that is, out of the air intake. In all engines there is a fairly long length of port, or duct, between the valve face and the outside of the cylinder head; this is mechanically unavoidable. The port has to be carefully shaped to reconcile the valve opening with these bends that are inevitable in the port and with the final shape of the carburetor port on the outside of the casting—all to give the necessary gas velocity and degree of turbulence.

Add to this the carburetor throttle body and mounting flange, and it will be seen that even with a conventional assembly, there is an appreciable length of intake ducting between the intake valve and the carburetor venturi; the minimum possible is dictated by straightforward mechanical requirements of the design. Fuel injection systems offer considerably greater design flexibility, especially when slide throttles are used. Also, of course, there is no venturi with an injection system.

10 / The Valve Gear

Valve Train Modifications

Modifications to the valve gear are aimed at reducing frictional and inertia losses and thus improving the mechanical efficiency—in other words, recovering power that is normally wasted and at the same time increasing reliability. If a production powerplant is to be used within its original rpm range, there is very little in most cases that can be done to the valve gear that will manifest itself in improved performance. There is, however, room for some gain in reliability. This is particularly so if heavier valve springs are being used on an engine with pressed-in rocker arm studs. Replacing the latter with screw-in studs may prevent an embarrassing failure.

In the case of pushrod engines, if the rocker arm shaft and the rocker arms are temporarily assembled in the same positions as when removed from the cylinder head, it will be possible to check how the rockers meet the valve stems, since the length of the valve stems exerts a large influence and is subject to alteration whenever the valves are ground or their seats recut. In addition, rocker arm wear at the point where the rocker arm contacts the valve cannot be tolerated in a competition engine. Less obvious to casual examination are worn cam followers, worn pushrods, and similar inaccuracies that can be measured in thousandths of an inch but that can add up to an appalling loss of valve lift and timing accuracy.

Except in "showroom stock" and "strictly stock" racing classes, the valve gear should have all new components. If possible, they should be special high-performance parts that are designed for use with a special camshaft. But even where modifications are proscribed, new components that are devoid of wear should certainly be given preference over high-mileage parts that have outlived their usefulness in competition terms.

Valve train geometry must be correct, but it can be upset by cylinder head milling, valve grinding, and careless selection of parts or haphazard assembly. The "lay" of the rocker arms in relation to the valve stems must be correct, both as to the angle vertically and with no bias to either side—unless it is intended. The rocker arms should normally contact the valve stems centrally. Some engines, such as the air-cooled VW, have the rocker arms offset so that they tend to rotate the valves. Whether you keep this feature for competition service is another matter.

Head milling is something that cannot be taken lightly. When the cylinder head has been machined for a compression ratio increase, it is necessary to install shims beneath the rocker arm pedestals or to use correspondingly shorter pushrods. Otherwise some valve lift will be lost and friction increased because of the increased angularity of the rocker arms. The thickness of the shims should be equal to the amount of metal removed from the head during milling. Pushrods need be shortened only by about half the amount that the head was milled. However, pushrod alterations can be made only if the pushrods have separate tips. Thus, using high-performance tubular aluminum alloy pushrods with hardened steel tips is the normal course.

Experienced (and successful) speed tuners spare no effort to obtain precise valve train geometry. In the case of engines with individually mounted ball-jointed rocker arms, the distances between the studs and those from the valve stems to the studs are corrected to the smallest increment measurable with a steel ruler. If necessary, the studs are bent slightly by tapping them with a soft-faced hammer. On engines with rocker arms shafts—and on overhead camshaft engines with rocker arms—

the contours of the part of the rocker arm that contacts the cam or the part that contacts the valve can be reground until the valve lift for each cylinder is uniform when measured with a dial indicator.

Rocker Action

The greatest room for error exists with rocker arms; overhead camshafts acting directly on the valves or cam followers can scarcely be faulted for accuracy. With rocker arms, the ideal is that, with the valve in the half-lift position, a line drawn through the rocker arm and making a right angle with a line drawn through the valve stem axis should run just across the tip of the valve stem. The rocker tip should contact the valve stem exactly at its central axis in plan. The tip of the rocker actually slides across the end of the valve stem during movement of the valve from open to closed and vice-versa, and this movement can set up an appreciable side thrust on the stem unless the dimension referred to is such that equal movement on either side of the central position is achieved. But with high-lift camshafts, the travel across the valve stem by the rocker arm face will be great. Hence, the proprietary rocker arms that are available for racing conversion of stock powerplants frequently have rollers at the tip of the rocker arm that contacts the valve.

In a comparatively little-used engine excessive valve stem length will not disrupt the valve train geometry, but in cases where the valve seats have been recut or otherwise altered, the valve stem lengths probably require correcting. (Many tuners would undoubtedly classify these cylinder heads as worn-out because the implied recessing of the valves is a definite hindrance to maximum output.) If the valve stem is too long, a little can be ground off the end—taking care to keep everything absolutely square. Alternatively, it may be possible to shim up the rocker arm shaft pedestals to increase the height of the rocker fulcrum, but this will lead to complications with the pushrod length if the pushrods are not subsequently replaced or modified as to length.

If the valve stem is too short, the rocker shaft standards

may be lowered to correspond by machining some metal off their lower faces to bring the rocker arms into the correct position. This may involve shortening the pushrods, but it is probable that the available adjustment at the pushrod end of the rocker will be sufficient to get everything correct.

Another method of compensating for a too-short valve stem is to fit a hardened steel cap over the end of the stem (which can also save a valve that has a lopsided wear pattern on the tip of the valve stem). This method (Fig. 10-1) means a loose component, which is undesirable, as well as added weight. Whenever possible, another approach to the problem should be found for the competition engine.

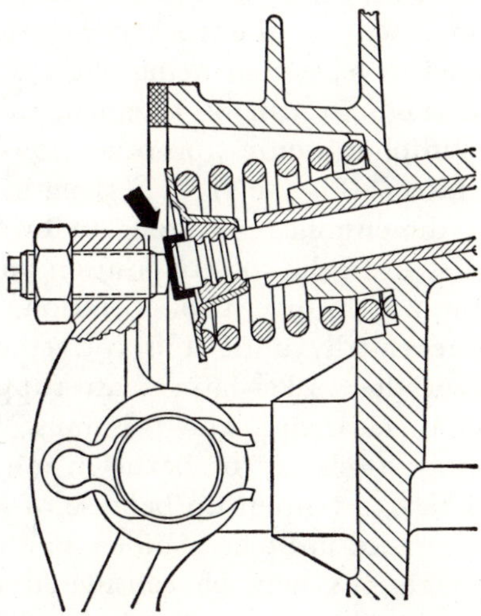

Fig. 10-1. Valve cap (arrow) on damaged end of valve stem.

Reduction of friction in the valve gear is well worthwhile. It is true that the power loss from this cause is insignificant compared with general frictional losses inside the engine. On the other hand the return motion of the valves is dependent on spring pressure and nothing else, and at high rpm there is not much margin of spring strength to overcome the inertia effect of the operating gear. This does not apply at lower rpm when inertia is small. But as the speed rises, there is progressively less margin of spring energy available to keep the valves following the cams. Thus any improvement that reduces frictional and inertia losses will mean that the valve timing will be adhered to over a higher range of rpm and reliability will be increased.

Rocker Arm Shaft and Components

The rocker arms should be quite free on their shaft (s) but without shake. The standard bushings (when bushings are used) should fully meet this requirement and last a long time with adequate lubrication and moderate loadings. Coil springs are frequently used as separators between the rocker arms; the slight stiffness imparted to the rocker movement by these springs is considered to have a silencing effect. If an unpredictable amount of extra noise can be accepted, these springs can be replaced by phosphor-bronze (or other low-friction alloy) tubular distance pieces (Fig. 10-2). The lengths of these must be very carefully dimensioned in order to bring each rocker arm over its valve stem in the correct position and to allow just sufficient end clearance to give free operation. The clearance should be tested repeatedly until it is correct because the act of tightening down the rocker arm shaft support bolts may cause a reduction in clearance, and this must be allowed for in the final assembly. It should be borne in mind that the distance pieces are likely to tighten up because of heat expansion. Too much end clearance must be avoided at all costs.

Lightening operations may be considered because of the mass of metal present in some kinds of rocker arms. The best course is to obtain new lightweight high-performance rocker arms from a speed equipment manufacturer. This, unfortunately, may not be possible on the engine that you are working with. If the stock rocker arms are lightened, the work must be carried out scientifically because the components are very highly stressed. Rigidity is the first requirement; undue flexing of the rocker arms will permit a valve timing very different from what the designer intended, not to mention the possibility of metal fatigue.

If the rocker is considered a beam that rocks about a ful-

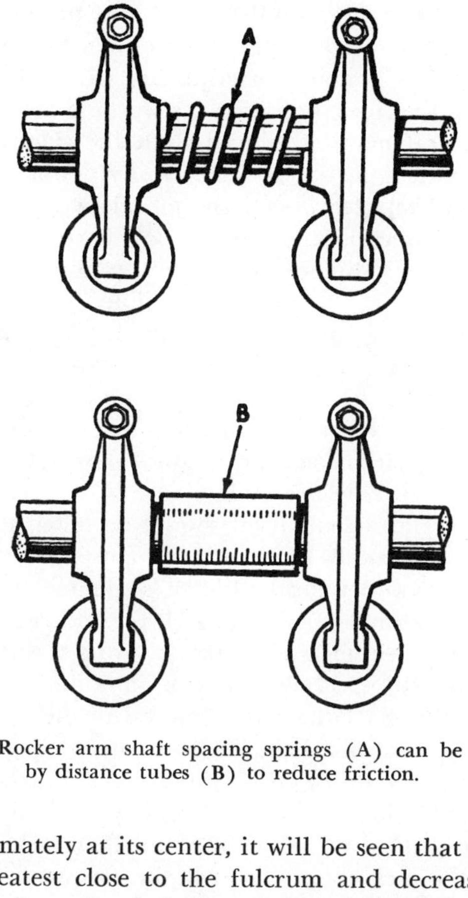

Fig. 10-2. Rocker arm shaft spacing springs (A) can be replaced by distance tubes (B) to reduce friction.

crum approximately at its center, it will be seen that the loading is roughly greatest close to the fulcrum and decreases progressively toward the ends of the beam. Many types of rocker arms lie at an angle to the rocker arm shaft in plain view so that the fulcrum is subjected to twisting as well as bending. Thus, the center of the rocker is obviously the part where liberties must not be taken. This part also moves at the lowest speed, and the lightly loaded extremities move at the highest speed. Lightening should, therefore, be concentrated at the points where stresses are low.

The length of the adjusting screw itself may be reduced where it projects above the rocker, or the locknut. In fact, a much thinner locknut or a hollow self-locking adjuster (as on the Ford V6) is quite in order. It is also practicable in some cases to grind an appreciable amount of metal off the rocker end that does not contain an adjusting screw. This should be removed from the sides, not from the working face. A very small amount of metal may be ground from the sides of the rocker throughout its length, but if there is any doubt about the wisdom of such a step, leave well enough alone.

The depth section should not be interfered with because it is primarily this dimension that resists the bending stresses. When the rockers are as light as possible, they should be polished all over using any means available. This polishing is a most valuable bar to fatigue failure, as can be a shot-peening to relieve internal stresses before the polishing operation. Of course, pressed or fabricated rocker arms should not be modified in any way; it is for these engines that light alloy replacements are most widely available.

The rocker arm shaft itself should be tested for parallelism and must be absolutely rigid in its supports, which are usually bolted to the cylinder head, though sometimes studs and nuts are used. It is normal practice to lock the rocker arm shaft bolts or nuts by thin steel locking plates or spring washers. A sounder method is to drill the bolt heads or nuts and lock them with safety wires. Wires facilitate the removal of the components and eliminate the need for hammer and punch operations on the plates, which may result in chips of metal breaking off. Locking compounds, such as Loctite®, widely used by speed tuners, have the advantage of taking up the inevitable wear on the threads that comes about through the frequent teardowns encountered by competition engines.

Rigidity

The amount of flexure or "spring" that occurs in the rocker arms and shaft assembly in normal running is accentuated under high-speed conditions. Normally the occasions when the engine operates at very high rpm may well be of only short duration, and the manufacturer faced with producing in a competitive market has to compromise on rigidity accordingly. It is not usually possible to go very far in improving the original design to give greater rigidity to the rocker shaft; but in cases where engines have been drastically modified, the improvement in high-speed performance by so doing can be considerable.

These modifications might take the form of a larger-diameter rocker arm shaft mounted in bearings carried in a rigidly secured rocker box; the box has a lid for access to the valve gear and takes the place of the usual light cover. In contrast, the rocker arm shaft in many standard four-cylinder engines is often carried on pedestals that leave the outermost rocker arms overhung on the shaft. Fig. 10-3 shows what a competition-oriented manufacturer is able to do to improve the common scheme. A decidedly rigid setup is used on the Toyota 1600 "Hemi" used in SCCA Showroom Stock Sedan racing. This assembly has very solid cross-braced pedestals that are held down by the cylinder head bolts (Fig. 10-4).

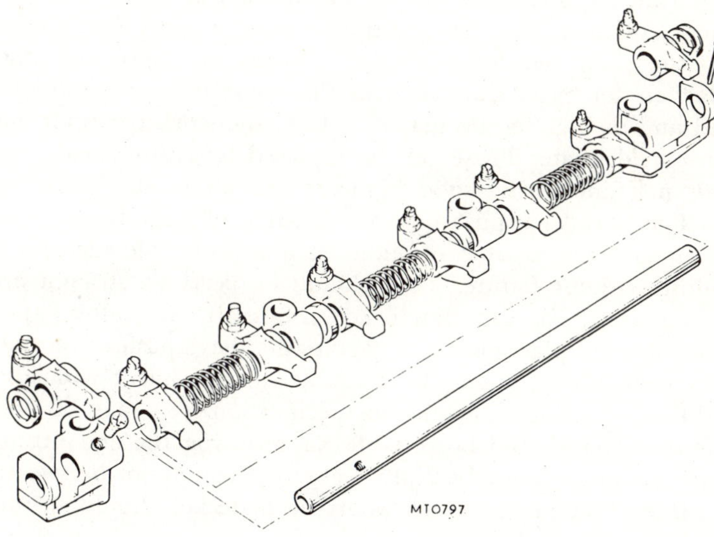

Fig. 10-3. Triumph Spitfire rocker arm assembly, showing unique pedestals that support shaft on both sides of outermost rocker arms.

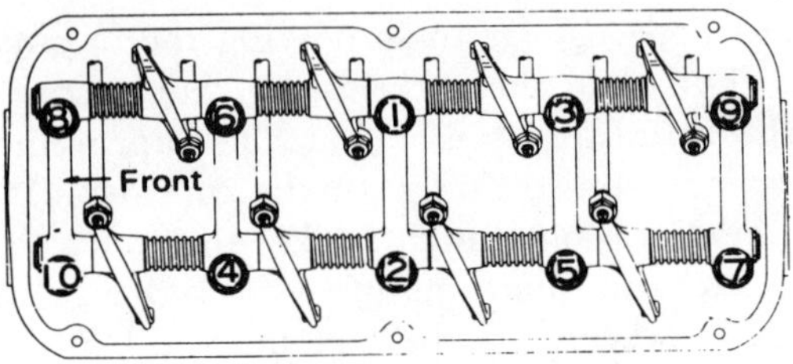

Fig. 10-4. Rocker arm assembly of Toyota Corolla 1600 "Hemi" engine. Notice cross-braces between pedestals, which are held in place by cylinder head bolts themselves.

It is sometimes possible to increase the shaft diameter by boring out the associated bushes and pedestal holes to suit; even an odd millimeter on the diameter will help, and the increase can be combined with very careful selection of material for the shaft. In some cases, additional support can be given to the shaft between adjacent rocker arms by taking support plates from suitably located studs on the head, but means must be found to ensure that the additional extra support does in fact grip the shaft tightly; otherwise it is useless. Also, any studs called on for extra duty such as this must be suitably lengthened and of high-tensile material. Last, the stock pedestals may prove inadequate. These can be replaced by custom-made pedestals machined from solid high-strength aluminum alloy.

The standard pushrods are usually of high-tensile steel tubing and are capable of standing a considerable increase in loading without failure. They should be perfectly straight and can be tested (as can the rocker arm shaft) by rolling them over a sheet of plate glass. On engines that have pushrods formed from steel tubing, it may be desirable to experiment with alloy pushrods; duralumin tubing to specification 4T4 may be used. The diameter should be suitable for receiving the end fittings, which obviously must be really tight, as they cannot be sweated in place. The pushrods, of whatever material, should be polished all over.

Titanium has also been used for valve gear reciprocating parts with excellent results in a number of sophisticated racing conversions. This material is comparable to alloy steel in strength and has a considerable weight advantage. Motorcycle engine designers who have specialized for years in obtaining extremely high powers and speeds from pushrod engines have used nearly every exotic space-age alloy available in their valve trains.

Cam Followers

The rocker-type cam followers used with some kinds of overhead camshaft engines can be dealt with by the same lightening, shot-peening, polishing, and other methods used for OHV rockers. Bucket-type OHC cam followers and the tappets or valve lifters of OHV engines can also be lightened by reducing the diameter at the center, leaving the full diameter at the top and the bottom only. These full-diameter areas provide the bearing surfaces.

A possible snag is that sometimes tappets may be ground externally to a final barrel shape as standard to allow for discrepancies in machining limits that might otherwise prejudice the accuracy of contact between the tappet foot and the cam lobe, thus leading to rapid wear. However, individual engines may show other possibilities for reducing cam follower weight, such as removal of metal from the inside, or enlargement of lightening holes, or the addition of lightening holes, suitably chamfered.

Valve Timing Modifications

Modifications to standard valve timing are a major item, and much can be done to enhance the performance of an engine without recourse to a special camshaft. It is usual for "sporty" production camshafts to employ valve timing that gives a good all-around performance fairly high up in the rpm range, possibly at the sacrifice of some torque at low rpm. This is generally evident by the degree of overlap at exhaust tdc.

To improve the low-speed torque, these essentially sporty camshafts are "detuned" by increasing the valve clearance. Thus, in stock cams that have considerable overlap, high-speed performance may be gained by using narrower valve clearances or low-speed power improved by using wider clearances. Though the technique is valuable to showroom stock and strictly stock class racers, the majority of competition engines do not race with the factory-installed camshaft. Hence, tuning the timing of a stock cam is not really within the scope of this book. Those interested in tuning unmodified engines for maximum performance are advised to obtain a copy of *Tuning for Speed and Tuning for Economy,* available from the publisher of this book at the address given on the title page.

Alternative Camshafts

Some makers provide alternative camshafts for racing, but in these days of emissions laws, they are available only through the manufacturers' competition departments to people who have cars used strictly for racing. The best alternative camshafts, however, come not from car manufacturers but from camshaft specialists that cater to the hot-rod and racing crowd.

Reground camshafts are based on the stock product and are widely available. Stock camshafts, being hardened, will stand having, say, up to .025 in. ground off the base, where the loading is negligible, and still leave sufficient wearing depth. This reduction in the base circle diameter (Fig. 10-5) will naturally increase the maximum lift and opening period with standard clearances. At the same time, by suitably blending the new base circle into the flanks of the cam, the opening period can be modified to almost any degree.

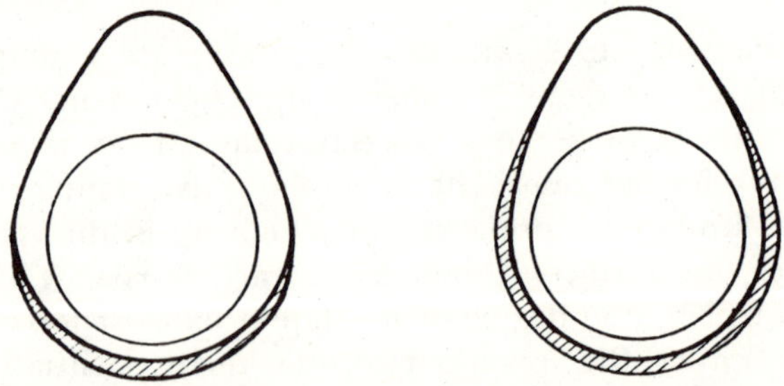

Fig. 10-5. Cam contours modified by removing metal (shown as shaded area) to increase both lift and opening period (left), and to increase lift only (right).

In planning the regrinding of a stock camshaft, the shape of the standard cams must first be obtained, together with the centerpoint. Then the base circle radius can be ascertained. If it is proposed to remove .025 in. from the base circle, the revised radius must be drawn inside the existing one. It will now be possible to see just what happens to the valve lifting and lowering points where the cam contours are blended smoothly into the new base circle; otherwise noise and undesirably heavy loading will be set up to no purpose. By comparing the existing timing diagram and cam contour with the proposed modified cam, it should be easy to find the best shape to give the required increase in lift, duration of opening, and overlap. There is obviously no point in going to the trouble of making a special camshaft unless fairly drastic alterations are made to the timing.

Assuming the use of an efficient type of exhaust system and with the emphasis on high rpm, it should be possible to increase the opening by a percentage amount equal to the percentage increase in maximum rpm. For example, if it is decided that the engine can peak at 6000 rpm instead of 5500 rpm, an increase in rpm of 9 percent, and the existing intake opening period is 250°, this should be also increased by 9 percent, making the new opening period 275°. The exhaust opening is dealt with in the same manner, and the overlap period is naturally increased automatically. Initially the increase is made equally, half on either side of the existing opening period. The work must be done accurately on a grinder with the correct types of attachments. The final blending is in some cases done by hand with an oilstone.

Camshafts that give valve timings for various forms of competition work are available for most production engines, and as dynamometer figures are often quoted with them in use, performances can be fairly assessed in advance. When considering the opening periods and overlap provided by these camshafts, bear in mind that no indication is given of the rate of lift of the valve. A long period, for example, might imply that the acceleration rate (or rate of lift to the fully open position) is moderate so that high loads are avoided; but a shorter period might give the same result in terms of total flow through the valve by the use of a quick-lift cam contour needing very robust valve gear. If the rules permit it, roller tappets can be used.

It is unusual for modern engines, which tend to run at high rpm, to employ cams having short periods and small overlap. If these cams are found in a stock engine, it can be assumed that the engine has been built to cater to a special purpose and is thus a type unlikely to lend itself to competition purposes. Modern engine design often demonstrates that a surprisingly large amount of overlap can be employed at quite moderate rpm with advantage to torque, and the same timing gives effective breathing at peak rpm also. If, in stock form, the engine's induction system imposes limitations on peak rpm, extended overlap timing cannot overcome these limitations, though it may possibly allow more mixture to be inhaled because of the longer time of valve duration. Even then, the benefit will be felt only over a very limited speed range at the top of the power curve; at lower speeds, the excessive overlap is likely to reduce torque by allowing backflow and charge mixing.

Because all tuning factors can be seen to work together, many tuners have advised their customers to apply "the three Cs" (cam, compression, and carburetion) in preparing an engine for competition. Small improvements in each of these three areas will usually produce a more rewarding power increase than a radical improvement in just one area. In fact, it will be readily noted that in the catalogs issued by various camshaft companies, most of the different available cam grinds will have comments attached to their descriptions. For example, the manufacturer will say, "This camshaft is excellent for longer courses, but should be used in conjunction with two four-barrel carburetors or with fuel injection." Reading what the camshaft grinding specialists have to say about their products and applications will be most informative.

11 / *Crankshaft, Cylinders, and Pistons*

General Condition

No engine should be put into competition service without a total disassembly of its working parts so that they can be inspected. Of course, new parts should be used in preparing an engine for racing, but used crankshafts and blocks can be retained if they are without flaws and can be remachined to like-new dimensions.

Crankshafts should be magnafluxed in order to detect invisible cracks. In some cases, you can obtain thoroughly prepared crankshafts for converting a production engine into a competition engine, but in other cases only stock components will be available. It will generally be found beneficial to have crankshafts (even forged ones) nitrided (or similarly heat treated) for additional strength and wearing ability. The heat treatment will cause some distortion, so the crankshaft must be straightened following this work. Then the bearings will need to be polished and the oil holes chamfered in order to avoid bearing damage and to promote better oil distribution. This work is best turned over to a crankshaft company that specializes in competition work.

The cylinder block must be free of cracks. The crankshaft and, where applicable, camshaft bearing bores must be checked for alignment and align bored if there are irregularities. The block should always be boiled out in a hot tank to remove any scale, casting sand, or other debris from the water jacket. During this operation, remove all the core plugs, including any for the oil passages. Otherwise the cleaning will not be thorough, and corrosive cleaner will find its way into the spaces between the plugs and the block and cause later leakage and other trouble.

The cylinder deck should be given at least a light cleanup milling even if inspection shows it to be exactly parallel to the crankshaft centerline. If it is not parallel, the cut will need to be deeper to correct the misalignment. The cylinders can then be bored to make them exactly 90° to the deck and the crankshaft and also to increase the bore to the maximum allowed by the racing class rules. Always turn the block over to an experienced speed tuner/machinist. Machine shops that specialize in passenger car repair work can seldom supply the care and precision that is necessary for winning races.

The pistons will have to match the reconditioned and modified cylinder block. Therefore, inspection of the pistons and rings is of importance only in tearing down a previously modified engine or a racing engine after a period of service. Though checking the pistons for collapsed skirts and acceptable clearance with the cylinders is important, your inspection must not stop here. In particular, check the piston ring side clearances. Top piston lands in particular may be distorted by racing service, causing the top compression ring to bind. In both new and used pistons, remove the rings (which on used pistons will probably be replaced anyway) and roll them in their grooves all around the piston—checking the clearance with a feeler gauge at all points on the piston's circumference. Minor lack of clearance can be corrected by a machine shop, but excessive clearance and tight clearances that prevent the insertion of the ring indicate that the piston should be replaced.

Of equal importance is checking the piston pin bores for uniform roundness and correct clearance. Also check the piston pin end clearance between the pin and the circlip on engines that are so equipped. This should not exceed .005 in. Excessive clearance can cause the pin to drive out one of the circlips.

Exacting Clearances

The cylinders and pistons together are the most important components of the engine. Piston friction accounts for a very large percentage of the total frictional loss, while the pistons themselves operate under conditions of high and varying speeds and temperatures. In addition to performing its main function of taking the expansion pressure and transmitting it to the connecting rod, the piston also acts as an efficient pump plunger and at the same time resists the connecting rod side thrust. Obviously, therefore, every component must be in perfect order in this department if the engine is to respond properly.

The maximum amount of wear normally occurs at the absolute top of the ring travel, where it is often evident in hard-worked engines as a top-cylinder ridge. This ridge makes measurement difficult with a micrometer, so a dial-indicator-type bore gauge is usually necessary. When the maximum bore wear is found at about the half-stroke position of the rings, it is a sign of dirty oil or oil dilution.

Near the bottoms of the strokes, the bores are little worn; but since this part may be unsupported by the main casting on some engines, slight bell-mouth or out-of-round distortion is possible. The best method of inspecting the bore for clearance determination is to remove the top-cylinder ridge with a ridge reamer and then measure the bore near that point. Wear in excess of the manufacturer's, your own, or your machinist's standard of precision will have to be corrected either by reboring or by installing new cylinders or cylinder liners.

Pistons that have been running with the correct clearance will show a perfectly even bearing surface over the thrust faces, without any highly polished areas that denote excessive rubbing contact caused by inadequate clearance. It is usual for speed tuners to allow about .001 in. of running clearance for each inch of bore diameter, at least on road-racing engines. But every kind of racing and every kind of engine creates special considerations, and small-block Chevrolet V8 engines used in drag racing often have clearances around .008 in.—about .002 in. for each inch of bore diameter.

The manufacturer's clearance can be used only as a starting point. In determining a good clearance for a particular purpose, it is advisable to read technical articles about the kind of engine being used and to talk to tuners and racers who use this kind of engine in the kind of racing that you intend to do. Loose-fitting and noisy pistons are usually not considered a disadvantage in a high-speed engine. However, the greater the clearance, the more sealing problems will arise as a result of increased ring wear. Therefore, the .001-in. per inch rule of thumb or the stock clearance plus .001 or .002 in. are good starting points.

Piston Rings

Piston ring fitting is treated in a rather casual manner in many repair shops, but for maximum efficiency, use a good deal of care when assembling and installing them. The main function of the rings is to curtail leakage past the piston. It is, of course, not possible to stop all leakage completely. In this case, the friction set up would be much worse in its effect on the mechanical efficiency than the power loss arising from the leakage. However, the leakage can be reduced to the unavoidable minimum by careful fitting. Three possible leakage paths are through the ring gap, around the back of the ring, and past the face of the ring.

Gap leakage was once thought to be the biggest offender, and many tuners went to great lengths to reduce the gap to very small amounts, often with poor results. Ring makers, for their part, produced rings with stepped and other peculiar gap shapes. Research has since shown that even quite large gaps allow little leakage. Regarding leakage around the back of the ring, it is necessary to allow a side clearance of about .002 in. to .003 in. so that the ring can spring quite freely even when the piston material has fully expanded because of heat. Thus, when the piston is moving upward on the exhaust stroke, the ring is forced against the lower face of the groove because of its own inertia and its friction against the cylinder. This close contact between the ring and the groove closes the leakage path around the back of the ring. In the same way, when the piston is moving down on the intake stroke, the ring is held against the upper face of the ring groove. During the compression and expansion strokes, the ring is held by gas pressure against the bottom face. Thus, the seal can be effective only if the sides of the ring and the groove faces are accurate and the ring-to-groove clearance is correct. Damage to the groove faces must be avoided when removing or installing rings; the contact should be similar to that of a valve and its seat.

The back clearance in the groove should also be checked. With the ring right back in the groove, there should be a minimum of .010 in. clearance between its outer face and the lands of the groove. The ring gap should not be less than .003 in. per inch of bore diameter. It is better to err on the large side because severe damage can be caused by the butting ring if heat expansion causes the gap to close completely. In an imperfect cylinder, gap variations can be caused by variations in cylinder diameter. Thus, the gap should be checked with the ring inserted into the bottom (least worn) part of the cylinder.

Gas Leakage

Leakage across the ring face increases with wear, particularly since bore wear is not truly circular, so that the radial pressure on the ring must be able to accommodate some variations. At the same time, excessive radial pressure will produce too high a frictional loss.

When a considerable increase in rpm is being considered, it is a good plan to consult the piston ring maker, the manufacturer of the racing pistons you are using, or an experienced speed tuner on the choice of rings, since it is possible for a sudden increase in blow-by to take place above a certain critical speed. Various theories have been put forward to account for this; one of the most feasible is that at very high speeds, the inertia of the ring is sufficient to lift it off its contact face with the side of the groove on the compression and exhaust strokes. This action allows the escape of pressure around the back of the ring so that gas pressure, which should be augmenting the natural spring of the ring in maintaining the seal against the cylinder wall, is relieved, and the ring collapses inward. Then leakage takes place across its face. If the ring gap is too small, actual ring breakage can result because the ends butt together.

A proper expanding tool should be used to remove or install piston rings. This will ensure that the ends of the rings are kept parallel and will prevent the piston skirt, lands, and grooves from becoming scratched. The side clearance must be present between the ring and the groove, and if it is necessary to increase the clearance, one side only of the ring can be rubbed down on a piece of plate glass coated with valve grinding compound. If much metal must be removed, say, up to .0015 in., begin by placing a sheet of fine emery cloth or 220-grit sandpaper on the glass. In either case, finish the job with metal polish applied directly to the glass. The work should be done to the top of the ring, and any ring thus treated should be installed with the ground-down side bearing against the upper face of the piston groove and the untouched face against the lower face of the groove. This will ensure that the best seal takes place on the compression and expansion strokes.

The usual rectangular-section ring acts to some extent as an oil control ring in addition to its main function of compression sealing, but it is not usually sufficiently effective in this capacity to obviate the use of oil control rings. The usual design of the ring—often called a scraper ring—relies on a suitable grading of its pressure against the bore to remove the desired amount of oil. The amount of pressure is determined by the contact area of the ring against the bore and is about 100 psi. The oil thus removed is transferred to the inside of the piston via suitable ducts. This is the general principle, but different designs vary slightly. Modern oil scraper rings are often assemblies made up of as many as five pieces, including the various expanding springs and support rings.

The oil scraper rings are placed below the compression rings, and it is important that they are sufficiently effective to prevent excessive oil consumption but not so drastic in action that they cause the bores to run dry. This would result in excessive friction and a high rate of wear. Many standard engines are excessively dry rather than the reverse, but for high-speed operation it is wise to err, if at all, on the side of underscraping by reducing the ring pressure.

Ring Sections

Although the traditional rectangular-section ring is still widely used, other sections are in general use, particularly in competition engines. Credit for much of their development must go to P. de K. Dykes, an engineer who has carried out extensive research in gas-sealing and oil control. Most piston manufacturers supply rings designed in accordance with the Dykes recommendations. Fig. 11-1 shows sections of some of the more usual types:

Type A is a normal plain compression ring used in the upper grooves. It also provides a good measure of oil control in this position.

Type B is a taper-faced ring, the shape of which promotes rapid seating—particularly on reringing overhauls where the bore is left unchanged.

Type C has an internal step so that when installed in the groove, the ring tends to twist. This gives the effect of a taper face on the periphery and in addition ensures rapid seating in the groove. Variations of this section, where the upper extension of the profile is pronounced, are typical of what are commonly known as Dykes rings—though Dykes was obviously concerned with many other types. If the upper extension goes all the way to the top of the piston, replacing the head land completely, it is known as a head land ring.

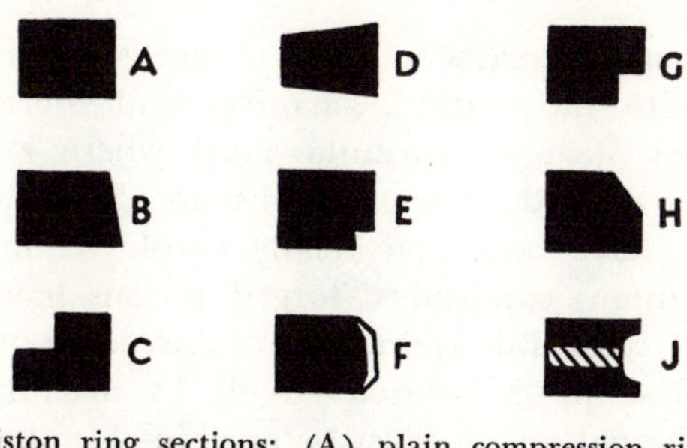

Fig. 11-1. Piston ring sections: (A) plain compression ring, (B) taper-faced compression ring, (C) internally stepped compression ring, (D) taper-sided compression ring, (E) oil-seal compression ring, (F) chromium-plated compression ring, (G) stepped scraper ring, (H) standard scraper ring. Correct positioning is shown, assuming piston to left of page.

Type D is a taper-sided compression ring. The taper may be on one or both sides, and the groove must be shaped to accommodate the taper. Ring sticking problems in high-output engines can be overcome by this design.

Type E is an oil-seal compression ring having a step on the periphery. It is usually installed in the second groove down and gives additional oil control, as well as a gas seal.

Type F is a compression ring for installation in the top groove. It has a thick layer of dense hard chromium plating on the working face that considerably prolongs both bore and ring life. It cannot be used with chromium-plated bores. The type is usually taper-faced as shown but can also be parallel-faced or stepped.

Type G is a stepped oil control ring.

Type H is another oil control ring with chamfered lands to provide greater oil control because of the narrow contacting area.

Type J is the standard type of slotted oil control ring, variations of which are universally used today.

Fig. 11-1 shows the actual positions in which the rings are installed in the grooves (the right way up, assuming the piston to be toward the left of the page).

Piston Design

For normal increases in performance, stock pistons can usually cope with the extra mechanical and thermal stresses in racing classes such as Formula Ford where extensive super-tuning is proscribed. These pistons are generally of the cast type, which have been universally employed for many years. For very arduous conditions, forged pistons have come to the fore in most cases. But apart from the expense of these pistons, the internal shape is dictated largely by machining considerations. The forged piston must be machined both externally and internally, at least in the smaller sizes, which are formed from solid bar material.

Although both forged and cast pistons are made from the

same sort of material and have essentially the same melting point, the forged pistons resist trouble better than the cast type. With cast material, the tensile strength of the metal is reduced with heat to a greater extent than is the case with forged material. A melted piston crown usually starts wih a crack, the edges of which fuse and accelerate the growth of the "hole". The forged piston is better able to resist cracking in the first place. However, in view of the strides made recently in cast piston manufacture, there is little likelihood of failure in blueprinted engines if clearances and installation are made approximately to the manufacturer's specifications.

Nevertheless, forged racing pistons are the rule today in any engine used in a racing class that permits nonstandard pistons. These pistons are available on a custom basis; the tuner can specify the height and offset of the piston pin bore, the shape of the crown, and a great many other design features. Often the pistons are supplied with a crown that can be machined by the tuner to obtain any desired compression ratio or cut with recesses for any valve diameter and lift. In addition, racing pistons suited to particular engines are often available from the car manufacturer's competition department.

Connecting Rod Alignment

Connecting rods are usually made of high-grade steel and are capable of withstanding a big load increase without failure, at least in the case of racing engines and production engines with a high-performance orientation to their designs. With ordinary passenger car powerplants, some care is necessary in deciding how far it is safe to go with respect to increased rpm and working pressures. If there is any doubt about the ability of the standard rods to stand up, it is advisable to investigate the possibility of obtaining replacement rods of higher-grade material. Here again, if the car manufacturer has a competition department, then high-performance rods can be obtained from this source.

Another profitable line of investigation is the use of lighter rods, since any reduction in reciprocating weight means less stress and greater reliability. It is not advisable to carry out any lightening operations on standard rods unless the risk of an expensive blowup is accepted. But it is possible to obtain for some engines—particularly American V8s—light alloy rods that provide a strength-to-weight ratio that is greater than that of the standard steel rods.

The tuner must keep a careful check on the condition of forged aluminum rods from one race to the next. If any rod has increased its big end to little end bore center length by .001 in. or more, it is usually cheaper to replace the entire expensive set of rods than to risk destroying the entire engine with a set of rods that may be stretching and losing strength. Connecting rod alignment can also be a problem because some aluminum alloys lose considerable strength if the rod is bent back to its original alignment; steel is very good in this respect.

The importance of parallelism of the big and little end bores cannot be overemphasized. This can be checked by the use of ground steel mandrels and V-blocks, together with a surface plate and a dial indicator, as shown in Fig. 11-2 and Fig. 11-3. While such checks and a bit of judicious bending can be carried out in your own garage, it is always best to leave the entire operation to a machine shop that specializes in competition engine preparation.

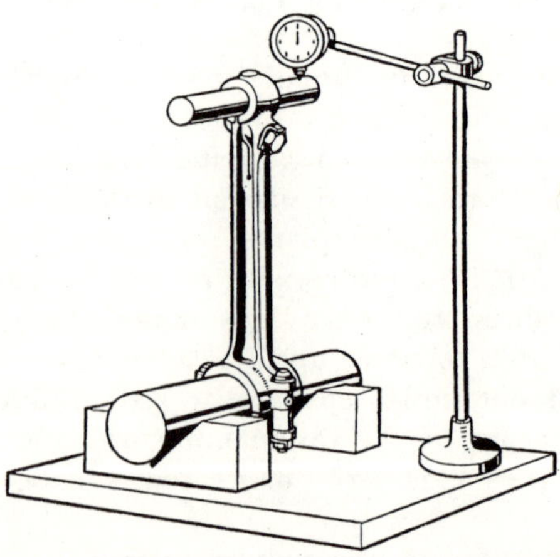

Fig. 11-2. Checking connecting rod for alignment of bearings.

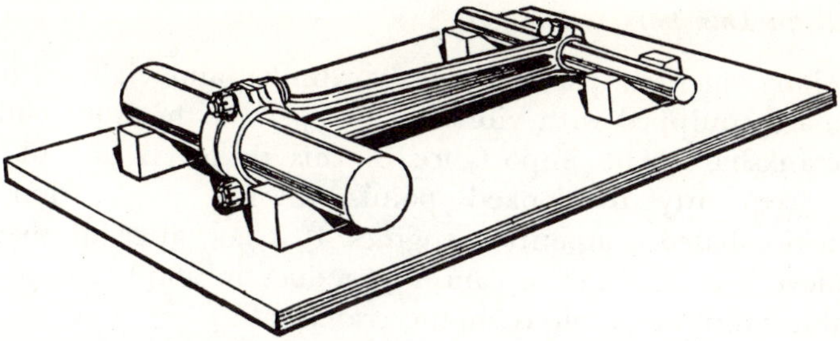

Fig. 11-3. Checking connecting rod for twist.

Balancing

In addition to machines that will quickly and accurately measure and correct connecting rod misalignment, a competition machine shop will have facilities for balancing the rotating and reciprocating parts of the engine. Balancing is an indispensable part of engine preparation for any racing class.

The crankshaft, flywheel, and clutch should be balanced both statically and dynamically as a unit. If the crankshaft has a pulley or a vibration damper, this component should be included in the balancing operation. The pistons and pins should be weighed individually and the heavier pistons machined internally until the weights of all the pistons are the same as that of the lightest piston.

The connecting rods should be balanced by the same method but not as a whole—that is, the little ends should be supported on a mandrel and the big end weighed. Then metal should be ground from the points provided on the big ends until all of the big ends have the same weights. The operation is repeated with the big ends supported on a mandrel and metal ground from the points provided on the little ends until all little ends have the same weight. If the work is done accurately, the entire rod will have the same total weight as the other rods in the engine. Because each end has been balanced separately, neither the rotating balance of the crankshaft nor the reciprocating masses of the pistons will be made uneven by installation of the connecting rods.

Vibration Dampers

Many engines, particularly six-cylinder and eight-cylinder types, are equipped with vibration dampers at the front end of the crankshaft. The importance of this device is one of the most frequently overlooked points in the preparation of production-based competition engines. It is not, after all, merely a pulley; it is designed to cancel or reduce in amplitude objectionable vibration periods in the crankshaft.

On many engines, the maximum rpm "redline" is determined by torsional vibrations in the crankshaft. If an rpm is reached where these vibrations have their natural frequency, the crankshaft will in all likelihood break. The crankshaft can also be broken at lower harmonics of this critical vibration frequency, and so a vibration damper is frequently installed to minimize the possibility. Thus, it does not pay to discard the vibration damper unless you know that the engine will never be operated at one of the critical rpm points. It is often worthwhile to contact the competition department of the car manufacturer to find out whether a competition-type vibration damper is available. These dampers sometimes make it possible to use higher rpm without crankshaft breakage or to improve acceleration if they are of lighter construction than the stock damper, which may have been designed primarily to limit noisy vibration periods at the slow speeds encountered in highway driving.

The viscous type of vibration damper consists of a circular casing that contains a flywheel rim, or *inertia ring*. The clearance between the casing and the ring is filled with viscous silicone fluid (Fig. 11-4). The casing is suitable for mounting on the front end of the crankshaft.

When no vibration is present, the casing and the inertia ring rotate as a unit; the ring is driven by the "stiction" of the silicone fluid, which requires considerable force to shear. When vibration starts and as the amplitude increases, the casing follows the crankshaft movement, but the inertia ring will tend to rotate uniformly because of its inertia. There is thus relative movement between the two components that is catered for by shearing action in the viscous fluid; the energy thus absorbed is dissipated as heat.

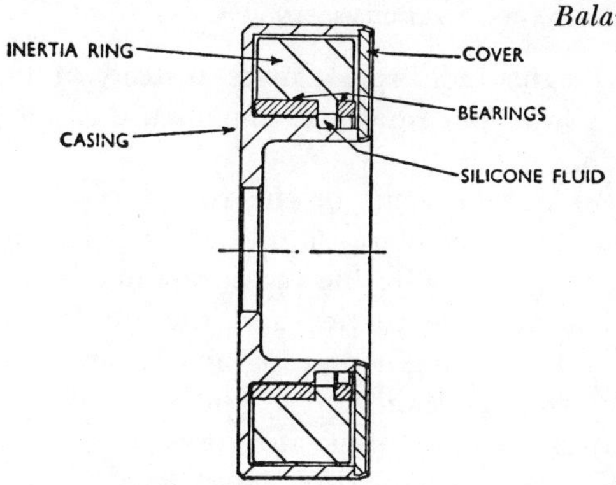

Fig. 11-4. Cross section of viscous-type vibration damper.

Another kind of device that is widely used is the bonded-rubber damper. In this case, the inertia ring and carrier are bonded to a rubber ring that forms an elastic connection between the two and that allows their relative movement; energy is dissipated in the rubber by molecular friction. Because the viscous type has an elastic element, it has a natural frequency of its own and can thus be regarded as a "tuned" absorber with one frequency of vibration. The frequency can be varied by altering the stiffness of the rubber ring or the inertia of the inertia ring (Fig. 11-5). It is by these means that competition vibration dampers can be developed that will damp out severe

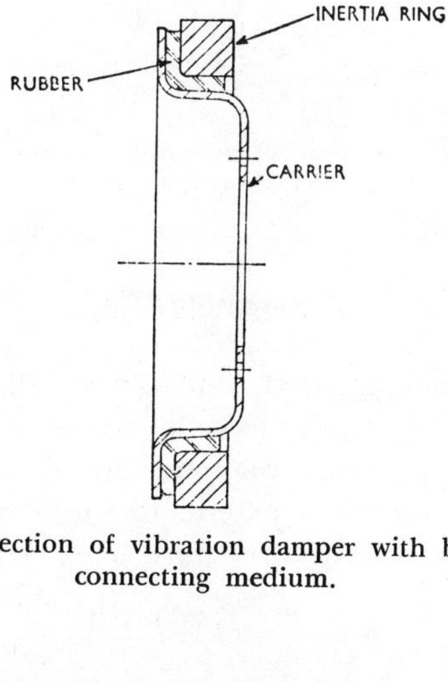

Fig. 11-5. Cross section of vibration damper with bonded rubber connecting medium.

vibrations in the high rpm ranges instead of handling vibrations in the low rpm range, as the stock damper usually is designed to do.

A recent development of the rubber-type damper has the rubber ring held in place simply by compression instead of being bonded (Fig. 11-6); the pressure alone is relied on to prevent slip between the carrier and the inertia ring. On most American cars, the engine timing marks are engraved on the inertia ring and, as shown, it is quite usual to incorporate a driving pulley for fan/alternator operation with the carrier. Nevertheless, these inertia rings have been known to shift, and so the competition engine builder usually makes it a point to find true tdc and establish his own timing marks rather than to depend on those located on the inertia ring.

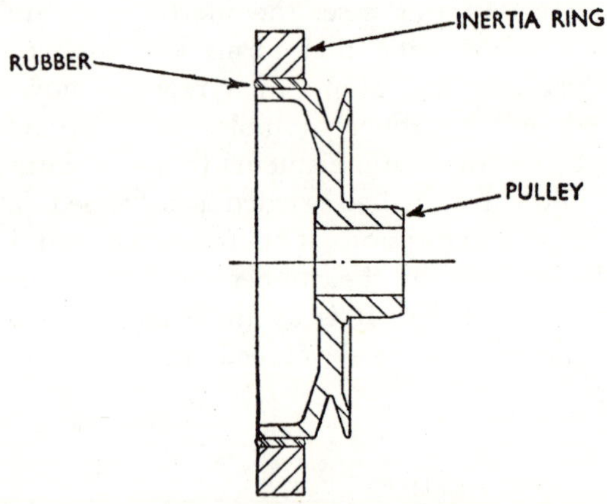

Fig. 11-6. Cross section of vibration damper with compressed rubber connecting medium.

Assembly Work

Any experienced speed tuner knows that it is not so much what you do to an engine as how you do it; correct assembly is of far greater importance than the brand of pistons or rings installed. Similar care must extend to the fitting of bearings, and regardless of how careful you have been with your micrometer, you should always check the bearing clearances with Plastigage® during assembly. Perhaps even more important is the attention you give to bolt torques.

Excessive bolt torques can cause cylinder block distortion. For this reason, competition machine shops install the main bearing caps and torque their bolts during cylinder boring operations. In addition, a heavy steel plate with cylinder-size holes in it may be torqued down in place of the cylinder head while the cylinders are being bored. During assembly, the same bolt torque should be used for the head and the main bearings so that the cylinder bores will remain as true as possible.

It is important to keep in mind that bolt torque is always second-hand information about bolt tension. The tension cannot be measured directly, but it is proportional to torque. Thus, the torque wrench readings should not reflect too much friction and not enough tension. Plated bolts require about 20 percent less torque than unplated bolts to achieve the same tension. Similarly bolts lubricated with graphite paste require about 20 percent less torque than bolts lubricated with engine oil. The same is true if Loctite® is applied to the threads instead of engine oil. Bolts normally must never be torqued dry unless specified because the friction may be so high and erratic that the torque wrench readings become meaningless.

To avoid bolt failure, the tension on the bolts must be at least equal to the loading placed on the bolts by engine operation. If the tension on a connecting rod bolt is 250 pounds and engine operation places a loading on the bolt of 300 pounds, there will be a fluctuating load of 50 pounds that will eventually lead to metal fatigue and bolt failure. Thus speed tuners sometimes increase the torque of bolts to above the factory specifications, but this can be dangerous if it is not done with care.

First, there is the possibility of distorting the parts. Second, excessive torque may cause the bolt tension to exceed the tensile strength of the bolt itself. The bolt will stretch during assembly and break during engine operation. Therefore, speed tuners should substitute higher-grade bolts in any application where the factory torque specifications will be exceeded.

The polishing of connecting rods will improve their resistance to metal fatigue. This work should always be done prior to balancing, or as a part of balancing, and should never be carried out in a way that will leave a canted surface for the big end bolt heads or nuts to bear against. This condition will place a bending strain on the bolts that can lead to their early failure.

12 / *Compression Ratio*

Selecting a Ratio

Because of emission control considerations, the oil shortage, and a general move toward lead-free gasolines, production car compression ratios are no longer as high as they were in the mid-1960s. The range is from about 8.0:1 to 9.0:1; fifteen years ago, 10.0:1 and 12.0:1 ratios were not uncommon. If one considers that in the 1950s many cars still had 6.5:1 compression ratios, today's figures sound more impressive.

For competition purposes, the compression ratios of production engines are increased whenever the racing class rules permit it. This has, of course, created problems since superpremium gasolines are no longer available. Consequently racers must frequently blend their own racing gasolines by adding an octane-improving fluid to ordinary premium-grade pump gasoline. Several fluids are on the market and can be obtained from speed shops or directly from companies that advertise in automotive periodicals. In racing classes where gasoline is required, the fuel is considered legal so long as the substance added does not change the fuel's specific gravity.

Where gasoline is not required, alcohol-based fuels are used with, in most cases, various percentages of nitromethane added. Greater power outputs are thus obtained and higher compression ratios can be used. Of course, with high-boost supercharging, lower ratios are used because the blower provides the additional pressure and a large combustion chamber space leaves room for more pressurized mixture to be burned on each expansion stroke.

Many car owners seem to arrive at a desired compression ratio figure by a process known only to themselves and then ask for details of the amount to be machined from the head in order to convert an existing ratio to, say, 10:1. Diplomatic questioning often reveals that the proposed new ratio has been selected because a popular competing engine with similar specifications uses a similar ratio.

We saw in chapter 1 the theory underlying the influence of compression ratio on thermal efficiency, but we have also noted that the theory assumes breathing conditions that do not obtain in practice. The absolute pressure at the end of the compression stroke depends not only on the compression ratio but also on the pressure existing at the start of the stroke. This in turn depends on how much mixture has been inhaled, that is, on the volumetric efficiency.

In an unsupercharged engine, the mixture at the end of the intake stroke may be "rarefied", and a high ratio will provide only the same final compression pressure as would be achieved with a well filled cylinder and a moderate ratio. Further, a combustion chamber design that does not promote rapid burning but leaves pockets of stagnant mixture may well stand a higher figure than one that gives clean and complete combustion. These facts should be apparent from foregoing chapters, but they are worth stressing.

Simple Arithmetic

The first thing to remember about compression ratios is that these are simple arithmetical dimensions, namely, the compression ratio is the difference between the cylinder-and-head contents *with the piston at tdc divided into the cylinder-and-head contents with the piston at bdc.* The ratio is thus very easy to calculate, and once the cubic content of the combustion chamber with the piston at tdc has been established (which may be a fairly complicated procedure on an engine with an irregularly shaped combustion chamber and domed pistons), the amount of metal required to be shaved from the gasket joint face of the head to increase the ratio to the desired amount can be ascertained. But, of course, the basic snag is in deciding just what the new ratio should be.

The arithmetical calculation ignores other factors influencing engine power, such as valve timing and ignition timing, combustion chamber shape, spark plug position, heat loss and gain, and the breathing efficiency of the engine, all of which have to be taken into account by the designer when he produces the original layout. While it is quite true that these factors do not necessarily have to be changed because of a moderate increase in compression ratio, they still cannot be ignored.

Even in the best designs, the volume inhaled per stroke inevitably varies with rpm because of compromises in the manifolding, the valve timing, and so on, which are based on the general requirements of use regarding the range of rpm over which high torque is desired. The effect of raising the ratio incautiously can thus be unpredictable unless the design as a whole is very carefully assessed.

Playing Safe

Production car manufacturers have tended over the years to choose a compression ratio that will enable the engine to run smoothly and to develop adequate torque over a reasonable speed range, even when driven poorly and possibly on the wrong fuel. That higher compressions are possible now than would have been acceptable twenty years ago is more a tribute to better transmissions and fuels than to any important advance in engine design. But because the car manufacturers tend to play it safe, both for the reason cited above and to meet legal air pollution limits, it is nearly always possible to raise the ratio somewhat without adverse consequences.

There is a reasonably reliable guide to the compression ratio that is feasible to use with fuel within the generally available range of octane: if the individual area of one piston does not exceed 8.5 square inches and the combustion chamber is of a good shape and of orthodox pattern, a compression ratio of 10 percent of the octane rating of the fuel will be about right. For example, with fuel of 82 octane, the compression ratio would be 8.2:1. If 95 octane is used, this can be raised to 9.5:1. For really excellent combustion chambers, or with even smaller cylinders, half a ratio higher is quite permissible.

The above piston area represents a bore of about 80 mm, which might be encountered on a modern 1500-cc "four" or a 3-liter V8. With larger cylinders, the acceptable compression ratio will be something less than 10 percent of the octane, though here also the combustion chamber characteristics may make a higher ratio practical in certain engines. All other things equal, it may be possible to use a higher compression ratio in a V12 than in an inline "four" of the same displacement.

Heating

Since a high compression ratio produces more piston thrust and therefore more power, it must also generate extra heat. Some heat must be dissipated over and above that normally dealt with by the cooling system. So long as the extra heat is not great, the standard exhaust valves, pistons, and so on will give no trouble, though more frequent inspection and, if necessary, adjustment of valves may be called for. Cooling and lubrication systems usually have sufficient margin, but it is wise to watch the temperatures of the coolant and the oil until conditions can be assessed. It may be necessary to add an oil cooler, increase the water pump rpm, or increase the radiator core area.

Also, when the engine is inhaling its maximum charge, as it does at the point of greatest torque, the increase in pressure will be at its maximum, and it may be necessary to retard the ignition somewhat within this range of rpm. The setting is determined by the automatic advance mechanism, which may have to be recalibrated using greater spring tension or weights that reduce the total advance—particularly if the compression ratio increase has been considerable.

Obtaining the Ratio

Several methods are available for increasing the compression ratio. A small increase is possible using a thinner head gasket. A more considerable increase will come about automatically by boring the cylinders oversize. For example, if a 1600-cc "four" with an 80-mm bore and a 79-mm stroke has a compression ratio of 9.0:1, increasing the bore to 81 mm will increase the compression ratio to 9.2:1. Of course, an increase in stroke will raise the ratio but not nearly so much as an increase in bore.

A reduction of combustion chamber volume will increase the ratio more than any possible remachining of the block or crankshaft. Therefore, any major increase in compression ratio must come about either by milling material off the cylinder head's gasket surface or by increasing the height of the piston crown. Domed pistons that will increase the compression ratio to almost any predetermined figure are readily available for all engines from the various racing piston manufacturers. Because these can in some cases interfere with the efficiency of the combustion chamber, it is often the practice to arrive at the racing ratio partly by milling the head and partly by higher-crowned pistons.

Machining Matters

Cylinder head milling has been carried to rather extreme limits in the past that cannot be duplicated with modern engines. Now, with the availability of high-grade iron and high-strength aluminum alloys, engine manufacturers are not so lavish in providing surplus metal thicknesses, and power-to-weight ratios have benefited accordingly. Generally there is still sufficient metal to allow a moderate amount of machining without weakening the casting structurally so long as there is no evidence of weakness in the water jackets (as might be caused by faulty casting or corrosion) and there is an adequate number of head bolts or studs. Nevertheless, today's thinner castings have made the domed piston more popular, and the wise machinist will mill a junk head and then saw it up for examination before he begins to work with a similar head that will go into the construction of a competition engine.

It is unlikely that a small amount of milling will have any bad influence on the combustion chamber shape, but points to be considered are the effect on free passage through the valves, around the periphery of the valve heads, and the proximity of the spark plug electrodes to the piston. If the electrodes are too close, local burning of the piston crown can occur. The effect of milling on the gasket seal should not be overlooked either because it is essential that the gasket be shielded from heat on its faces and not exposed as an intruding "flange" in any way.

The valves in a normal head are tucked up out of harm's way so they cannot foul anything. But in most cases, the chamber overlaps the cylinder bore top to some extent, so that when fully open, at least one of the valves has its edge above the flat face of the block.

The clearance at this point allows some margin for valve float on overrevving when the valve may open more than usual. Obviously, machining the head reduces this safety clearance (Fig. 12-1). Thus, there is a possibility that in similar circumstances the valve may contact the block and bend the valve stem. The remedy is to increase the clearance at the appropriate point on the block top by grinding a relief into the upper part of the cylinder. This kind of work is illustrated throughout the book, for example, in chapter 18. The block may also need to be relieved because a camshaft is used that increases valve lift or because, after head milling, the valve would otherwise strike the block even at cranking rpm. Reliefs, of course, reduce the compression ratio slightly. So will any cleaning up or polishing of the combustion chambers that involves removing appreciable metal.

Because the cylinder/cylinder head joint is recessed into the cylinder heads of Porsche and VW air-cooled engines, the compression ratio cannot be increased by milling. Instead, the cylinder sealing surface inside the head is machined deeper into the head by flycutting. The flycutter is a relatively simple machine tool attachment that can be used in an ordinary drill press or, better yet, a milling machine.

The rules of many classes state that metal can be removed

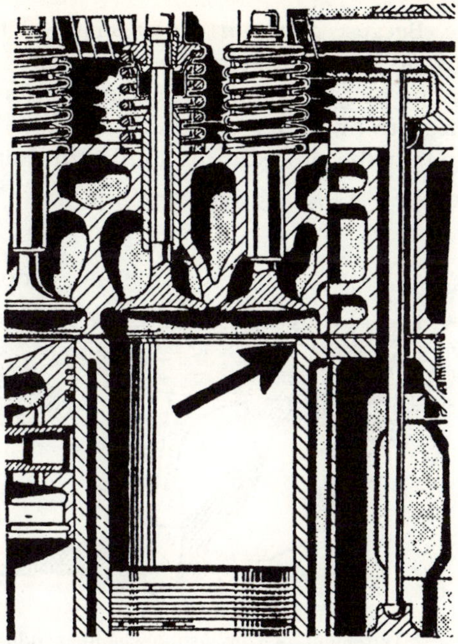

Fig. 12-1. With reduction of head depth by milling face, valve clearance to block at point indicated may be insufficient for safety.

from, but not added to, the production cylinder head. Nevertheless, where it is permitted, the combustion chamber can be reshaped by the addition of welded metal. This is, of course, the only way in which the compression ratio can be raised on a Wankel engine—by partially filling the recesses in the rotor face. Great precision is necessary to obtain three uniform combustion chambers per rotor, and, of course, the rotors must be carefully balanced and checked for distortion afterward. In some cases, high-compression rotors are available from the manufacturer's competition department.

Obtaining Head Volume

In order to determine the amount of metal to be removed from the head to obtain any desired ratio, it is necessary to measure the volume of the combustion chamber when the piston is at tdc. Because of the irregular shape of the combustion chamber, any attempt to determine the volume by linear measurement is out of the question. However, it is not a difficult matter to measure by means of liquid introduced into the combustion chamber; the procedure is similar to that mentioned in chapter 9 (Fig. 12-2).

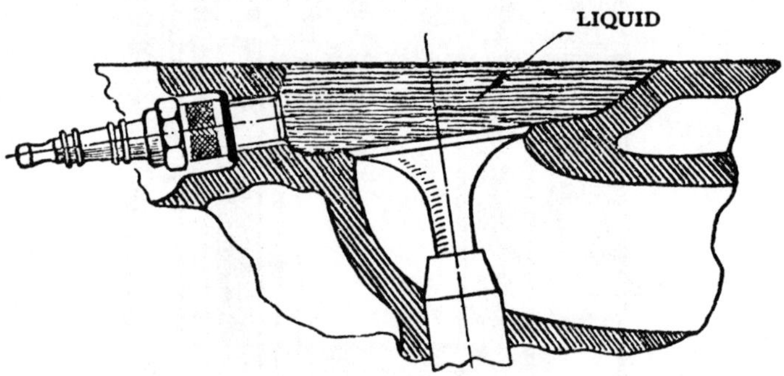

Fig. 12-2. Measurement of combustion chamber volume by quantity of liquid contained in space.

To avoid having any one chamber that exceeds the compression ratio specified by the rules in showroom stock or strictly stock classes, the volumes of all the chambers should be measured and then be made as uniform as possible. Variations can be accomplished by machining the valve seats deeper or by raising the valves slightly by using new valves instead of valves that have been ground. Performance will, of course, be sacrificed if the intake valves are recessed very much.

With all the chamber volumes uniform, the head can be milled or flycut to achieve the desired ratio or the blueprint ratio in every cylinder. For the purpose of finding how much the head must be milled to raise the compression ratio, it is easier to refer to the *displacement ratio.* The displacement ratio is the displacement of one cylinder divided by the combustion chamber volume and is always 1.0 less than the compression ratio. The displacement ratio for a cylinder having an 8.6:1 compression ratio would be 7.6:1. If you want to raise the compression ratio from 8.6:1 to 10:1 you must therefore mill enough off the head to change the displacement ratio from 7.6:1 to 9:1.

Let us assume that we are working with a cylinder that has a 9.1 centimeter bore and a 7.7 centimeter stroke. It would take a combustion chamber volume of 65.5 cm^3 to give an 8.6:1 compression ratio or a displacement ratio of 7.6:1. To determine the amount of head milling required to change the displacement ratio to 9:1, do the following:

1. Subtract the present displacement ratio (7.6) from the desired one (9). 1.4 is the answer.
2. Multiply the desired displacement ratio by the present one ($9 \times 7.6 = 68.4$).
3. Divide 1.4 by 68.4 (obtained in step 1)—*not the other way around!* This will give .0204.
4. Multiply .0204 by the stroke in millimeters ($.0204 \times 77 = 1.5708$ mm). 1.5708 mm (about .062 in.) is the amount that must be milled from the head.

Expressed as an equation, the procedure would look like this:

$$\text{Milled Amount} = \frac{\text{New Disp. Ratio} - \text{Old Disp. Ratio}}{\text{New Disp. Ratio} \times \text{Old Disp. Ratio}} \times \text{Stroke in mm.}$$

If you are working with an unmodified stock head and want to mill to achieve the blueprint compression ratio, the matter is simplified because the manufacturer usually specifies the combustion chamber volume. Simply run this volume of kerosene/ATF into the combustion chamber that has the least volume and then carefully scribe the chamber wall to show the high-water mark. The head can then be milled by that amount. If any measurements appear doubtful—for example, if the level seems to call for an unduly large amount of metal to be removed for the specified ratio—recheck carefully. The head must be absolutely level during these operations.

Gasket Volume

The head gasket volume and the piston head land volume must be taken into account if you are trying to approach a specified compression ratio with total accuracy. For example, in SCCA Formula Ford racing, the rules are very explicit concerning the maximum ratio. But to simplify matters, a standard figure of 4.75 cm^3 is always used for head gasket volume when the engine is checked by the officials. Similarly, the space between the piston head land and the cylinder wall above the top compression ring is measured and included in the unswept volume. Thus, the simple formula

$$\text{Compression Ratio} = \frac{\text{Swept Volume} + \text{Unswept Volume}}{\text{Unswept Volume}}$$

becomes

$$\text{Compression Ratio} = \frac{\text{Swept Volume} + \text{Gasket Volume} + \text{Head Land Volume}}{\text{Swept Volume}}$$

Before you begin building an engine, consult the rules of the sanctioning organization of the racing class.

The gasket volume, of course, remains static when the compression ratio is modified, and thus the figure can in some cases be subtracted from the calculated chamber volume to determine how much you should mill the head. But here again, check the rules. In some cases, a constant figure is applied; in others, the actual gasket thickness is measured. Some engine makers state the thickness of the compressed gasket and perhaps the gasket volume. This may be the figure applied to all engines in a blueprinting class by the sanctioning body, regardless of whether a standard gasket is used.

If the rules are vague, it is wise to err on the safe side, and with some modern embossed steel gaskets the volume is very small. It is, however, not difficult to allow for the extra volume contained by the gasket and which can be added to the combustion chamber volume in making the calculation for the standard engine. When the cylinder bore apertures are approximately circular, the calculation is simple, involving merely the area of the circle (the radius $\times$ the radius $\times$ 3.14) times the gasket thickness.

If the hole is irregular, divide up this shape into a series of thin rectangles (Fig. 12-3). Obtain the area of each rectangle, add the areas together, and then multiply by the gasket thickness. In either case, the gasket thickness should be that of the gasket when it is compressed. Embossed steel or "shim" gaskets can be measured with a micrometer because the metal does not compress much when the cylinder head bolts are tightened. With sandwich-type gaskets, a well-worn one should be used for measuring with the micrometer firmly tightened onto it.

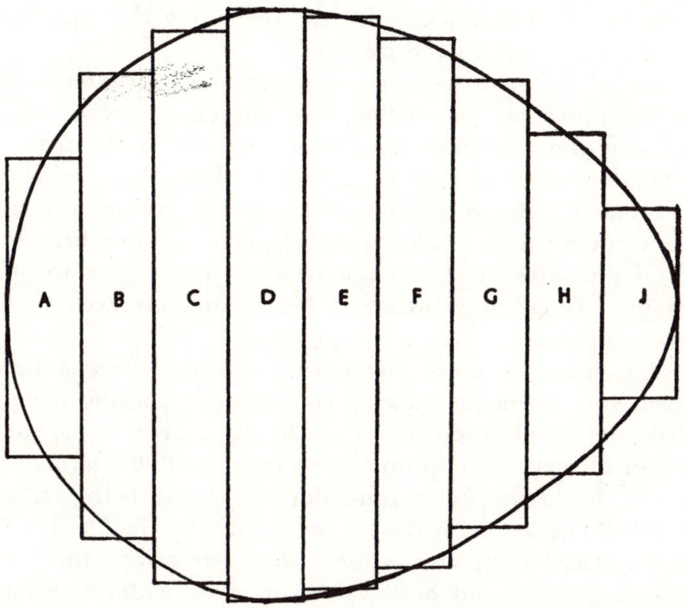

Fig. 12-3. Method of obtaining area of gasket opening. Narrow sides of each rectangle are bisected by combustion chamber profile. It is not necessary that all rectangles be the same width. The more rectangles used, the more accurate the result.

Domed Pistons

If the piston has a convex crown, its volume must be subtracted from the unswept volume in computing the compression ratio. This can be arrived at in several ways—all awkward. If a large enough graduate can be obtained with calibrations in cubic centimeters, the graduate can be filled with kerosene/ATF and the piston crown lowered into the liquid until the dome is immersed. The rise in fluid level indicates the volume of the dome.

Another method is to make a female cast of the piston crown in clay. This is then measured by running in liquid from the burette. Modeling clay is also useful for determining piston-to-valve clearances. By placing the clay atop the piston of the engine during trial assembly and then turning the crankshaft through two revolutions, indentations will be left in the clay by the valves. The thickness of the remaining clay can then be measured to ascertain the clearance.

In engines with bowl-in-piston combustion chambers, compression ratios can be changed only by changing pistons, and valve-to-piston clearances can become a problem. As in the case of domed pistons, it may be necessary to cut valve reliefs into the piston crowns to avoid interference. Of course, any engine with vertical valves has less valve clearance trouble because the face of the valve head and not its edge would have to project into the piston's way, which is highly unlikely even with extreme camshafts.

To check for valve clearance problems before a head is milled, you can make up a jig comprising a packing piece that enters the combustion chamber to an extent equal to the amount of metal it is proposed to remove. (This is mainly in the case of flat-topped pistons; domed pistons, being irregular, are better checked with clay as previously described.) The head is then placed right side up on a flat surface with the packing piece in position and both valves in place, without springs or other fittings. The valves are then allowed to drop until their heads touch the packing piece; this position approximates that obtained with the valves touching the piston. Measure the distance from the end of the valve stem to the top of the guide in this position, and measure from the same point with the valve on its seat. The difference should not be less than the valve lift plus a safety margin of 3/32 to 1/8 in.

Other Machining Operations

In some cases, the spark plug hole is fairly near the head surface that is to be milled, and machining may actually bring it right down to the joint. This does not matter, providing the gasket is well back and does not protrude into the combustion chamber when the plug hole has been faired off by filing as indicated in Fig. 12-4. There should be at least 3/32 in. of flat face between the gasket edge and the plug hole edge. The bottom threads of the hole must not be left ragged after machining the face but should be filed and blended into the surrounding metal. To keep the plug threads from projecting into the combustion chamber, it may be necessary to install a thick solid copper gasket between the normal spark plug gasket and the head. These spacer gaskets are readily available for just this purpose from spark plug companies, such as Champion, that maintain a racing department.

After milling, you will have to accommodate a greater length of cylinder head bolt in the block or, if the engine has studs, place thick steel washers between the head nuts and the head. Bolt holes should be tapped deeper, if this does not involve drilling into the water jacket, or the bolts can be shortened or somewhat shorter bolts obtained. The valve train geometry will be upset by head milling on pushrod engines, and after milling the head either the pedestals must be shimmed higher or shorter pushrods must be used.

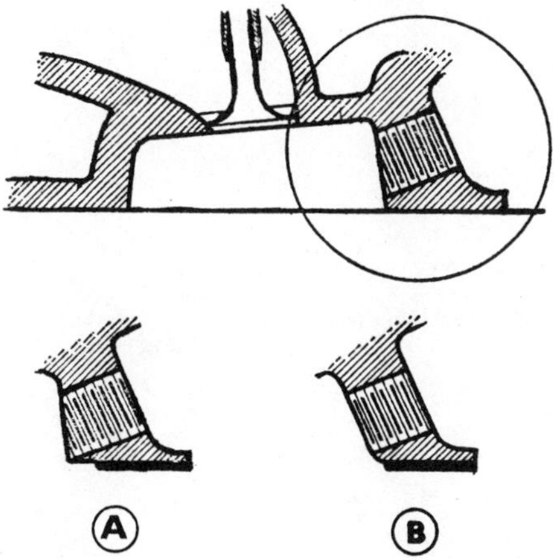

Fig. 12-4. Smoothing contours. Any sharp edges or corners resulting from matching head face (A) must be removed (B).

On overhead camshaft engines, head milling will place slack in the timing chain or the camshaft drive belt. This must be accommodated in the correct manner if trouble is to be avoided. Some comments on this appear in chapter 18.

Though modern head gaskets are very good, O-rings around the cylinder bores are better and are used in almost all racing engines today. O-rings are particularly worthwhile if very high compression ratios are used. Gas-filled O-rings, a recent development, are certainly worth making room for in engines that will be used in rigorous professional racing events. At present, lapped joints—without gaskets or O-rings—are found only on air-cooled engines.

13 / *The Engine in Operation*

The Engine Installed

When the engine is installed in the car, it is subjected to conditions and influences that can reduce the amount of power available at the flywheel. Usually some power is inevitably required for driving auxiliaries, and this loss must be accepted (although we shall examine in this chapter some ways of reducing the power drain to a minimum). There are other power losses that may simply be the consequence of faulty installation or a disregard of the basic requirements for obtaining full efficiency.

For example, close confinement in a compact engine compartment means that the heat liberated during engine operation greatly increases the temperature of the air that surrounds the engine. Much of this air has already been warmed by entering through the radiator. Adequate means for getting the hot air out of the engine compartment is therefore a primary necessity if engine reliability is not to be impaired. In most production cars there is a fairly free flow down and back, the movement of the car helping to some extent in the extraction effect. Normally this flow will keep the temperatures of the block casting and the oil pan within bounds.

However, there is still likely to be a concentration of warm air toward the top of the hood, particularly in the region of the exhaust headers or manifolds. Radiation from these exhaust system components is added to the heat coming from the radiator and from the engine itself. Because it is from this same upper under-hood area that the intake air is drawn, there is a probability of power loss in comparison with dynamometer figures, where normal atmospheric temperatures and pressures apply.

Present-day emission-controlled cars have, in addition, an intake air preheating system, which is taken into account in selecting the standard carburetor jetting and ignition timing. So, in the quest for cold air during the conversion of a production car into a competition car, the introduction of cool outside air will almost always make richer carburetor jets a necessity rather than a nicety and may even call for some basic alterations to the spark advance curve. It is possible to obtain a good idea of this influence that cool air has by considering what has just been said on the basis of Charles's law, as given in chapter 1.

Air Temperature and Volume

According to Charles, the volume of a gas such as air becomes zero at −273° (or 0°K). The effect of a reduction in air intake temperature from 30° to 15°C (or from 86° to 60°F) is equal to a reduction from 303° to 288°K, or about 5 percent. This will increase the air density by the same percentage. The maximum temperature, however, is only that of a normal hot day, and in practice air-intake temperatures of over 75°C have been recorded after a couple of flat-out miles on a standard sports car. This being 348°K, a reduction to the normal under-hood temperature of 303°K with the engine warm would equal 15 percent, while the result of bringing it down to an average outside-air ambient temperature of about 15°C would make the difference over 20 percent.

The improvement on engine power output that will be obtained by increasing the weight of air drawn in (because a given volume of cold air is heavier than the same volume of warm air) can be calculated theoretically by a standard formula. The horsepower of the engine is multiplied by the square root of the number obtained when the existing intake temperature (absolute) is divided by the proposed intake temperature. The following example will show the procedure (it is assumed that the temperatures already quoted apply).

Imagine an engine with 80 bhp. The existing intake air temperature with the engine installed in the car is, say, 348°K, and modification of the intake air system will reduce this to the atmospheric ambient, or 288°K.

$$\begin{array}{ll} \text{Existing } T = 348 & \dfrac{348}{288} = 1.21 \\ \text{Proposed } T = 288 & \end{array}$$

$$\sqrt{1.21} = 1.1.$$

Therefore, engine power will be increased to $80 \times 1.1 = 88$ bhp.

The example suggests an impressive power increase by what seems to be a simple modification, but it is possible to take full advantage of the increase only on a racing car that never needs to cope with winter cold and is not required to produce flexible power for highway driving. Also, richer jets would be necessary to keep the same fuel/air ratio.

Simple cool-air grilles or scoops, such as the one shown in Fig. 13-1, are very common on all competition cars where the rules permit them. Where body changes are proscribed, a cool-air duct from the region of air intake to a point behind the grille but ahead of the radiator works very well. But bear in mind that the pavement can be very hot, and so the intake duct must not be placed too near the road.

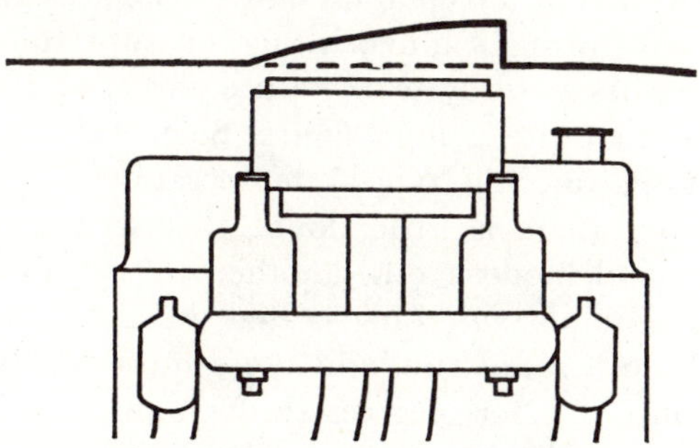

Fig. 13-1. Additional air intake adjacent to carburetors for intake air temperature reduction.

Pressure Ducting

If a large intake is located at the front of the car low down, in the high pressure area but well above the road, a good volume of cool air will be forced into the engine compartment or air intake plenum chamber, particularly at high speeds; the pressure buildup in such an intake may be about 0.5 psi at 100 mph. The feasibility of using this pressure for supercharging has already been considered in chapter 9. Generally, however, it is preferable to have a free outlet around the actual carburetor or injection system intake, the flow from the duct being guided past the intakes rather than forced into them. A plenum chamber with an open outlet to the rear fulfills this requirement very well.

Much can be done also to reduce heat convection from the exhaust manifold or header on engines where the latter is mounted just below the carburetors. A shield laminated from a sheet of asbestos sandwiched between two sheets of aluminum, which can often be fitted beneath the carburetor or injection intake, will help considerably. This kind of shield can also be used beneath air intakes that, because of engine design or installation, are very near the road, the radiator, or some other heated part.

Oil Temperature

With an engine in good condition, the oil temperature is usually kept within acceptable limits when extra power has been achieved through blueprinting or supertuning. So long as the oil reaches a steady temperature and stays there, no harm will come from quite a high figure. Some car makers still advocate caution, quoting typical temperatures as 75° to 80°C maximum, but there is little doubt that with modern engine construction and modern oils, another 10° to 20°C should do no harm. The important thing is that a sudden burst of power must not cause a rapid rise, and there must not be a slow but steady buildup of a few degrees throughout the time of engine operation; eventually the danger point will be exceeded.

Oil coolers are available that can be coupled into the lubrication system to augment the normal means of cooling. The oil cooler is most effective when it handles the entire oil flow, being connected in the circuit between the oil pump and the main oil gallery. The bypass type of cooler, connected in an auxiliary circuit such as can be combined with a bypass oil filter or with the pressure gauge line, helps to reduce the temperature to a small extent and is useful when overload conditions are of short duration. A typical oil cooler and its installation hardware are shown in Fig. 13-2.

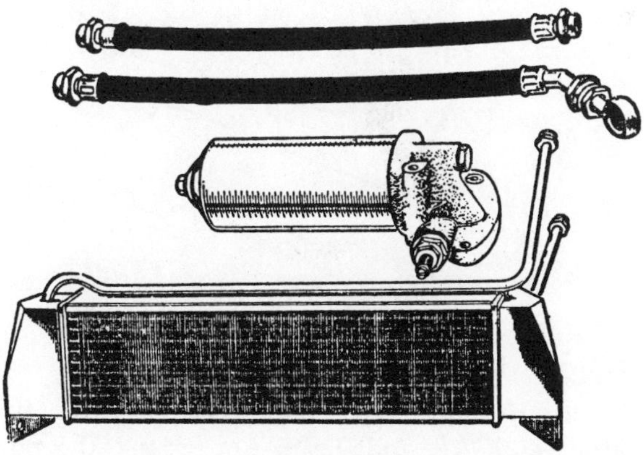

Fig. 13-2. Components of a typical oil cooler for installation on a competition-prepared production engine.

An air scoop (Fig. 13-3) can in some cases be fitted to assist in directing cool air over the oil pan. But if the rules do not prohibit it, an oil cooler is always best and will be even more efficient if it is mounted at the front of the car. This will require long hoses or pipelines on many cars. It is of vital importance that the hoses or pipes are properly chosen for the duty, resistant to vibration, not in a position to be caught in moving parts, and firmly secured to the car. Aircraft-type plumbing is almost universally used for applications such as this in professional racing.

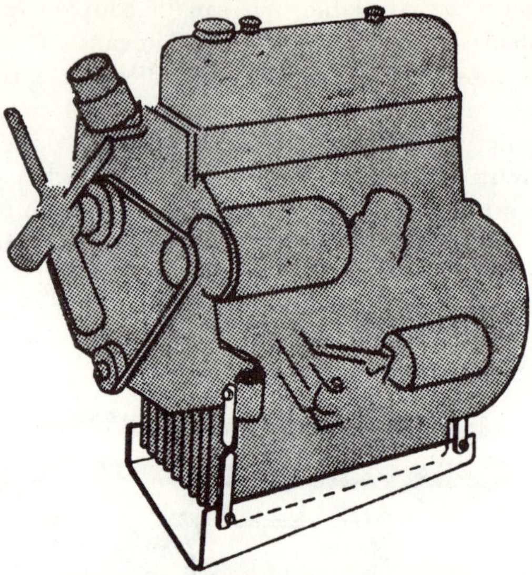

Fig. 13-3. Airflow over sump augmented by a simple design of scoop.

Mechanical Losses

The alternator, the water pump, the fan, the fuel pump, and, on large NASCAR and Trans-Am sedans, the power steering pump all represent auxiliaries that take power to drive. But it is possible to cut these losses considerably so long as they do not lead to engine overheating, a dead battery, or excessive driver fatigue. Electric fuel pumps are invariably used on competition cars, except in showroom stock and strictly stock classes. They will improve fuel system reliability while reducing the load on the engine.

Pumps for dry sump lubricating systems also drain power from the engine, but balancing this is the reduction of mechanical loss caused by oil drag. The dry sump also lowers oil temperatures and reduces problems of oil sloshing, oil starvation, and oil loss. Any power taken by the pump should be considered well spent.

Various friction-reducing oil additives are marketed, and,

judging from the decals on racing cars, one would think that these are being used in every engine. In truth, racers usually do use the additive that they advertise, but the actual quantity may be exceedingly small—just enough to obtain sponsorship money from the additive manufacturer. A few team managers use additives liberally and swear by them as agents for reducing internal friction losses.

Generators and Alternators

Some formula cars dispense with the alternator or generator for races that last only a few miles, which is frequently the case in club racing. Other tuners choose to put a very large pulley on the alternator or generator in order to reduce its speed and hence the power absorbed in driving it. Of course, no charging system and, in many cases, no battery is used in drag racing machines that run for only a few seconds at a time. Oval track race cars are push-started and have magneto ignitions, so no battery, generator, or alternator is required.

In addition to taking power from the engine, the generator or alternator can fail during a race. If it does, the water pump or power steering pump may be affected so that the car must be withdrawn from competition. Therefore, some tuners drive the alternator or generator from a point other than the engine. An installation such as the one shown in Fig. 13-4 has at least three advantages. First, a failure will not interfere with engine operation; second, because it is driven off the axle, the rpm is low and less power is absorbed; and third, it is nearer the battery, which has been placed inside the trunk in order to balance a nose-heavy car.

Fan Output

The air output of the radiator cooling fan varies in proportion to speed that is something between the square and the cube of its rpm. The power needed to drive the fan is approximately in direct proportion to the weight of air moved. Thus, a fan delivering 10 cubic feet of air per minute at 1000 rpm might deliver up to 1000 cubic feet per minute at double this

Fig. 13-4. Alternator, belt driven from rear axle, installed in trunk of racing Chevrolet Camaro.

speed. So if we assume that the engine idles at 700 rpm, crawls at 1000 rpm, and is revved to 2000 or so rpm for bursts of traffic acceleration, it might be thought that an average engine speed of 1200 to 1400 rpm is suitable as a basis for designing the fan to cool the engine under the worst conditions.

However, the fan, driven directly from the engine, has to vary its speed over a range of five or six to one; this is where the waste of power comes in. Even if we assume that the fan is reasonably efficient over a good part of its speed range (and this is usually not the case), at any speed above the one designed for, the fan will be moving far more air than is necessary and taking a lot of engine power to do so.

Multibladed fans are quieter than two-blade fans; sometimes multiple blades are set at irregular angles to one another to limit further the possibilities for resonant frequencies that passengers might find objectionable. In a competition car, therefore, it is often possible to reduce fan drive losses by using a fan with fewer blades or a smaller diameter. For example, the two-blade fan of the Ford Pinto will readily fit in place of the multibladed fan of the Ford Capri.

Viscous fan drives, which are a kind of fluid clutch, are widely used today even on low-cost production cars. The clutch has enough drag in its viscous fluid to turn the fan at a 1:1 ratio at low rpm, where fan cooling is needed. At higher speeds, the clutch allows slippage so that the fan's maximum rpm remains constant and proportional to the quantity of air it is moving, while the engine rpm goes to very high figures during acceleration and high-speed driving. The saving offered by a fan clutch is shown in Fig. 13-5.

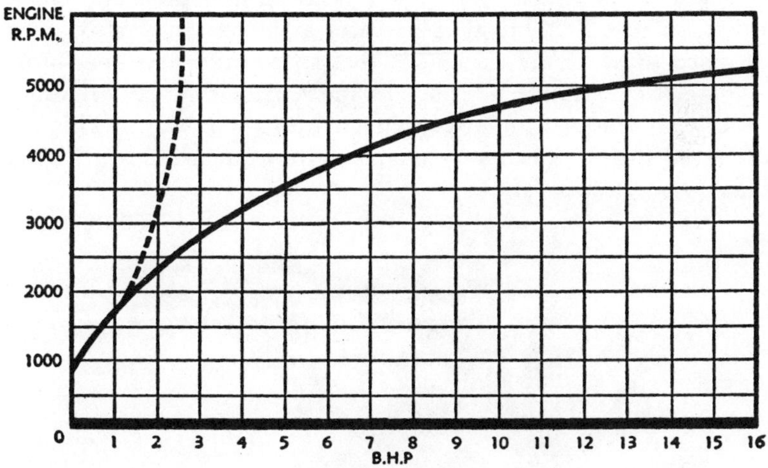

Fig. 13-5. Fan power absorption on 3.4-liter engine. Dotted line indicates drive with clutch running disengaged.

Freewheeling fan drives are sometimes incorporated in the clutch assembly in order to reduce inertia loadings on the V-belt. The use of a larger pulley for the fan is not desirable, as it may be for an alternator, because the fan is usually mounted on the water pump shaft, and a reduction in fan speed would also mean a reduction in coolant flow, which could lead to overheating.

The Electric Fan

Electric fans are used in many production cars, particularly VWs, Audis, MGs, and Triumphs. These can be transplanted to other cars, thus eliminating the mechanically driven fan. An electric fan not only can be of constant-speed type, thus minimizing the current needed to drive it, but also can be thermostatically controlled so that the motor runs only when the coolant temperature in the radiator exceeds a predetermined level.

The Lucas installation on the E-type Jaguar takes about 7 amps at 12 volts; a typical proprietary unit (Fig. 13-6) takes 6 amps. These figures represent less than 100 watts, or 0.13 horsepower; of course, this power is unvarying since the fan is either on or off. A fan that operates at a constant rpm can be designed to be highly efficient at that speed only; fans that must operate over a range of rpm are inefficient at nearly all times because of the compromises that go into their shape. This explains why the losses to an engine-driven fan are much higher than to an electric fan.

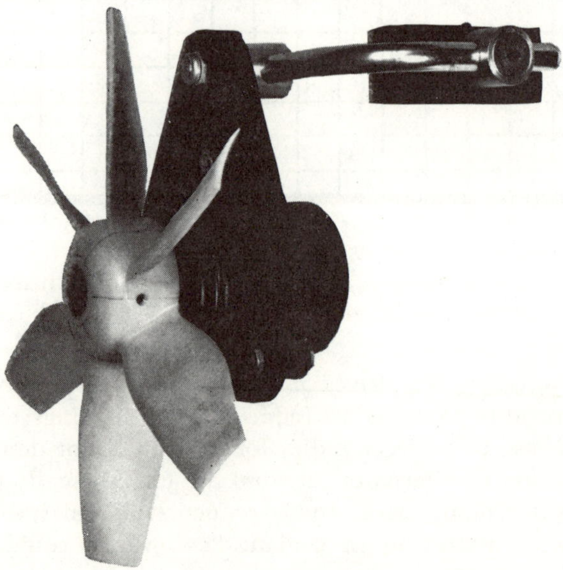

Fig. 13-6. Radiator cooling fan with electric motor drive.

It may seem that power wastage figures such as those quoted earlier for engine-driven fans are incorrect because the thin-section fan belts used today are not designed for transmitting such power. But despite its small size, a modern V-belt can actually transmit greater power (with less frictional loss in bending over the pulleys) than the old-style heavy belts. The average 3/8-in. belt on pulleys of suitable diameter will transmit 5 to 7 bhp at 5000 feet per minute of belt travel with no trouble at all.

Hardware and Plumbing

The hardware and plumbing used to install the engine in the car can have a life-or-death influence on the survival of the engine and, in some cases, the driver. If flexible engine mountings are used, they should be modified with catch straps so that the engine will not radically shift its position if a rubber mounting breaks. Only high-grade bolts should be used for solid engine mountings, and the best available scatter shield should be installed over the clutch/flywheel bellhousing. Most racing classes have strict rules governing this equipment, and while compliance may not help your car win races, it may keep the driver in one piece.

Aircraft plumbing, if not required, is still desirable. The hoses, for example, can be obtained in both flame-resistant and aromatic-resistant materials. Some hoses have both qualities and may also be self-sealing. No matter how well an engine is designed and built, it will not win races if there is an oil pressure loss, cooling loss, fuel pressure loss, or under-hood fire resulting from a burst hose.

Aircraft quality metallic lines should also be used with aircraft quality fittings. Plumbing designated AC (Air Corps) is obsolete and becoming hard to obtain. Choose AN (Army/Navy) standard fittings. The cost need not be high because often this material is available as military surplus. Several companies that advertise in racing enthusiast publications offer these kinds of surplus goods.

Remember that any kind of makeshift plumbing or hardware is a potential weak link in the car. The car may be good and the engine may be good, but if the engine is not carefully installed in the car, neither will be worth the trip to the race track.

14 / *Supercharging*

Historical Background

A supercharger is a kind of air pump that is used to deliver air (diesel engines) or a fuel/air mixture (spark-ignition engines) to the cylinders under pressure instead of having the mixture enter under atmospheric pressure as a result of the depression (partial vacuum) caused by downward movement of the piston on the intake stroke. In regard to supercharger design, a normal cylinder and piston can be used for blowing, as in large industrial and marine diesels. For small high-speed engines it has the inherent disadvantages of reciprocating action and high frictional loss. It was in all probability last used in the small gasoline engine field on the early DKW racing motorcycle engines; it was subsequently superseded on these by a rotary-type blower, and this is now the universal type for competition use (Fig. 14-1).

In past years, the centrifugal supercharger has come in for a great deal of bad press in automotive technical books. The assumption, of course, was that the impeller was driven mechanically by the engine, which was the case years ago. The high point for engine-driven centrifugal blowers was certainly the Miller and Duesenberg race cars of tne 1920s and 1930s. While the design combines simplicity and small size with a high out-

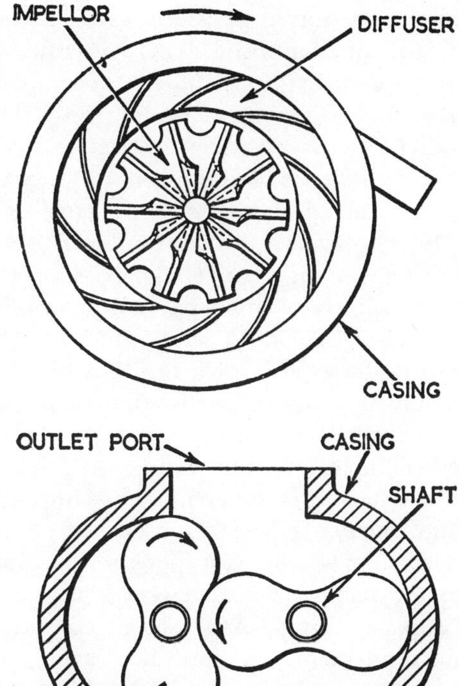

Fig. 14-1. Three basic types of compressors used for supercharging: (top) centrifugal impeller; (center) Roots with synchronized rotors; (bottom) sliding vane and eccentrically mounted drum.

put of mixture, the output pressure varies approximately as the square of the impeller speed. This is because the operation of the unit, as implied by "centrifugal", depends on the kinetic energy imparted to a whirling mass of air by its inertia—akin to the energy that tends to "scatter" an overspeeded clutch or flywheel. Thus, if an impeller is designed to supercharge at 8 psi at 5000 rpm, it would give only 2 psi at 2500 rpm.

On the Duesenberg and Miller cars there was no problem because races were on high-banked board tracks or at the Indianapois Motor Speedway and rpm was maintained in a narrow band. Supercharging disappeared when fires and the Great Depression ended the board track era. On dirt tracks the centrifugal supercharger was a costly toy that wore out in the abrasive dust and provided low boost at the time when high boost was most needed—in leaving the turns.

Though a centrifugal supercharger with a turbine operated by engine exhaust gases was used with great success on certain kinds of aircraft, its vast potential is only now being realized in motor racing. Turbocharging, as it is called, is the only kind of supercharging being used or considered on today's oval track and road racing engines. The revolution might have come about sooner were it not for the bad things that had been said about centrifugal blowers in general.

Despite one or two attempts to reintroduce centrifugal supercharging in Championship oval track cars during the 1950s, the most widely known failure for centrifugal blowers was the 1.5-liter V-16 BRM Grand Prix engine of that period. Countless Britons contributed money to the building and development of the first BRM, a car they hoped would redeem England's honor in international racing. When the car failed, many journalists rushed to criticize, particularly the choice of the centrifugal supercharger. At low speeds, the car could not get out of its own way; and when the boost from the blower finally came in, it was with a force that frequently destroyed the engine. Thus we can forget about mechanically driven centrifugal blowers. The successful superchargers of the mechanically driven variety have been either the Roots or the vane-type.

The Roots supercharger was used on the all-conquering Mercedes Benz Grand Prix race cars of the 1930s and on the Alfa Romeo Grand Prix cars that dominated Formula I racing in the late 1940s and early 1950s. The Auto Union Grand Prix cars that outdid even the Mercedes during some seasons of the 1930s were equipped with vane-type blowers. All of the last types of these three racing marques used two-stage supercharging, that is, a large supercharger compressed the mixture into a smaller supercharger, which compressed the mixture further and crammed it into the combustion chambers. At the end of its development, the Alfa produced in excess of 420 bhp from 1500 cm^3 (91 $in.^3$) of piston displacement.

Roots Blowers

The operating principle of the Roots blower (developed in America before the Civil War for foundry blowing) is similar to that of a gear-type oil pump; two rotors carrying lobes are synchronized by external gearing to run with extremely close clearances between themselves and the housing (Fig. 14-2). When there are two lobes on each rotor, there is some impulsiveness

Fig. 14-2. Components of a Roots-type supercharger.

in the delivery and suction, which becomes of no consequence at higher speeds and is virtually eliminated by using three-lobe rotors. Skewed ports or spiral rotors also reduce the effect, but, of course, these features in no way affect the operating principle.

With a Roots blower there is no change of total volume throughout the pumping cycle and thus no internal compression. After entering the blower, the mixture is expelled from the delivery port against whatever backpressure is encountered in the manifold, which depends on the ratio of blower delivery volume to the swept volume of the engine cylinders. The vane-type blower, on the other hand, has its own internal compression which is significant and will be discussed later.

Roots blowers are used today only on drag racing engines. This is mainly because Roots-type GMC blowers from diesel engines are very widely available in the United States, and a number of experts have become proficient in setting these units up for drag racers. Normally the GMC blower is mounted atop a big V8, and a special fuel injection unit is installed atop the blower.

For drag racing, the GMC unit should be carefully examined and modified before it is put on the engine. First of all, heavy duty end plates must be fitted because the stock plates tend to crack near the bearing bosses if there is a backfire in the induction system. Also, the rotors, which are aluminum, must be pinned to keep them from slipping on the steel stub shafts as a result of backfiring. This is done by drilling through the rotors near their ends and through the steel stub shafts. Steel pins are then driven in with an interference fit.

In addition, a visual check should be made for obvious cracks, nicks, and marks that show interference between the rotor lobes. The clearances between the lobes must never be zero, and this clearance can vary if there is excessive end play because of the spiral of the lobes. The clearance between the rotors and the housing should be .005 to .007 in., the clearance between the rotors and the front of the housing should be .008 to .010 in., and the clearance between the rotors and the rear of the housing should be .016 to .018 in. These clearances can be checked with a feeler gauge and must not vary from the specifications because of end play.

Though multiple V-belts were formerly used to drive GMC

blowers used on drag engines, the Gilmer belt is now used universally; it eliminates all slippage and reduces the weight of the belts. With many setups this means that there is no way of driving the water pump (although on most of the cars there is no radiator). The Roots blower, being virtually positive displacement, starts working right away and is ideal for drag racing. As we shall see, there is a slight time lag involved in getting a turbocharger up to its efficient rpm.

Vane-type Blowers

The vane-type blower (Fig. 14-3) has a drumlike rotor mounted in bearings so that it is eccentric to a circular casing, leaving a crescent-shaped chamber formed between the two, with circular end plates. A number of vanes pass through slots in the rotor, with very close clearances between their tips and the casing. These vanes divide the casing into separate portions —usually four.

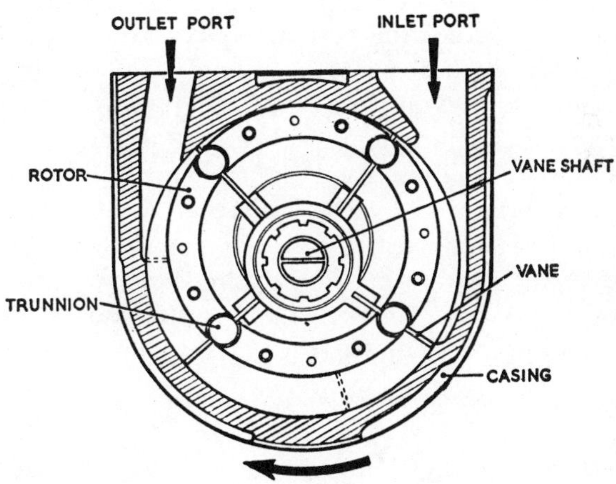

Fig. 14-3. End elevation of vane-type supercharger, showing disposition of ports.

The vanes, which pass through the rotor and have very fine clearances between their extremities and the casing and the end plates, virtually subdivide the crescent-shaped chamber into four separate portions. The intake port on the casing is positioned so that as one of the portions receives its full volume of mixture, the adjacent portion, on the intake side of the unit, is increasing in volume and creating a vacuum at the intake port as the vanes revolve. As soon as the vanes have reached a position where the portion of the chamber between them contains its maximum volume, the volume between the vanes begins to diminish, since the space between the rotor and the casing becomes progressively less as the delivery port is approached. Thus the mixture is compressed inside the casing before it is released to the intake manifold.

The vane-type blower therefore has its own internal compression, which is adiabatic. This means that the internal pressure will increase with the speed of compression, that is, with an increase in rpm. The result is that as the weight of mixture delivered increases, the power required to pump it decreases for any given unit weight. The delivery temperature also decreases in relation to the quantity. This feature has made the vane-type blower attractive to designers at various times. But despite careful engineering and sturdy construction (Fig. 14-4), racing applications have suffered from mechanical unreliability.

A number of add-on vane-type blowers have been marketed, and some of these, such as the Shorrock unit shown in the illustrations, had well-designed lubricating arrangements and thus held up very well by comparison to some older units. As Fig. 14-5 shows, the construction was robust enough to be reliable under all conditions. At present, however, the only add-on superchargers being marketed for production cars are turbochargers, and, as we shall see, there are some very good reasons for this.

Pressures

All types of superchargers are limited in the maximum pressures that they can produce, but the centrifugal impeller is capable of producing pressures that are in excess of anything yet seen in a car. For example, obtaining even a 30 psi boost with a positive-displacement blower would require at

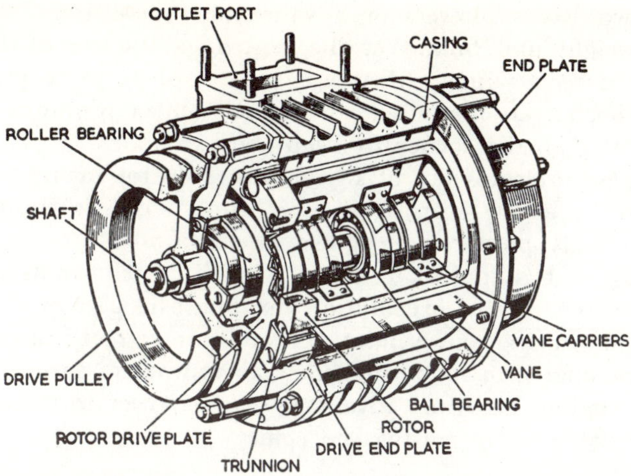

Fig. 14-4. Internal construction of Shorrock vane-type supercharger.

Fig. 14-5. Shorrock vane-type supercharger with end cover removed, showing vane carriers.

least two Roots blowers or a vane-type supercharger of very great weight and bulk. Yet the centrifugal blower of the ill-starred sixteen-cylinder BRM was capable of 65 to 70 psi, and the turbochargers used on Indy cars can deliver pressures above 80 psi, though they generally keep boost to about 45 psi.

These pressures are of significance only for racing because a blower pressure even 5 psi over atmospheric is enough to "cook" many production engines. About a 7 to 8 psi boost is the most that is applied to a stock engine, which points up the strength and careful preparation that must be part of any turbocharged racing powerplant. In addition, street conversions keep the engine in its normal rpm range, whereas turbocharged racing engines take full advantage of the power increases that are available by increasing the rpm.

Turbocharging

The turbocharger is not a complicated device. It consists of a centrifugal supercharger, with a rotor and housing of lightweight alloy, connected by a short enclosed shaft to a turbine wheel only three to four inches in diameter. The turbine wheel and its housing are of heat-resistant steel alloy, and the housing is designed to direct the engine's exhaust gas stream directly over the turbine wheel. The fuel/air mixture is drawn in at the center of the centrifugal rotor and exits at the periphery, whereas the exhaust gases enter at the periphery of the turbine wheel and exit from its center.

An internal combustion engine wastes a great deal of energy, which is thrown out the exhaust pipe in the forms of heat, pressure, and gas velocity. A turbocharger puts this waste energy to work, using it to drive a supercharger. There are no gears, chains, or belts to wear out, and the supercharger and its carburetor or injection system can be located out in the cool air. The duct to the engine's intake manifold can be designed to have an efficient shape, with no worry about mechanical drive complexities.

It might seem that the turbine would create exhaust back pressure and thus reduce the engine's power output. However, in a normally aspirated engine, back pressure is a problem only because it hinders the entry of fresh fuel/air charges into the cylinders. Because the intake manifold is under pressure, there is no problem in getting the exhaust gas out and a fresh charge in.

Blowing Down

The technique of blow-down scavenging, or *blowing down,* is not new to supercharging and is illustrated in Fig. 14-6 with a mechanically driven vane-type supercharger. Because the induction system is under high pressure, if a very large valve overlap and early intake valve opening are adopted, the whole combustion chamber can be blown through by fresh cool mixture for quite an appreciable period. During this time, all traces of burned gas are ejected, and the valves, the plug, the piston, and the cylinder head are subjected to a cooling blast; the heat is transferred to the mixture, which goes on its way down the exhaust pipe—burning on its way and helping to drive the turbine with increasing vigor.

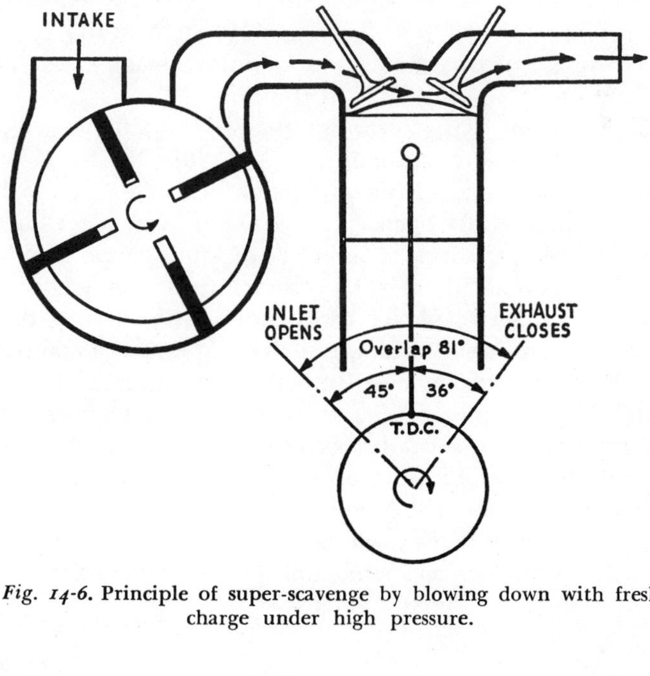

Fig. 14-6. **Principle of super-scavenge by blowing down with fresh charge under high pressure.**

High-pressure charging of the cylinder begins as soon as the exhaust valve closes, prior to which the combustion chamber is already well packed with fresh mixture. The success of blowing down is highly dependent on valve timing, and the use of a turbocharger must be anticipated in designing the camshaft (s). But even in cases where turbochargers are added to engines that were designed for normal aspiration, exhaust backpressure is not a source of loss because the supercharger output can be increased easily to compensate. By comparison, at maximum rpm the GMC Roots-type supercharger on a blown nitro-fuel dragster requires about 100 bhp from the engine's crankshaft output. While turbocharging is not totally "for free" or "something for nothing", it is about as close as can be attained.

The Catch

This is not to say that turbochargers will make as great an impact on drag racing as they have on road racing and oval track racing. For drag racing, the positive-displacement Roots blower is certainly better at present. With any positive-displacement, mechanically driven blower, the boost is available immediately, which is important for obtaining the low-speed power that is necessary coming "out of the hole".

The problem with turbocharging is the infamous "turbo lag" that Indy drivers complain about. At idle, the turbocharger is hardly turning over. Even at part throttle, there is no boost so long as the throttle opening is too small to increase the pressure in the compressor to a level above atmospheric. But when the throttle is opened fully, even at relatively low engine rpm, the flow of exhaust gases increases enormously, causing the turbine to speed up and begin pressurizing the intake manifold.

This takes time because from the virtual standstill at idle the turbocharger must accelerate to a speed as high as 120,000 rpm. In a production car turbocharged to perhaps 8 psi, some boost will be available immediately at speeds of, say, 50 mph or above. It does not take long for the blower to reach this pressure. But on an Indy car, where full boost may be more than 40 psi, the turbine must "wind up" higher to achieve the pressure; thus the time lag is greater.

Despite this shortcoming, the turbocharger has a compensating virtue: it supplies boost more on the basis of engine demand than on engine speed. So despite the time lag, full boost is possible at full throttle and low engine speeds. This can be very decisive in road racing, where a mechanically driven blower might never get up to full boost before the next turn. Another failing of the positive displacement blower is that it takes up excessive space (turbochargers are exceedingly compact) and needlessly heats up the mixture, as the blower is actually relatively inefficient as a compressor.

Finally, it might be supposed that mechanically driven blowers work all the time and are therefore supplying boost all the time. Actually, they are only going through the motions. With the throttle partially open, there is inadequate mixture for the blower to compress. Under these conditions the turbocharger is not drawing power from the crankshaft, but the mechanically driven blower is, flailing away in a partial vacuum and costing bhp.

Output Characteristics

To achieve high pressures with Roots blowers, designers have sometimes resorted to two-stage supercharging. Turbochargers have no such limitation; the design problem presented is frequently that of too much output rather than too little. A turbocharger that is designed to give the desired output in the mid-range of engine rpm will produce a boost pressure that is much too high at maximum engine speed. In addition, inertia keeps the turbine spinning at relatively high speed even after the throttle is closed and, upon reopening the throttle, the engine rpm may have changed to a speed that does not match the blower pressure. Therefore, several methods have been developed to control the turbocharger's output.

Perhaps the simplest is a compressor blow-off valve, or *pop valve.* When a predetermined pressure is reached in the intake manifold, the valve is forced open so that the pressure does not exceed this level—exactly the principle of the pop valves or safety valves used on the boilers of old-time steam locomotives. On an engine with port-type fuel injection, the pop valve presents no problems because the compressor is handling air only. If the carburetor or fuel injection is on the atmospheric side of the compressor, then the pop valve would be releasing fuel/air mixture. Aside from the fuel waste and air pollution, this also presents an obvious fire hazard.

Another simple system is to restrict the turbine's outlet. Although this reduces the available turbine power at high speeds, it also tends to increase exhaust backpressure to a point in excess of that which can be overcome by the blower. Thus, although it is a safe and inexpensive system, it is not suited to competition engines.

Waste Gate

The most successful system for controlling turbocharger output is the waste gate. Despite its immoral sounding name, the only thing being wasted is exhaust gas, which in any piston engine other than a turbocharged one would be totally wasted anyway. In its simple form, illustrated in Fig. 14-7, there is a diaphragm valve that opens the waste gate whenever the pressure in the intake manifold exceeds the predetermined maximum. With the valve open, a portion of the exhaust gas is diverted to the atmosphere, preventing the turbine speed from rising above the rpm needed for maximum boost. More complicated applications of the waste gate principle, which give a boost rate that rises or falls proportionally to engine speed and load, are in use.

The Turbo Porsche (Fig. 14-8) has a highly developed waste gate turbocharger system. By incorporating a compressor recirculation circuit, the Porsche Turbo reduces turbo lag to almost nothing under conditions where the throttle is opened wide after a long period of coasting. In the Porsche system, the turbocharger handles air only; fuel is supplied by a special kind of Bosch K Jetronic fuel injection.

Add-on Turbochargers

There are a number of turbocharger kits on the market that can be installed on cars that formerly had normally as-

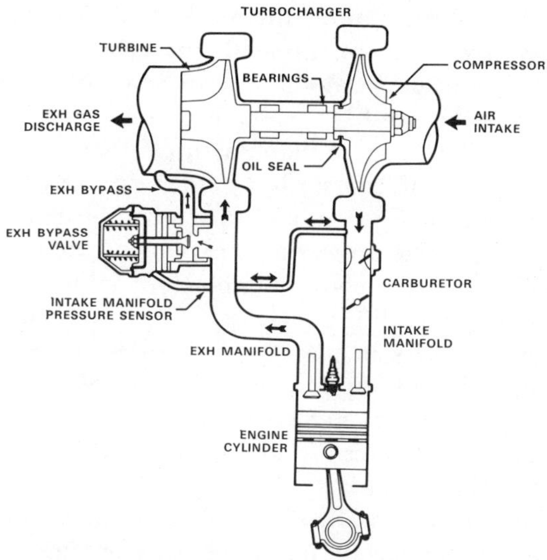

Fig. 14-7. Simplified operation of waste gate.

Fig. 14-8. Waste gate on competition-prepared Porsche Turbo allows surplus exhaust to exit through a separate, smaller exhaust pipe, reducing quantity of exhaust directed into turbine.

pirated engines. In addition, there are turbochargers available from AiResearch (which supplied turbochargers to General Motors for Corvairs, Oldsmobiles, and Pontiacs in the 1960s), Rajay (these seem readily available), and Schwitzer (which would rather sell to engine manufacturers than to individuals).

The question, of course, is whether one can add a turbo to a Toyota and then go out and dice with the Porsche Turbos on the race track. The answer at this point is probably no. The add-on kits supply low boost suitable for highway applications with otherwise unmodified engines. These kits are attractive because they generally increase performance without increasing emissions or fuel consumption. But even if you obtain a Rajay and design your own racing system, you probably will not be able to duplicate the results obtained by a factory-developed system such as the Porsche.

Carburetion is a particular problem. Aside from the fact that a broader-range carburetor is necessary because of the wider band of horsepower, there is the problem of where to install the carburetor in relation to the blower. Should the supercharger supply compressed air to the carburetor?

Fuel injection does not solve the problem easily. Indy cars, for example, frequently use two fuel injection systems—one for low speeds and the other for full boost—because of the wide horsepower band. Also, the question still remains, should the injection take place before compression, or should it be into the intake ports where the fuel pressure must be higher because of the blower pressure in the induction system?

The GM system, and the setup used with most add-on systems, places the carburetor ahead of the compressor to produce the fewest problems with carburetion. But when the throttle valve is closed, the compressor is subjected to full engine vacuum. This makes necessary a very stiff seal on the compressor shaft to keep lubricating oil from being sucked out of the turbocharger's bearing section. The resulting seal friction tends to interfere with turbocharger response.

Placing the carburetor downstream creates many problems, and almost certainly will require a special carburetor and fuel pump. Even with injection there is the problem of compressor surge. This occurs when the supercharger pumps a reverse flow out of its inlet when the outlet is closed by the throttle at high speeds. Indy cars get away with it because their throttles are closed for only short periods of time, but serious problems can arise in developing a homemade system for racing.

There are signs, however, that the SCCA and other race sanctioning organizations may liberalize their positions on turbocharging. Once it becomes legal to add a turbocharger for racing, it will not be long before development begins. Nevertheless, it will still be many years before the average tuner can hope to equal the results that have been obtained by Porsche on the race track.

PART III

Engines and Applications

15 / *The American V8 Engine*

Expert Tuners

Some of the finest tuners in America have been consulted in the preparation of this chapter and later ones. Each is an expert in preparing the particular engines covered by that chapter. Whenever possible, their names are mentioned. A reader who needs further advice or information concerning the preparation of a particular engine for competition should contact one of these experts directly.

The present chapter, which is concerned with American V8 engines, is one exception. Information for it comes from many sources. Probably in every reader's immediate area there are shops that devote themselves to the tuning of American engines for drag racing, oval track racing, or road racing. Therefore, these facilities and speed shops are the best places to go for any advice or machine shop services that may be required. Many of the most respected racing engine builders use outside machinists for the preparation of cylinder heads, crankshafts, and other components that are virtually remanufactured for racing.

Standard Design

Since the mid-1950s, the V8 engine has been the standard American engine (Fig. 15-1). Rarely has it been a small engine.

The idea of a 2.5-liter or 3.5-liter V8 seems foreign (no pun intended) to the American designer, and for vaguely incomprehensible reasons, six-cylinder and four-cylinder designs are resorted to by the American auto industry for anything smaller than about 4.5 to 5.0 liters. Really small V8s are excellent powerplants. But probably because Americans have associated "big" and "more" with "better" and "expensive" for so many years, the U.S. car makers assume that anything "small" and "less" must automatically be "worse" and "cheap".

Fig. 15-1. The Chrysler/Plymouth Hemi, one of the greatest of American V8s in stock car racing and in drag racing.

For well over twenty years, big OHV V8s have been the cornerstone of racing in America. This is likely to be the case for some time, although that time will eventually come to an end.

Government regulations that demand reasonable levels of fuel economy for each manufacturer's line of cars are causing the American auto industry to phase out the biggest V8s and limit the use of others. Instead of building smaller, more economical V8s, the trend is toward V6s and inline fours.

But even though some fairly large V8s will be with us for a number of years in the bigger, more expensive sedans, these powerplants may not be usable for competition. For example, many factory-installed components on today's federally regulated engines are simply unsuitable for high performance use. Emission laws have resulted in lower compression ratios, and the car makers have taken advantage of the reduced power outputs by using cheaper crankshafts and connecting rods, which lack adequate overstrength for racing. Currently the manufacturers show some interest in providing forged crankshafts and connecting rods for some engines, so that these parts can be installed by racers for particular classes that require factory components. But this situation may not last very long because there is little to be gained in the way of publicity by racing V8s when V8 cars no longer comprise a significant part of the manufacturer's output.

Since we are paying higher car prices because of emission controls, it does not seem altogether right that we should be given cheaper engines, which are made possible by the emission controls. Still, that's Detroit. And we are fortunate indeed that there is still a great deal of speed equipment available for the good V8s manufactured in more carefree times. Because millions of V8s are on the road, we would not be likely to run out even if the Big Four stopped manufacturing them today. After all, Ford flatheads and Buick straight-eights are still seen occasionally in drag racing and hot rod competition, though neither has been built since the early 1950s.

Classes

In road racing, American V8s figure prominently in Can-Am cars, the big GT classes, in sedans, and in large production sports car classes. Although other makers have enjoyed an occasional win, Chevrolet engines are at present the only V8s actually racing on road courses in a competitive manner. Partially this is because the Chevrolet Camaro and Corvette cars are capable of being made into good-handling road racing cars and because the Chevrolet small block V8 is the lightest and most compact 5-liter block available for use in formula cars and sports racing cars.

In drag racing, the situation is more democratic. While Chrysler and Chevrolet engines predominate, there are plenty of good Ford and American Motors engines on hand to keep the competition keen. Oval track racing, both with sprint cars and dirt-track championship cars, is very much Chevy-dominated. Compact size, a low center of gravity, and extensive development know-how are probably responsible for this, and some truly magnificent designs based on the Chevy engine have been achieved (Fig. 15-2). In NASCAR, every brand of American V8 is well represented.

Fig. 15-2. The Moser racing engine, based on a Chevy small block. Double overhead camshafts are spur gear driven from a big Gilmer belt sprocket on each head.

This chapter will show what can be done to obtain the maximum power from these engines. Whether these modifications are applicable in part or at all to certain kinds of racing is something that can be determined only by studying the rules for the particular racing class in which an engine is to be used. We will not spend a great deal of time discussing speed equipment because there is so much equipment available. Instead we will concentrate on the cylinder heads, cylinder block, and engine moving parts. These may differ radically from one engine design to another, but the supertuning principles are the same.

Cylinder Heads

Fig. 15-3 is a cross-section view of a small block Chevrolet engine, showing the well designed ports and the classic wedge combustion chamber with its extensive squish area. It is obvious from examining the top end of all American V8s that the de-

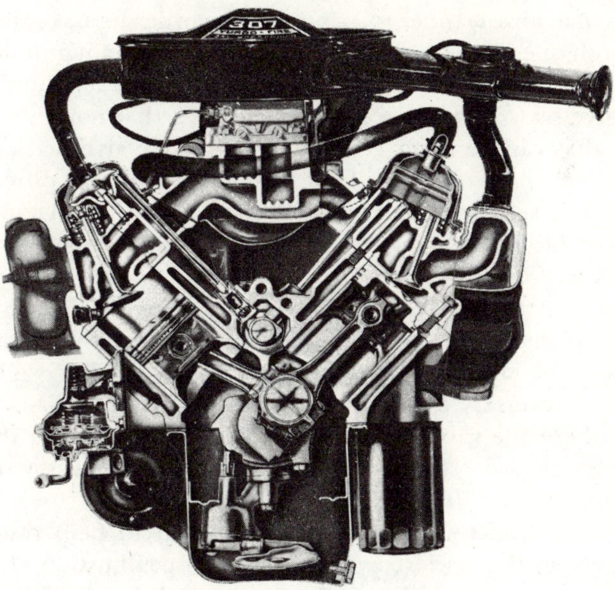

Fig. 15-3. End elevation of a Chevrolet small block V8, showing ports and combustion chambers.

signers have achieved exceptional volumetric efficiency without resorting to overhead camshafts. The big block Chevrolet, the Chrysler Hemi, and several other large-displacement designs do have staggered valves, but these engines still represent a great accomplishment for fundamental design techniques that were decades old when the first post-World War II OHV V8s appeared in the 1940s.

In some blueprinting classes, it is permissible to enlarge the ports so long as the valve diameters are not changed. Usually this is called a *competition valve grind.* The procedure is to ream the ports larger right at the valve seats, moving the seats out to the extreme outer periphery of the valve. This can normally be done without a loss of gas velocity at the valve because the valves are already rather undersize for the cylinder volumes.

When a tuner is under no rule restrictions, it is entirely practical to increase the valve diameters on all American V8 engines. The possibility that this may need to be done seems to have been taken into account by the designers; there is usually ample metal at the valve seats and at critical parts of the ports for the machinist/tuner to grind away in order to make the course a smooth one all the way to the bigger valve. Keeping in mind the axiom that the narrowest point in an intake tract should be at the valve seat, it can be seen from Fig. 15-4 that not only is there metal that can be removed from around the valve seats, but the taper from the intake manifold flange to the valve is pronounced. Thus, it is seldom a problem to maintain a gradual taper from the port face to the valve even when the largest possible valves have been fitted.

Most American V8s have wedge-type combustion chambers. It is thus inevitable that increasing the valve sizes will cause the valves to be shrouded by the combustion chamber walls. This makes it necessary to relieve the chamber all around its deepest part, where the valves are located. Domed pistons are therefore necessary in order to achieve competition compression ratios in modified engines (Fig. 15-5).

The cylinder heads are usually milled to help restore the compression that is lost by enlarging the combustion chambers. However, milling causes the intake ports to be misaligned with the manifold flange. This is taken care of either by reshaping

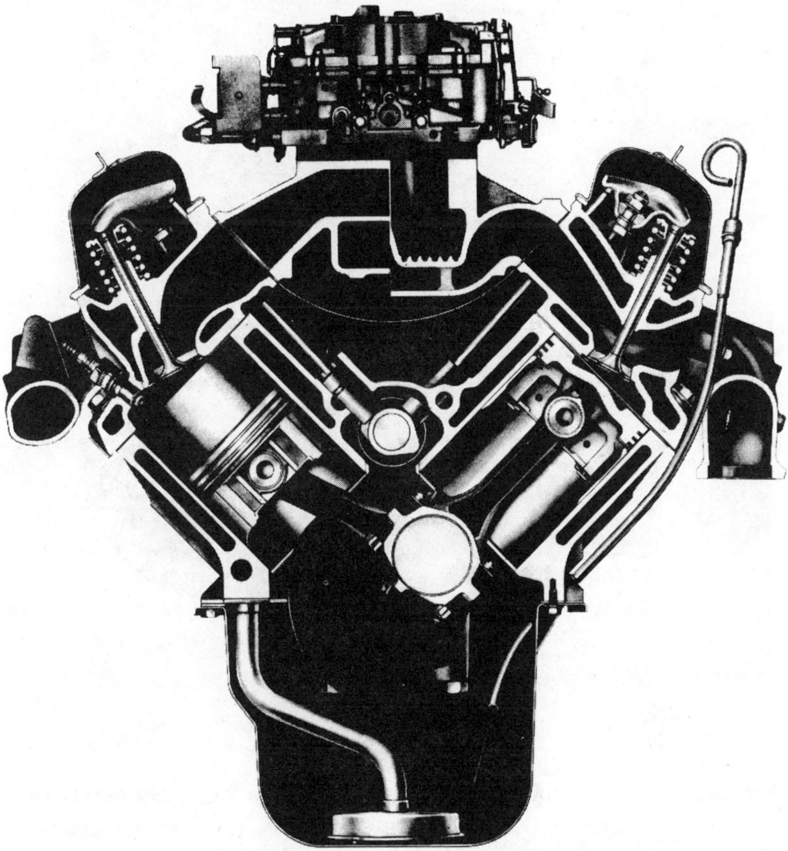

Fig. 15-4. Cross-section of American Motors V8. Notice the pronounced taper in the stock intake ports and the thick area of metal around the valve seat.

the ports or by using a proprietary manifold that has been designed to help compensate for these problems; usually the manifold has been designed so that it can easily be milled to fit.

No American V8 has machined combustion chambers, and so merely polishing the surfaces and rounding off sharp edges can eliminate hot spots and make higher compression ratios practical. Chevrolet small block cylinder heads and American

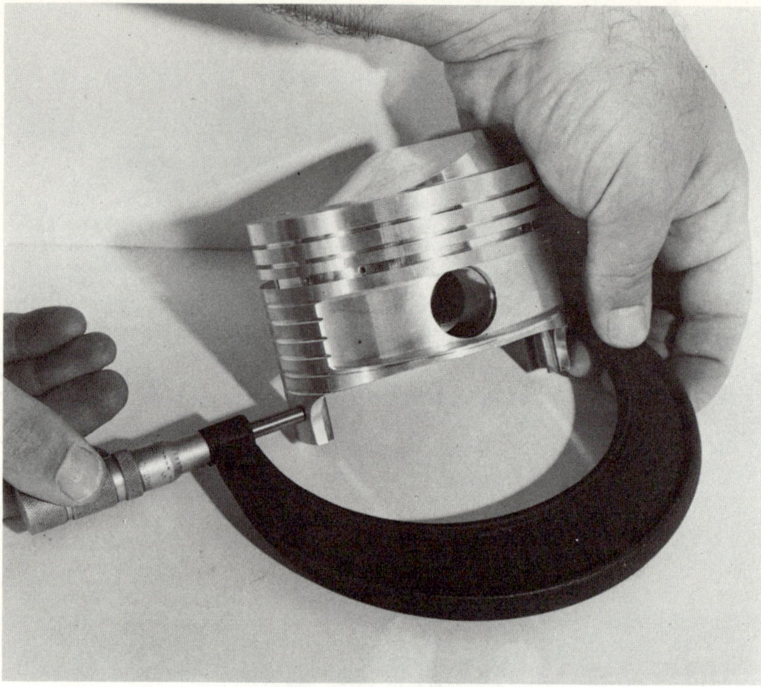

Fig. 15-5. Skirt diameter being checked on typical domed piston.

Motors heads, to name two, have dime-size depressions in the combustion chamber roofs. In most heads, the depressions are located adjacent to the spark plug, but in some American Motors heads they are on the side of the chamber that is opposite the spark plug. In cleaning up or reshaping the combustion chambers, these depressions should merely be smoothed and polished; under no circumstances should they be removed because the turbulence that they create has a beneficial influence on combustion.

Valve Gear

Most American V8 rocker arms are mounted on individual studs with ball joints. The studs are a press fit in the cylinder head casting and in some cases extend into the water jacket (Fig. 15-6). The stock studs must be withdrawn so that the holes in the head can be tapped and screw-in studs installed—using waterproof sealer if the holes extend into the water jacket. Competition camshafts and higher rpm make stiffer valve springs a necessity. These are usually dual or triple springs, and the increased loading would cause the stock rocker arm studs to pull out during engine operation.

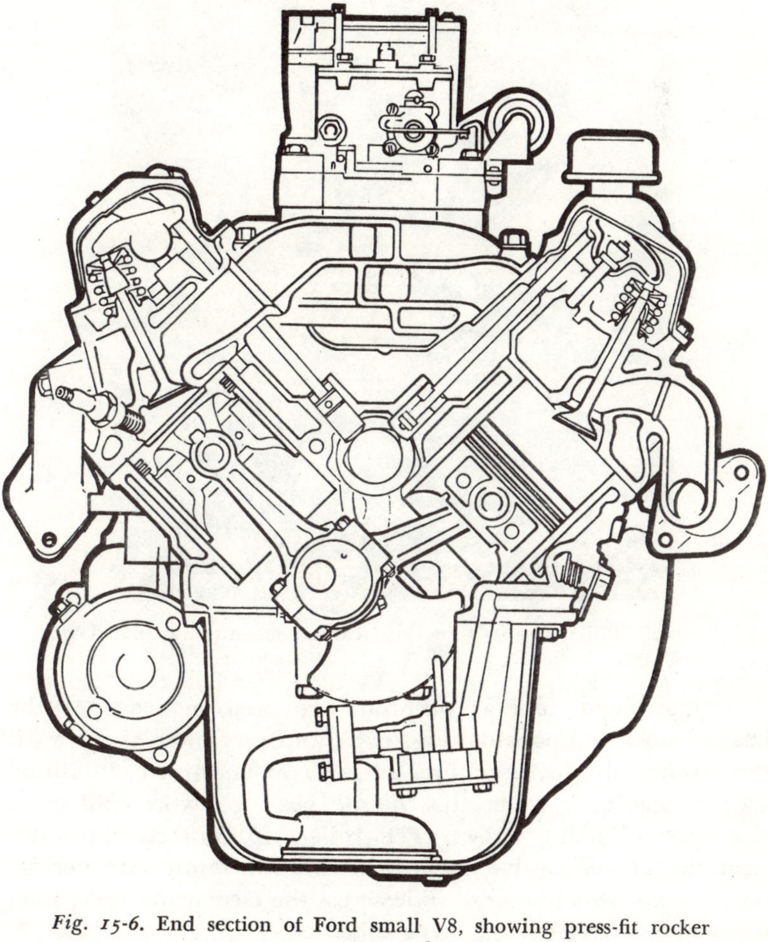

Fig. 15-6. End section of Ford small V8, showing press-fit rocker arm studs.

Hydraulic valve lifters are another almost universal feature of American engines. These are heavy and tend to "pump up" at high speeds, holding the valves open. Therefore, solid lifters are used in supertuned engines. If the cam profiles are exceptionally radical, roller tappets are used (Fig. 15-7). Roller tappets reduce cam and lifter wear almost entirely and permit valve timing that would otherwise be obtained only with overhead camshafts.

Fig. 15-7. Roller tappets being installed in competition-prepared V8.

Lightweight tubular pushrods are used in place of the heavier stock components (Fig. 15-8), and very often the pressed steel rocker arms are scrapped in favor of machined aluminum rockers that have roller tips, needle bearing pivots, and valve clearance adjusting screws. The roller tip contacts the valve stem and prevents valve guide wear that can result with normal rocker arms, which tend to slide across the face of the valve stem when high valve lifts are being obtained.

Link-belt "silent" timing chains (Fig. 15-9) are used except on a few rare high-performance models that have double-row roller chains. For racing, double-row roller chains are sometimes

Fig. 15-8. Road racing Camaro engine with intake manifold removed. Lightweight tubular aluminum pushrods can be seen.

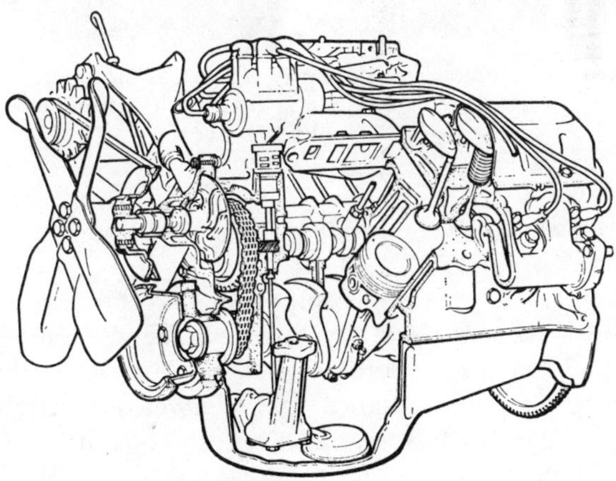

Fig. 15-9. Side section of Ford V8, showing typical silent-type timing chain.

substituted on engines that did not originally have them. Alternatively, a spur gear camshaft drive can be installed (Fig. 15-10). The gear drive is strong and provides accurate valve timing. However, noise is considerable because of the straight-cut teeth.

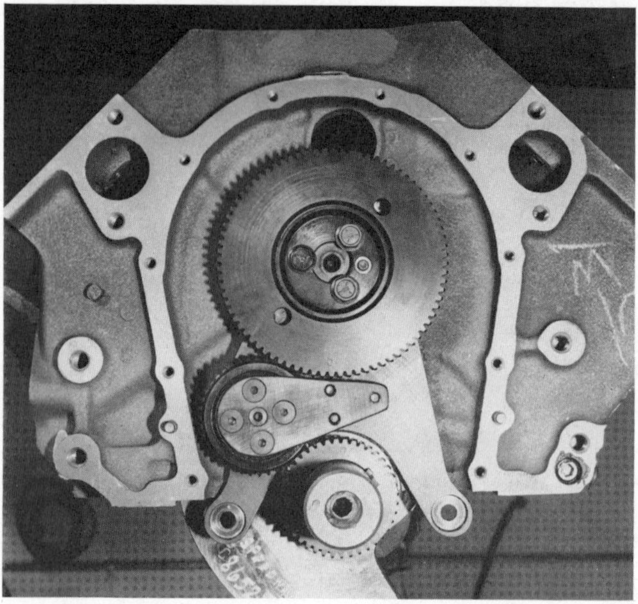

Fig. 15-10. Spur gear camshaft drive installed on big block Chevy. Idler gear is necessary to keep camshaft rotation in same direction as in chain driven setups.

Crankshaft, Connecting Rods, and Pistons

A number of crankshaft suppliers can provide excellent competition crankshafts for any American V8. Furthermore, for some engines, the car manufacturers have forged crankshafts available for high-performance and competition applications. As delivered by the factory, however, these crankshafts leave something to be desired.

First of all, the factory crankshaft must be magnafluxed. If it is a cast shaft, X-ray examination is sometimes used to detect internal flaws and pockets that have resulted from bubbles in the molten metal or from irregular flow into the mold. Unless the car manufacturer is known to provide good heat treatment, the crankshaft should be nitrided or Tuftrided and then straightened to correct any distortion that has resulted from the heat treatment. In some instances, crankshafts are shot-peened to relieve stresses.

Crankshafts that are not machined may be lightly polished to inhibit further the formation of fatigue cracks. The shaft is always dynamically balanced with electronic equipment, first by itself and then in combination with its flywheel and clutch. Oil holes are chamfered (Fig. 15-11) and carefully radiused (Fig. 15-12). Finally, the journals are given a very high polish. In the case of drag racing engines that burn nitromethane fuels, the crankshaft journals are sometimes chrome plated in order to avoid corrosion caused by caustic fuel residues in the lubricating oil.

Fig. 15-11. **Oil holes being chamfered with abrasive stone in high-speed grinder.**

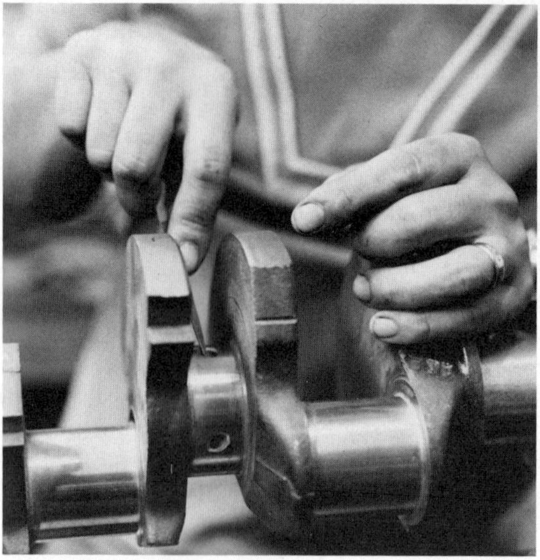

Fig. 15-12. Chamfered oil holes being radiused and smoothed with fine file.

Stock connecting rods can be used for mildly supertuned engines, but for anything approaching a full-race modification, something stronger must be used. Sometimes stock rods are "boxed", which consists of welding steel covers over the I-beam parts of the rods. Boxed rods are stronger, but they are also heavier. For most racing applications, special rods are used (Fig. 15-13). The forged-aluminum rods are widely employed in drag racing and oval track engines, but often they don't last as long as steel rods do. The Carillo forged-steel rods have been known to go as long as two racing seasons without attention or replacement. These are widely used in V8s prepared for road racing.

The connecting rods must be carefully balanced. Not only should each rod weigh the same amount as every other rod in the engine, but each end of the rod should weigh the same amount as the corresponding end of any of the other rods (Fig. 15-14). If there are pins in the big end bores to keep the bearing shells from spinning, the tuner must remember to install the pins when the rods are balanced (Fig. 15-15).

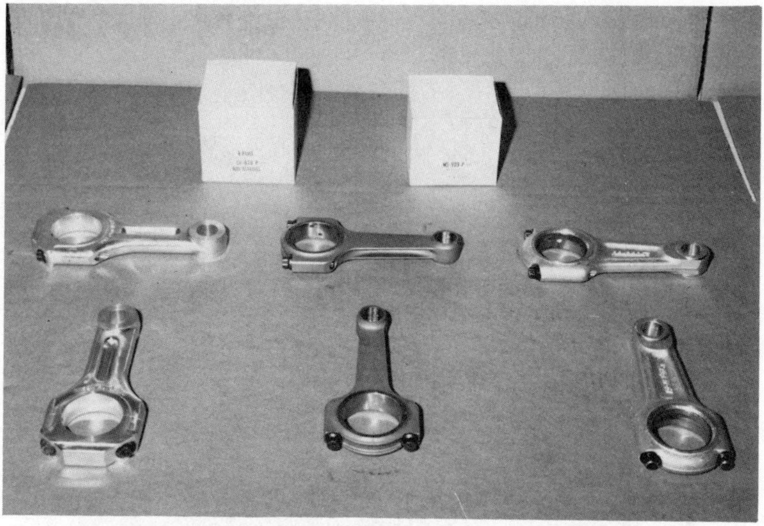

Fig. 15-13. Three kinds of high-performance connecting rods for V8 engines. At left is forged-aluminum rod with large pad for balancing. Center rod is a Carillo forged-steel rod. Forged-aluminum rod on right is contoured for better aerodynamics at high rpm.

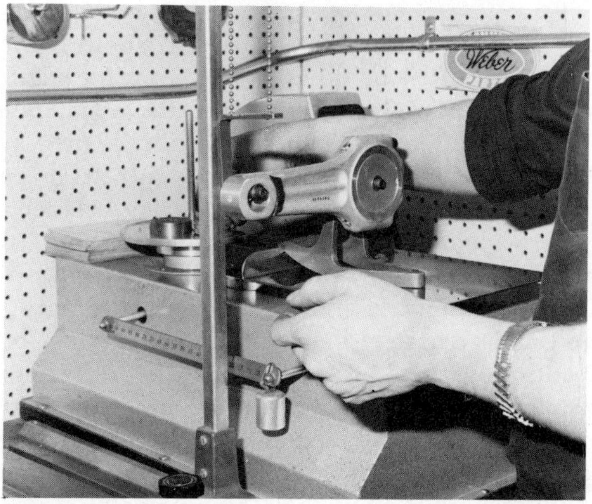

Fig. 15-14. Big end of forged-aluminum rod being weighed for balancing. Notice that little end is supported.

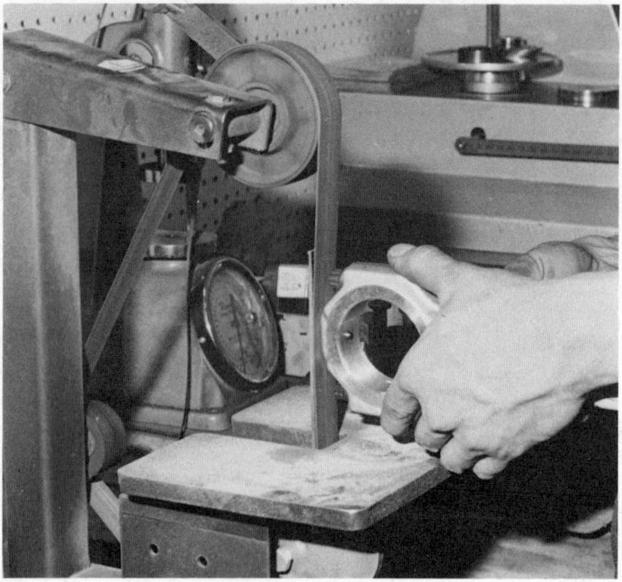

Fig. 15-15. Metal being polished from balancing pad on forged-aluminum rod. Bearing antirotation pin can be seen installed in big end bore.

If stock rods are used, it is usually not safe to lighten them greatly, which can often be done on less powerful engines. It is possible to polish the rods and to reduce the weight somewhat, for example, by grinding the flash off the sides of the I-beam section. If the stock connecting rods have rather large pads that provide places for metal to be removed during balancing, all of the rods can be lightened by removing almost all of the pads before balancing the rods. But no matter what kind of rod is used, it must be absolutely straight and the bores must be perfectly round and accurately aligned with one another.

Competition pistons for American V8s are available in a vast number of designs. Whatever a tuner needs for a particular purpose can be readily supplied from stock or can be custom made (Fig. 15-16). Full-floating piston pins are universally used, and it is extremely important that the pins be fitted to the pistons and connecting rods with precision. Racing pistons are expensive, and if a pin seizes in the piston because of inadequate oil clearance, the pin bore will gall and the piston will be ruined. Excessive clearance will hammer the bores out-of-round and can even break the piston or the rod. There have been "backyard engineers" who have tried to fit piston pins using brake cylinder hones and who have installed piston pins that had almost a full 1/16 in. of clearance in the rod little end. In every case the consequences were far more costly than it would have been to take the pistons, pins, and rods to a competition engine builder for precision fitting on a special machine.

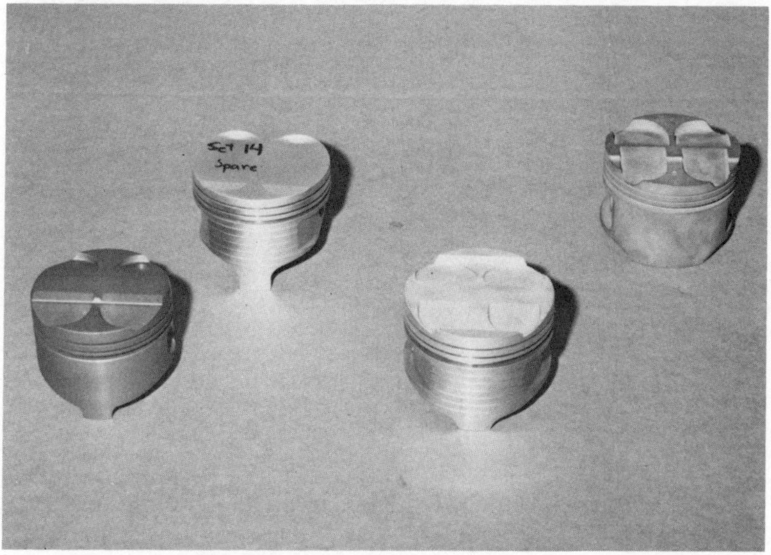

Fig. 15-16. Various kinds of racing pistons used in the Moser racing engine. Different finishes and crown shapes that give various compression ratios can be seen.

Cylinder Block

In addition to the usual and very necessary blueprinting operations, such as align boring, deck milling, and cylinder boring, V8 blocks receive a wide variety of special modifications that match them to their particular applications. For example, the Chevrolet block in Fig. 15-17 has been equipped with studs for installation of the Moser double overhead camshaft cylinder heads that are used to convert Chevy small blocks into Moser racing engines. This is a four-valve-per-cylinder conversion, and so it is not necessary to relieve the block for valve clearance, as it is when two big valves are employed.

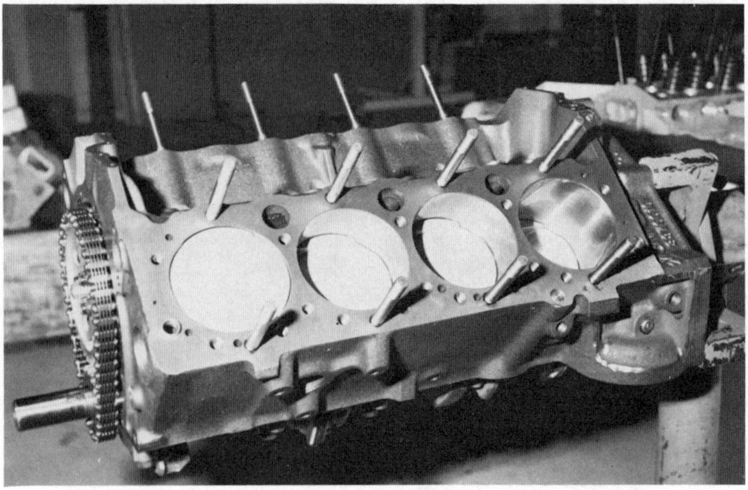

Fig. 15-17. Chevrolet 327 block prepared for installation of Moser double overhead camshaft conversion. These engines are competitive with DOHC Offy and Ford engines in USAC racing.

Valve reliefs are mainly necessary on the intake sides (Fig. 15-18). Another vital step in block preparation is to chamfer the bolt holes. This operation reduces stresses at the junction of the hole and the block surface and also prevents thread pull-up from interfering with the compression of the gaskets. Some tuners have been known to polish the lifter and pushrod gallery of the Vee in order to get better oil drain-down. Main bearing saddles (Fig. 15-19) are frequently used to strengthen the crankshaft end of the block—especially in high-output drag racing engines (supercharged or unsupercharged). Several high-performance V8s have been built with four-bolt main bearing caps, such as those shown in Fig. 15-19. In the Chrysler Hemi, the main bearing caps are crossbolted.

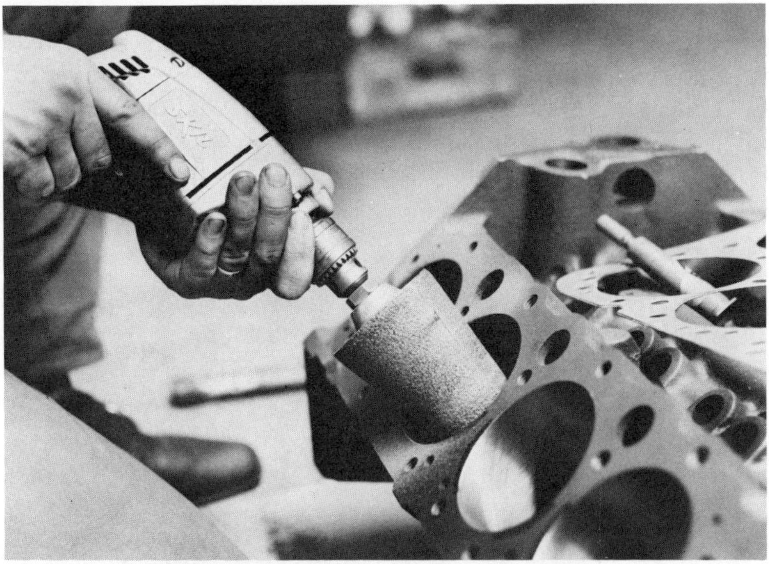

Fig. 15-18. Valve reliefs being ground into V8 block.

Fig. 15-19. Additional main bearing rigidity achieved through installation of support saddle.

For road racing and oval track racing, a dry sump system is nearly always necessary (Fig. 15-20). There are, however, baffled oil pans available for several popular V8s, and some Corvettes and Camaros use wet sumps in road racing. The high banking and left-turn-only conditions that prevail in stock racing make it possible for these engines to run with modified wet sump systems.

Fig. 15-20. Scavenge pump for dry sump system on Moser racing engine. Pump is driven by small Gilmer belt. Notice aircraft-type oil lines that connect DOHC head to sump.

Tuning

The setup of induction systems depends greatly on how the engine will be used. A large variety of different carburetors, injection systems, and supercharger installations is used in drag racing (Fig. 15-21). Road racing engines are generally equipped with multiple Weber carburetors, but fuel injection is also used in some installations. Fuel injection is always used on oval track race cars, and stock cars are in most cases limited by the rules to a single carburetor.

Fig. 15-21. Roots supercharged drag racing V8 with "bugcatcher" type fuel-injection unit. Throttle valves can be seen in three ports in air intake scoop atop blower.

Dynamometer testing is usually employed for selecting the correct carburetor or fuel injection jets, and so forth. However, drag racing engines—particularly the big supercharged nitro-burners—cannot be tested on a dynamometer. These engines simply will not survive a run of more than a few seconds at full throttle, and so any tuning that is not learned on the drag strip is not learned at all.

The trend is toward electronic ignition in all forms of racing. But V8s with magnetos and with conventional coil and battery ignitions are still very common. As is the case with carburetor jetting, optimum ignition timing is something that can best be arrived at on a dynometer. Dynamometer testing is, of course, the only way to determine the correct advance curve for the distributor. After the ideal degree of advance has been determined on the dyno for every racing rpm, the distributor is installed on a distributor testing machine and calibrated to deliver the correct advance at each speed (Fig. 15-22).

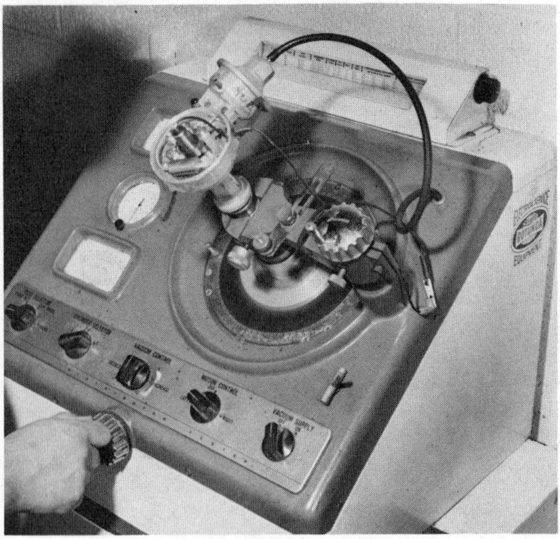

Fig. 15-22. Distributor spark advance curve being determined on a distributor testing machine.

In an age when gasoline is becoming scarce and emissions are measured in tons per mile, the future of the big American V8 as a passenger car engine is in doubt. Its future as a racing engine is not in doubt, however; there is no cheaper way to obtain vast amounts of power. Even if the supply of gasoline ran out tomorrow, it would not restrict the big V8s that use alcohol/nitromethane fuels. These fuels, of course, can be used for any kind of racing and are a renewable resource. So if the age of the big-inch V8 comes to an end, it will not be because the oil ran out; it will be because we have run out of parts, and that sad day fortunately lies many years in the future.

16 / *Formula Ford*

The Engine

Formula Ford racing started in England in about 1967 and arrived in America shortly thereafter. So far, at least one world champion driver began by racing Formula Fords, and undoubtedly many more will follow. Though the modern Formula Ford chassis is fairly sophisticated and expensive, the engine itself is not. A majority of the engines found in Formula Ford racers were rescued in part from junkyards. And although a fully prepared Formula Ford powerplant costs $2500 to $2600, this is still cheaper than a competitive Formula Vee engine. (see chapter 19.)

Two kinds of Ford engines are permitted by the rules, and both are versions of the English Ford "Kent" engine. One of these is referred to as the *Cortina GT*, which, as its name implies, comes from a 1968, 1969, or 1970 English Ford Cortina GT car. The other one, the *uprated,* is the engine used in 1971, 1972, and 1973 Pinto 1600 cars. See Fig. 16-1 and Fig. 16-2.

The Cortina GT engine has a 10.0:1 compression ratio, and the uprated has a 9.3:1 ratio, but the uprated engine is more powerful. This power difference is compensated for by the rules, which require that cars with uprated engines carry an additional 50-lb. handicap. A good uprated will produce 111 to 112 bhp, but some tuners are beginning to approach this figure with the Cortina GT. So, while the majority of tuners are still working with uprated engines, there has been a recent trend back to the Cortina GT. Fifty lbs. is a considerable weight penalty for cars in this class.

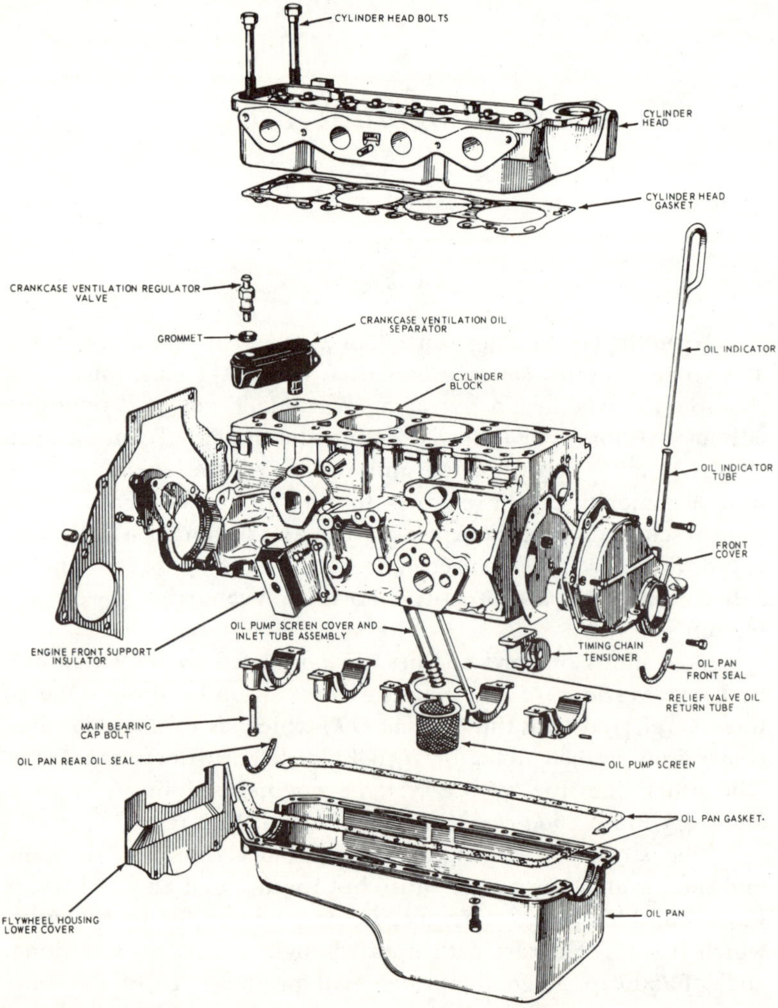

Fig. 16-1. Exploded view of main castings of Ford uprated engine.

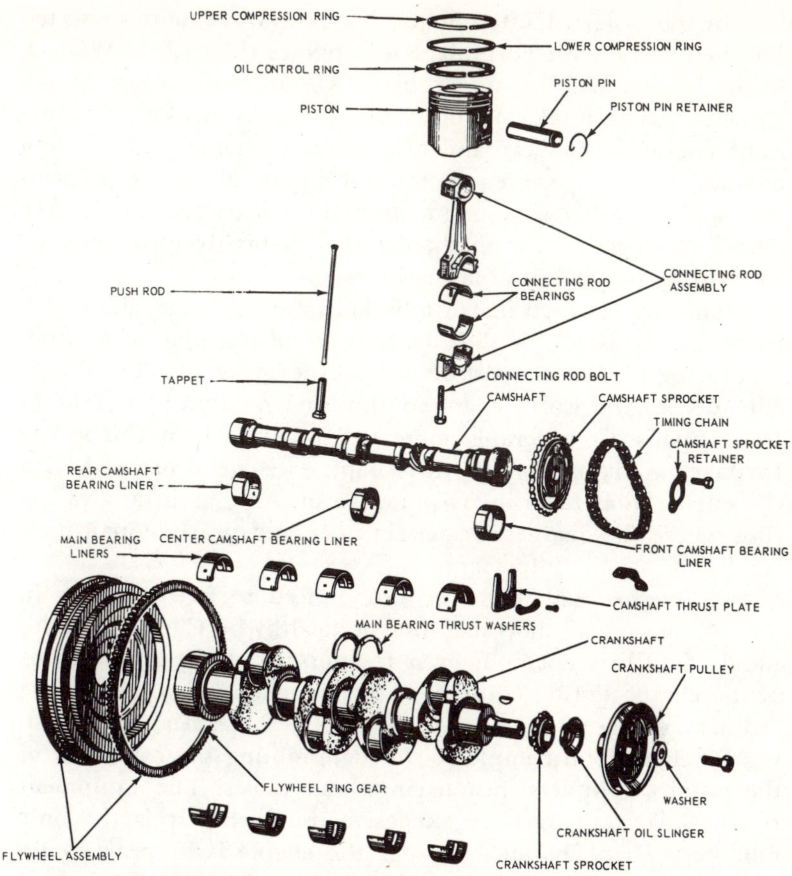

Fig. 16-2. **Moving parts housed in block of Ford uprated engine.**

Precision Assembly

Whereas Formula Vee is definitely an engine tuners' class—tuners frequently have more to do with which car wins than do the drivers—Formula Ford rules are such that actual tuning requires neither a great deal of machine shop work nor a search for rare production components. Precision assembly has more to do with whether a Formula Ford engine wins or blows up than any esoteric supertuning trick.

In the eastern United States, few people are more respected for their work with Formula Ford engines than Chris Wallach of the Marblehead Racing Group (MRG) in Marblehead, Massachusetts. Chris totally rebuilds twenty to twenty-five Formula Ford engines each year and also works on many other racing engines. He has mastered all the fine points of engine preparation and assembly (not to mention the tricks) and conducts a school for formula car mechanics that is highly recommended for anyone new to formula machines and engines.

The engines used in Formula Ford are, of course, descended from the "80-bore" English Ford powerplants that revolutionized racing during the past decade. In the 1970s, the "bloodline" extended to the world championship and finally to the pole at Indianapolis. Considerably more will be said about this racing heritage in chapter 17. In this chapter we are concerned with the engine as it races in its purest form. And, as in any racing class where the engines are nearly stock and nearly equal, their conditions are all important.

An engine that might be pronounced perfectly healthy in the average repair shop may not be healthy by Chris Wallach's standards. Chris does not even recognize the existence of some of the engine testing equipment that is taken for gospel by the ordinary garage mechanic. At the MRG shops, leak-down testing is relied upon completely for determining the condition of the pistons, cylinders, piston rings, and valves. The equipment required is comparatively expensive, but using it is the only reliable way to find out whether the engine is in peak condition. Peak condition is what wins races in Formula Ford.

Leak-down testing differs vastly from an ordinary compression check. The tester itself is an aircraft engine tool. Very often it will turn up burgeoning troubles that no compression tester could ever find, such as a cylinder head gasket that has developed a weak spot. When one considers the problems that a blown gasket can cause during a race, the extra precision involved in leak-down testing seems well worth the effort.

A compression tester can be used only with the engine turning. Consequently variations in cranking rpm, throttle opening, and similar factors can have considerable influence on the gauge reading. Furthermore, exceedingly minor leakage will not register at all. Leakage, of course, means lost power, or, as in the case of the head gasket, a probable dnf. Even if a compression gauge does indicate a dubious compression pressure for a cylinder, it is impossible to determine the exact site of the leakage with any precision.

With a leak-down tester, the engine is tested while its moving parts are stationary. The pressure pump on the test equipment is used to raise the pressure in the cylinder to a predetermined point. If there is leakage, the gauge will descend noticeably within a given time period. More important, the site of a leak can be determined precisely by the sound of escaping air. If air is heard escaping at the carburetor intake, for example, an intake valve is leaking. Similar sounds from the exhaust pipe indicate a leaking exhaust valve. Hissing sounds from the crankcase point to faulty piston rings, and a gradual descent in the pressure without audible leakage should prompt the tuner to check for bubbles in the coolant or to remove the spark plugs and listen for leakage to adjacent cylinders.

Assembly techniques are outside the scope of this book and are better left to the MRG school. In this chapter we will concentrate on what must be done to the Ford pushrod 1600 engine in order to make it into a Formula Ford competition engine (Fig. 16-3). It must be emphasized that, while virtually all of the

Fig. 16-3. Formula Ford engine with valve cover removed.

parts in the engine are stock Ford components, they are in most cases modified. But though you cannot take an engine straight from the junkyard to the race track, neither can you do anything to it that you please. Every modification is closely governed by the rules of the Sports Car Club of America. And no work should be carried out on a Formula Ford until the General Competition Rules (GCR Book) and recent issues of *Sports Car,* the SCCA journal, have been carefully studied.

Cylinder Head

Cylinder head preparation is the most important part of Formula Ford engine building. Though the combustion chamber is in the piston, only the uprated head has a perfectly flat surface; the earlier Cortina head has a slight combustion chamber that shrouds the valves somewhat. Because no milling of the Cortina GT head is permitted by the rules, this design feature accounts for some of the power handicap.

Ports can be reshaped by the removal of metal so long as the GCR sizes are not exceeded (Fig. 16-4). In 1977, the intake ports could be no larger than 1.42 in. in diameter and the exhaust ports could be no more than 1.16 in. in diameter when measured at the manifold faces of the head. These dimensions are very near the blueprint diameter. So, keeping in mind that the narrowest point in any induction system should be at the valve, it is not effective to enlarge the seat so that the valve is seated at its extreme periphery, as described in chapter 15 in connection with American V8 engines. On the contrary, in a Formula Ford engine the valve is seated at the innermost part of the valve facing so that when it is closed most of the valve head stands proud of the cylinder head surface.

This valve seating, in addition to keeping the port diameter correct for best gas velocity, also produces the optimum gasflow. The valves, being almost completely unshrouded, even by the seat, allow maximum gasflow almost from the moment that they begin to open. The effect is nearly the same as that of increasing the cam duration (which is illegal), and because this high valve seating accounts for much of the power that can be derived, every effort is made to preserve it.

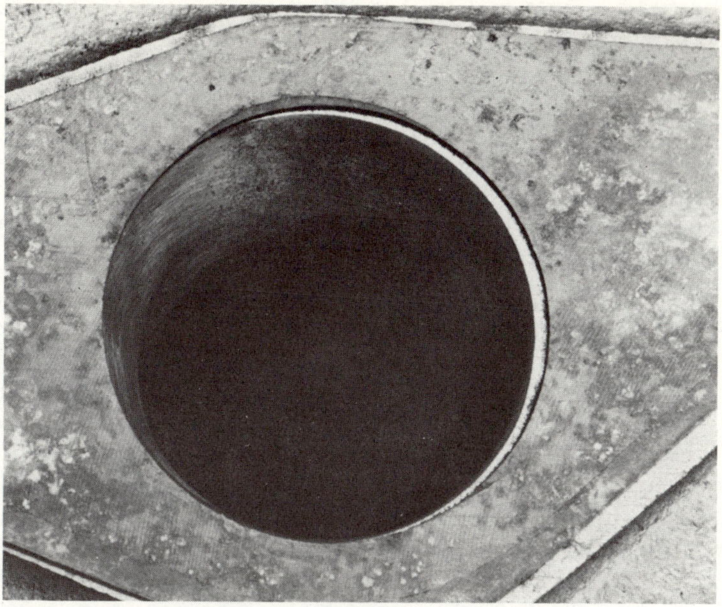

Fig. 16-4. Polished, slightly enlarged intake port of Formula Ford engine.

The uprated head can be milled as part of each valve grind so that the high valve seating is maintained. But because the Cortina GT head cannot be legally milled, the cylinder head must be discarded if much metal has been removed from the seats during reconditioning. On either kind of head, it is extremely important to remove as little metal as possible when cutting the seats. This is especially true in the case of the uprated head; each milling reduces the gasflow because of the gradual change of the port shape. Ultimately, uprated heads must be discarded.

The standard head gasket is specified because its thickness has an influence on the compression ratio and because the rules assign a fixed value to gasket volume when compression ratios are being checked. The only additional modification is that it is permissible to ream the integral valve guides for Ford valves with oversize stems (which are heavier) or to install press-fit valve guides in place of the integral guides. Guide replacement is necessary sooner or later because of an important valve gear modification that greatly increases valve guide wear.

Intake and Exhaust

The intake ports and the exhaust ports can be matched to the intake manifold and to the exhaust headers, but this work must not enlarge the port diameters beyond the previously given dimensions. The exhaust system itself is not limited in any way by the rules, and though a great deal of development has been carried out in this department, there is room for further experimentation.

The carburetor flange of the intake manifold must be milled so that the carburetor is level (Fig. 16-5); in a formula car, the flywheel end of the engine is not lower as it is in a Cortina or Pinto sedan. Many tuners apply this rule with a vengeance, milling as much from the manifold as they can in an attempt to get the carburetor venturis closer to the valves. The power gain is mainly imaginary.

Fig. 16-5. Carburetor and intake manifold. Notice how top of manifold is flattened by milling, thus lowering the carburetor as much as possible.

A number of other tricks have come and gone because they too offered gains that were purely imaginary. One of these was a kit that could be installed on the carburetor so that both throttle valves would open simultaneously. This modification, though legal, produced no measurable power on the dynamometer and had the disadvantage of making the car impossible to drive from the paddock to the track. The moment the throttles were opened at low rpm, the engine simply died. Another fad was velocity stacks for the carburetor. Some of these were ingenious in design, with two throats of different lengths to "compensate" for the uneven lengths of the intake manifold branches. For several years, every engine had them (Fig. 16-6); by 1977 they were out of favor and rarely ever seen.

Fig. 16-6. Formula Ford engine equipped with velocity stacks on carburetor. Notice cast-aluminum valve cover.

The inside of the intake manifold can be cleaned up and enlarged so long as the GCR measurements are not exceeded. In addition, the water-heated intake manifold feature is scrapped, for obvious reasons, and the water holes in the manifold are plugged. The carburetor itself can be rejetted in any way, though the venturis cannot be changed or modified. This work should be done on a dynamometer.

Valves and Valve Gear

The camshaft must be the standard Ford camshaft for the engine, and it cannot be reworked in any way. The rules on this are very strictly enforced. However, the valve timing can legally be retarded up to about 2° for better high rpm power by means of an eccentric pin used for locating the timing chain sprocket on the camshaft. One of the few nonstandard Ford parts is the timing cover shown in Fig. 16-7, which permits a mechanical tachometer to be driven from the end of the camshaft.

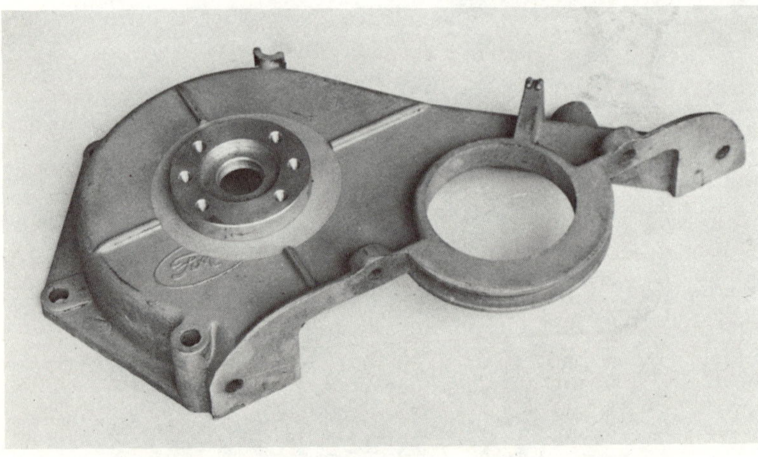

Fig. 16-7. Timing cover with hole for tachometer drive. Flange, atop cover, actually goes inside cover and is bolted to end of camshaft.

The valves are usually replaced instead of refaced. There are two very good reasons for this. First, grinding metal from the valve facings causes them to seat deeper in the cylinder head and this must be avoided at all costs. The second reason is that the Ford valves have a circumferential ridge just inside the narrow-diameter part of the valve facing. Though this ridge is probably no more than 1/64 in. high, it disrupts the gasflow. Nevertheless, the rules forbid that it be ground off, and refacing the valves can remove all or part of the ridge—thus causing a rules infraction that can disqualify a car.

Intake valve lift can legally be increased to .356 in. and exhaust lift to .358 in.—measured at the spring retainer with the valve clearance adjusted to zero. This modest increase is obtained by modifying the contour of the part of the rocker arm that contacts the valve stem. This modification, called profiling, causes the rocker arm to contact the valve off center and increases the distance that the rocker arm slides across the valve stem as the valve is forced open. The result is rapid valve guide wear, and many tuners choose to install press-fit guides from the very first so that replacement will be simpler when the wear becomes excessive. This price is a small one to pay because profiling is a vital source of increased top-end power.

All of the pushrods in the engine can be reduced to the minimum weight specified. Otherwise the valve gear components must be stock, aside from profiling the rockers. For racing, the valve clearances are usually set near .010 in. for intake valves and .015 in. for exhaust valves, with the engine hot.

Crankshaft, Connecting Rods, Cylinders, and Pistons

According to Chris Wallach, there is no real power to be found in the bottom end of the Ford engine (Fig. 16-8). However, a great deal can be done to improve reliability, and reliability can win races that speed will not. Crankshafts have been a definite trouble area. The rules now permit a steel center main bearing cap to be used on Cortina engines in place of the stock cast cap; uprated engines do not need this, and it is not permitted by the rules. Crankshaft breakage occurring at the crankcase end of the rear main bearing journal has been the main problem on both engines. When this happens at full throttle, as is usual, the engine instantly overspeeds and is destroyed.

Fig. 16-8. Bottom end of Formula Ford engine.

Crankshaft reliability is improved by skillful preparation, careful assembly, and limiting the crankshaft's service life to no more than fifteen races. The first step in preparation is magnafluxing, followed by Tuftriding and straightening. The crankshaft is indexed, together with the rods, so that the piston strokes will be uniform in length and at exactly 180° intervals. Both the crank and the rods are then polished sufficiently to relieve stresses (Fig. 16-9). Finally everything—crankshaft/flywheel/clutch, pistons, and connecting rods—is carefully balanced, usually in conjunction with the polishing operation.

The crankshaft and the connecting rods can be lightened to the minimum weight specified by the rules. The flywheel cannot be lightened, but it can be drilled to accept a Formula 3 clutch (Fig. 16-10), which saves four pounds and moves mass toward the center of the flywheel where it has less inertia. In 1976, flywheel lightening was seriously considered as a rules change to help reduce the threat of crankshaft breakage. But it

Fig. 16-9. Rod big end and crankthrow. Notice light polishing used for balancing and relieving stresses.

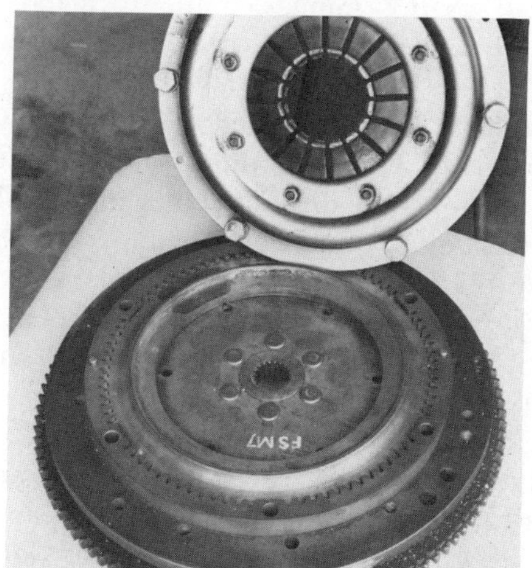

Fig. 16-10. Formula Ford flywheel equipped with small-diameter Formula 3 clutch.

was decided, at least for the present, that crankshaft reliability was not an insurmountable problem if the engine were correctly prepared and if the driver did not use techniques that placed abnormal acceleration and deceleration loads on the crankshaft.

The main bearing and connecting rod big end clearances are .0025 in. to .0030 in. in a newly prepared Formula Ford engine. Chris Wallach has long preferred Vandervell bearings because they seem to be more forgiving of foreign matter, which, if not absorbed into the bearing shell, can increase friction and cause score marks on the journals. Recently, however, Teflon bearings have begun to come into favor. Because of their low friction, the bearing clearance can be reduced. Therefore, less power is taken from the engine to drive the oil pump.

The English Ford engines are blessed with an external oil pump so that a wide variety of pump designs can be accommodated easily. A dry sump system is used for Formula Ford racing, making this very compact engine even more diminutive and ideally suited to formula cars. A pressure pump is installed in the normal oil pump location (Fig. 16-11), and an additional scavenge pump is fitted—often driven by a small Gilmer belt, as is the water pump. (Some competitors have used electric water pumps, but here again the power gained is probably imaginary.)

An oil radiator is always used (Fig. 16-12), which helps to preserve bearing life. Other modifications include plugging the crankcase ventilation and dipstick holes in the block, installing a ventilation hose and connection on the rocker arm cover, and cutting off the oil filler and welding a plate in its place. The oil pressure is maintained at 40 to 50 psi hot, in fact, *very* hot—90° to 95°C. Hot oil results in less oil drag and consequently more power.

Standard Ford pistons are required by the rules. The rings can be of any manufacture, so long as two compression rings and one oil scraper are used and the pistons are not reworked in any way. The cylinder blocks used are the Cortina GT 1968–1970 and the uprated block with the part number DIFZ-6010-C. The uprated block cannot be bored for oversize pistons; the Cortina block can be bored .030 in. oversize. If the cylinders wear beyond this diameter, either the block must be replaced

Fig. 16-11. Oil pump used with dry sump system, installed in place of stock pump.

Fig. 16-12. Oil radiator mounted at rear of Lola Formula Ford. Notice modifications to valve cover.

or dry-type liners must be installed and bored to the prescribed diameters. Either block can be honed as required in order to obtain the optimum .006-in. piston clearance with stock pistons. Blocks are always align bored, and this operation may include the camshaft bearings also.

Tuning and Maintenance

The valves should be ground every third race and, on the uprated engine, the cylinder head must be shaved as described earlier. Piston rings last 12 to 15 hours, and the crankshaft should be discarded after 15 races.

The flywheel bolts tend to shear after long service—especially if the driver uses harsh clutch engagement with the engine speed and the transaxle mainshaft speed not carefully matched. Shifting techniques are becoming rather artless in today's formula cars. The kinds of gears and synchronizers used make it possible to change gears without fully disengaging the clutch or without using the clutch at all.

The modern technique should not be confused with traditional no-clutch changes made by matching engine and mainshaft rpm; the common procedure now is to move the gear lever as quickly as one can and rely on the machinery to match the cogs. The whole thing is as easy to learn as it is hard on machinery. Or, as D. B. Tubbs said in another context, in the Barron-Tubbs book *Vintage Cars,* "Great efforts were made to offer a foolproof gear-change, but, as always, this produced only a new kind of fool."

The stock Ford flywheel bolts are entirely satisfactory. However, they should be discarded and replaced by brand new bolts at every tear-down—or whenever the transaxle or engine is removed, making the bolts readily accessible. New bolts should, of course, be used during initial preparation, as well as in the course of maintenance. Bolts that have seen service on the highway are no better than those that have had several hours on the race track.

The normal racing rpm range is from 5000 to 6400 rpm (only an inept driver will attempt to race below 5000) . Consequently, it is in this 1400 rpm range that the ignition timing must be absolutely correct; at lower speeds it can be approximate. The 6400 rpm limit can be exceeded without valve float; but no more power is obtained, and a change in gearing is more in order under these conditions than any splendid efforts by the engine tuner.

A stock Ford distributor must be used, and the best reliable power is obtained with 41° to 42° of total ignition advance. For an important professional race or in the SCCA annual run-offs, the total advance may be increased to as much as 45°. This much advance burns things—pistons, for example—but it does offer somewhat better high-rpm power.

The distributor's centrifugal advance should be calibrated to produce the correct advance for best power at 5000 rpm and at every rpm up to the 6400 rpm redline. This requires stiffer advance springs, and the advance must take place smoothly throughout the 1400 rpm racing range, without sharp changes in the curve.

Formula Ford engines will run on good pump gasoline without preignition or detonation. Nevertheless, wise competitors use an octane improver, which is legal so long as it does not alter the specific gravity of the fuel. Higher octane not only is insurance against detonation induced by overly enthusiastic spark timing, but also is preventive medicine should hard cornering or foreign matter in the carburetor cause the mixture to lean-out during a race.

There is probably no better or cheaper way to begin a professional racing career than in the Formula Ford class. Secondhand cars are not costly; brand-new cars in the most competitive designs are, although the engine is relatively cheap and has a very long lifetime when it is expertly maintained. Formula Ford is a class for real race cars, with professional events for those who have exceptional skill.

17 / *Cosworth*

Choice of Champions

Ken Duclos, president of Kay-Dee Automotive Engineering in Westford, Massachusetts, speaks almost with reverence when he talks about Cosworth engines—with good reason. Duclos has been SCCA N.E. Division Formula B Champion from 1969 through 1976, with the exception of 1971 when he did not race. He won these championships (plus two national championships) using a Kay-Dee Engineering Cosworth BDA. In 1975 he won the national championship by a full three seconds, leaving the rest of the field behind.

Other drivers have won championships with Kay-Dee engines, which proves that their success is not dependent on who is driving. Nor are Kay-Dee powerplants the only Cosworths found among the leaders. Cosworth engines are excellent pieces of raw material, and whether they are refined into merely good Cosworths or into excellent ones depends a great deal on who has assembled and tuned them.

Many drivers have bought Cosworth BDAs in kit form and assembled them in their own garages, at a savings of $1000 to $2000 over what a similar engine would cost after it has been professionally prepared. These home-built engines, provided the amateur tuner knows what he or she is doing, can easily be winners—until they come up against a trick powerhouse from a professional shop.

History

If you take the cylinder head, the moving parts, and the accessories away from a Cosworth BDA (Fig. 17-1), you will find the Cortina GT block used in Formula Ford racing. There is a reason for this. The BDA is directly descended from the "80-bore" English Ford engines that first appeared at the end of the 1950s and were, by the mid-1960s, outracing the other small engines on the race track.

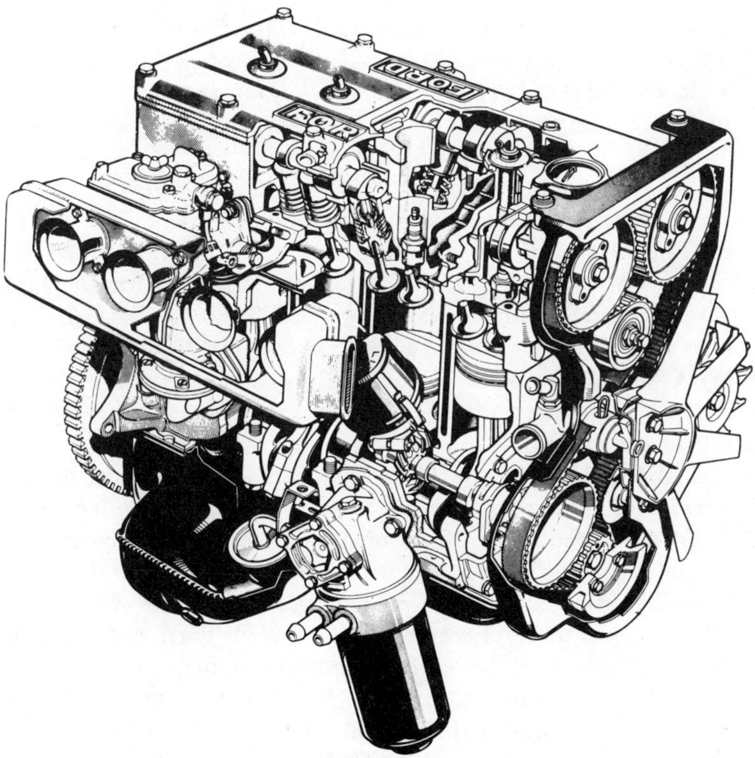

Fig. 17-1. The Ford BDA DOHC engine as installed in the Escort RS 1600 car. Cosworth version has dry sump, detail differences.

The earliest Cosworth development of the Ford 80-bore was the 1-liter MAE, which retained pushrod valve operation. This engine seems to have provided Ford with the inspiration to develop its own high-performance pushrod units, which culminated in the Cortina GT of Formula Ford fame. Along with the MAE came the Cosworth SCA. The "SC" stands for "single cam," and this SOHC unit of 1-liter capacity is still active in American club racing.

The Lotus Twin-cam (Fig. 17-2) was designed by Lotus for the Elan and executed by Cosworth. Primarily this is a single-port-face Cortina GT engine, bored to 82.55 mm in order to obtain a displacement of 1558 cm^3 and equipped with a twin-cam cylinder head. It soon replaced the Cortina GT as the competition engine in Cortina sedans. The resulting cars were known first as Lotus 28s and later as Ford Lotus Cortinas. The fully developed Lotus Twin-cam with about 140 bhp was soon a dominant engine in small formula cars and in sports racing cars.

The MAE, SCA, and Lotus Twin-cam have been rendered obsolete by the present Cosworth four-valve-per-cylinder engines. The first of these was the FVA. ("FV" stands for "four valves.") This engine, which dominated Formula 2 in the late 1960s and early 1970s, displaces 1594 cm^3 and has gear-driven double overhead camshafts. The Cosworth DFV (double four valve) Grand Prix engine is essentially two FVAs siamesed. (Of course the crankcases for these engines are not standard Ford passenger car parts.)

At about the time the FVA was developed, the Cortina engine was undergoing its redesign for the crossflow cylinder head. Also, Ford was tooling up for a new car that was slightly smaller than the original Cortina, the latter having grown in Mk. III form to almost compact car proportions. The new car, called the Escort, eventually became the standard Ford competition model, and some of the early high-performance Escort sedans had the Lotus Twin-cam engines. But the lure of the FVA's performance potential led Ford to approach Cosworth for an economical version of the successful four-valve engine for use in the Escort RS 1600.

The new engine was built around the Ford Kent crossflow

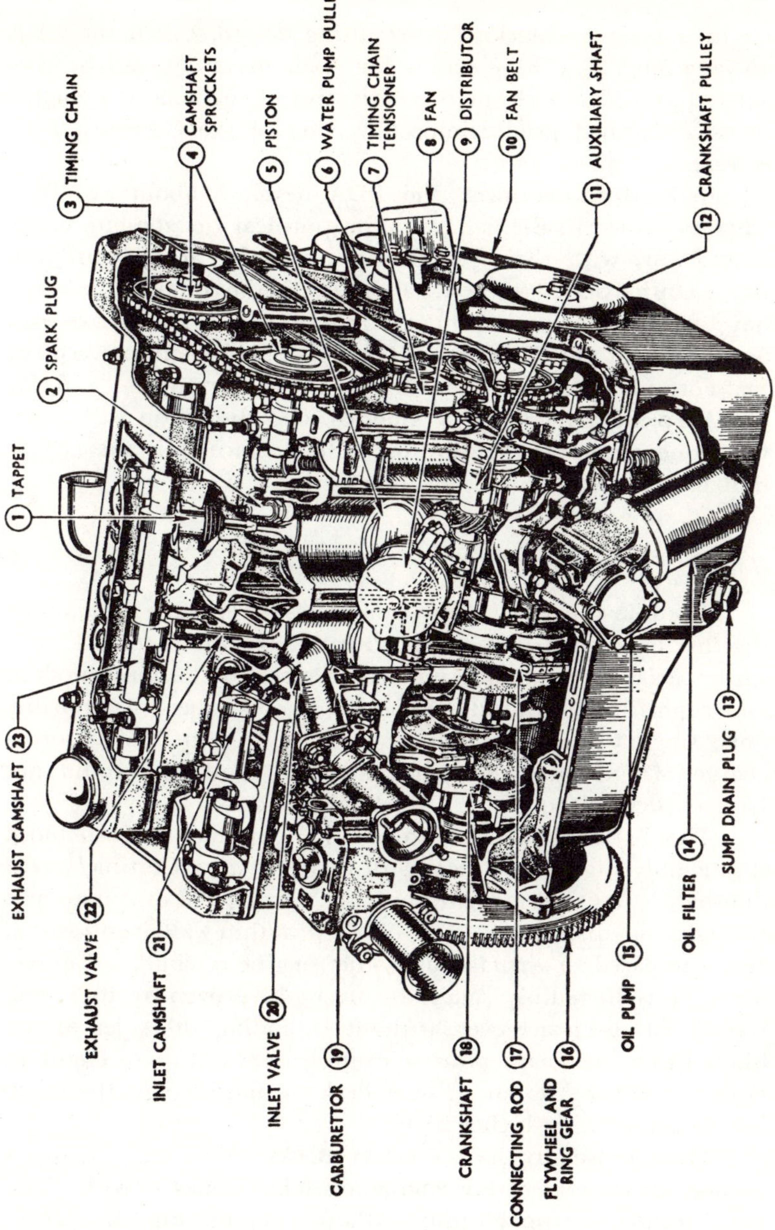

Fig. 17-2. Ford Cortina Lotus engine, in part section.

engine's cylinder block. The resulting design, called the BDA 16-valve engine ("BD" stands for "belt driven"), can be considered as a high-performance derivative of the 1600 GT engine or as a detuned productionized version of the Cosworth FVA engine.

In Ford Escort form, the BDA develops about 120 bhp, whereas Cosworth BDAs have an output near the 200 bhp mark. The FVAs, with their gear-driven camshafts, are seldom seen in the United States (there are about five in the country), and they have fallen from their pinnacle in Formula 2 racing overseas. So at present the Cosworth BDA is the dominant powerplant for Formula Atlantic (no fuel injection permitted) and for SCCA Formula B racing in America. Our discussion will therefore concentrate on the Cosworth BDA, since it is the engine that most readers will be dealing with.

Racing Thoroughbred

The four-cylinder English Ford engines represent a bloodline that has extended, since the late 1950s, from the humblest backyard specials ever to win important races to the most modern and sophisticated Grand Prix cars. Though there is probably not a single Cortina or Escort part in a Cosworth DFV or turbocharged DFX, these successful Cosworth Ford V8s would not have existed without the little English Fords.

The BDA, in addition to having a production Ford block that is only slightly modified, still carries a stock Cortina/Escort camshaft as an intermediate shaft for driving the oil pump and the ignition distributor. The block is a thin-wall iron casting, and, compared to virtually any other engine of similar displacement, it is incredibly small. Its strength, proven by its racing record, has been achieved without extending the sides of the block below the main bearing centerlines—certainly a credit to Ford engineering. From cylinder deck to sump flange, the block is only about nine inches high.

Though usually spoken of as BDAs, there are actually a number of versions of this engine available from Cosworth. The most popular is the Formula Atlantic engine, known in Cosworth catalogs as the BDD. It is equipped with Weber carburetors and retains the standard Ford Escort valve diameters (Fig. 17-3). The BDG is an approximately 300-bhp fuel-injected version that uses bigger valves and has a special aluminum alloy block that gives it a 2-liter piston displacement.

Fig. 17-3. Fully assembled Cosworth BDA ready to be installed in car.

The BDH is a 1300-cm³ version with the Ford iron block, Escort valve sizes, and carburetors. It is used in sports racing cars, such as in the SCCA DSR class. An 1100-cm³ engine, designated the BDJ, is available for Formula C, and this unit has appeared both with carburetors and with fuel injection. Returning briefly to the Cosworths with gear-driven camshafts, there is, in addition to the 1594-cm³ FVA, the 1790-cm³ FVC. Cosworth, however, has found that the belt drive is in many ways superior to the gear drive. In addition to being simpler, cheaper, quieter, lighter, and more quickly assembled, it is equally reliable and less subject to valve timing irregularities caused by wear.

The BDA needs very little maintenance. It will, barring disaster, go five or six racing hours between teardowns. It is fully content at 9400 to 9500 rpm and never gives serious problems so long as it is assembled in clean surroundings and to Cosworth specifications. Ken Duclos insists that the engines should always be run in under load on the dynamometer before they are exposed to the irregularities of racing. If nothing else, this ensures that adequate cooling will be available during the vital break-in period.

The older Lotus Twin-cam engines had some cylinder head sealing problems in highly tuned form. The present Ford blocks have a stiffer deck, which, together with the design of the BDA cylinder head, has eliminated that kind of trouble. Thus, the BDA has no chronic weaknesses, and being a Cosworth product it could justifiably be called the finest engine in the world for its particular kind of racing. The Cosworth specifications have proven to be entirely correct for the engine, and the only tuners who get into trouble are those who try to outguess the engineers.

Cylinder Head

The construction of the cylinder head is interesting (Fig. 17-4). It is designed as a three-layer sandwich to facilitate casting and machining and to minimize replacement costs. It also makes valve adjustments easier. The lowest layer, an LM 8 aluminum casting, contains the combustion chambers and ports and houses the valves with their single springs (Escort RS) or double springs (Cosworth BDA). The second layer, or cam tray, is also cast in LM 8 and provides bearings for the two camshafts and machined bores for the bucket-type cam followers. The top layer is the single-piece cam tray cover, a magnesium alloy casting. Though the designs of the FVA and DFV/DFX cylinder heads are similar, these engines have removable bearing caps and bearing shells for the camshafts; on the BDA the camshafts are inserted from the end and run directly in the cam tray casting.

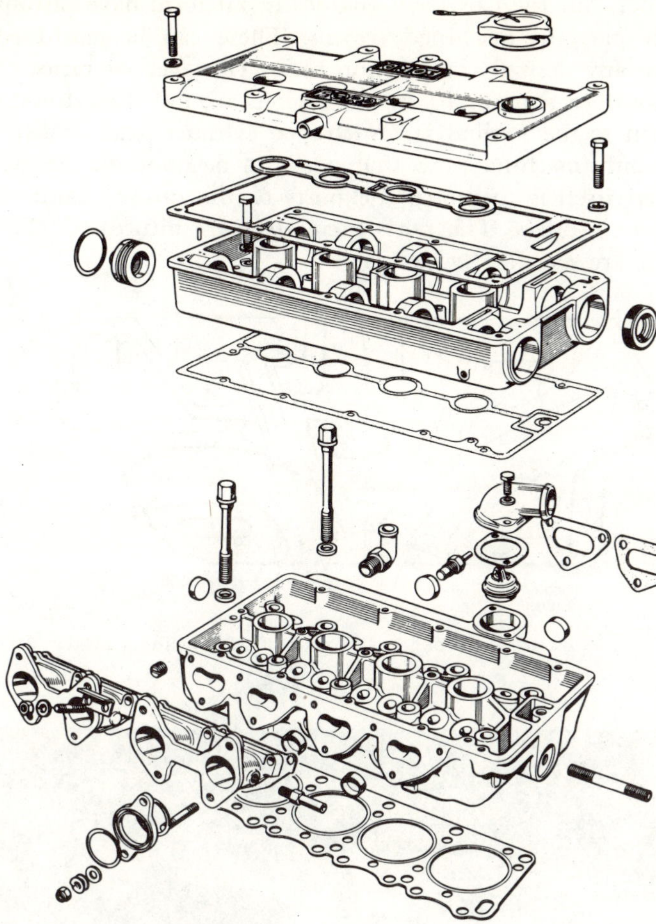

Fig. 17-4. Exploded view of BDA cylinder head assembly. Notice separate cam carrier, or cam tray, with its full-ring camshaft bearings.

The valves (Fig. 17-5) are of 1.22-in. (intake) and 1.0 in. (exhaust) head diameter arranged at a 40° included angle, with vertical spark plugs mounted centrally in each pent-roof combustion chamber. With standard Escort pistons the compression ratio is 10:1, but the Cosworth engines in kit form have pistons with only partially machined crowns. These can be machined to achieve any desired compression ratio. However, at ratios of 12:1 or above, it is necessary to use O-rings around the cylinders in addition to the normal sandwich-type cylinder head gasket.

The only machine work that needs to be done on the cylinder head itself is to match the ports to the intake manifold and exhaust headers. The ports themselves are bifurcated (Fig. 17-6) and are excellently designed.

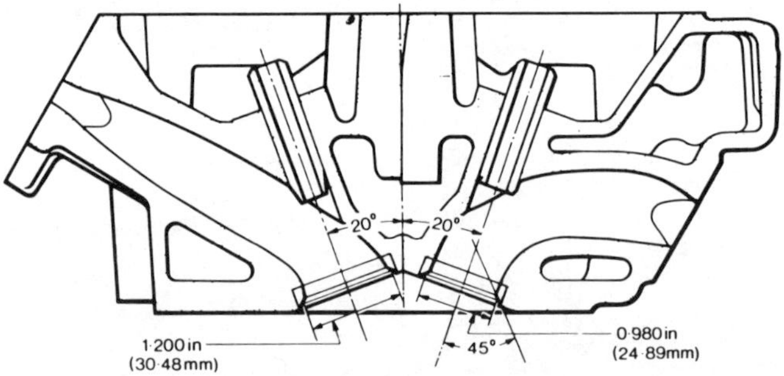

Fig. 17-5. Cross-section through valve seats of BDA cylinder head.

Fig. 17-6. Bifurcated ports in Cosworth sixteen-valve cylinder head.

Valve adjustment is accomplished by means of shims inserted between the cam followers and the valve stems. On the Lotus Twin-cam engines, the same system was used. However, it was necessary to remove the camshafts before the cam followers could be lifted out for access to the shims (Fig. 17-7). On the BDA, the cam tray (Fig. 17-8) can be lifted off with the camshafts installed in it for access to the adjustment shims. If the engine is on an assembly stand, the head can be placed upside down so that the cam followers will not fall out of the cam tray. If the engine is installed, a package of sixteen bar magnets is available for holding the cam followers in place as the cam tray is lifted off.

Fig. 17-7. Partially disassembled Lotus head, showing how camshafts must be removed before cam followers can be taken out.

Fig. 17-8. Cosworth BDA cam tray, showing machined bearings and cam follower bores.

Cylinder Block

It is nearly impossible to tell a Cosworth BDA block from that of a stock Cortina GT, although the cylinders, the deck, and the front surface of the Cosworth block have a much finer finish and are perfectly uniform. The main bearing bores and the intermediate shaft bearing bores are machined to Cosworth standards, and the cylinder bores are at right angles to both the deck and the main bearing centerline.

On engines raced with higher compression ratios, O-ring grooves are cut around each cylinder bore (Fig. 17-9). O-rings eliminate high compression sealing problems, which are always a possibility when an aluminum head is used on an iron block, or vice versa. Ken Duclos O-rings his engines with copper wire first cut to length and then inserted in the grooves. The wire is compressed as the cylinder head is torqued down, augmenting the sealing capability of the regular head gasket.

The glass-fiber-reinforced camshaft drive belt engages no fewer than four sprockets: crankshaft, two overhead camshafts, and the semiredundant pushrod-engine camshaft. In addition, two idler pulleys bear on the back of the belt, one to provide tension

Fig. 17-9. O-ring grooves around cylinder bores in Cosworth block.

adjustment and both to provide adequate wrap-around on the sprockets. Additional studs are set in the front of the cylinder block for mounting the idler pulleys, and a new cylinder block front cover is used to convert the engine from a pushrod unit with a roller-chain camshaft drive to a belt-driven DOHC.

Crankshaft, Connecting Rods, and Pistons

The production Escort RS engine has a Tuftrided nodular cast-iron crankshaft. The Cosworth BDA has a Laystall crankshaft machined from a forged steel billet (Fig. 17-10). This component is strong and is held to the flywheel by twelve high-strength socket-head bolts. The flywheel and crankshaft require no reworking by the tuner for competition purposes.

Fig. 17-10. Laystall crankshaft for Cosworth BDA.

The Cosworth engine uses forged-steel connecting rods, one of which is shown in Fig. 17-11. These have a pressed-in bushing at the little end, as on production Ford engines. The rods themselves, however, are several times as strong as stock Ford rods, and high-strength 12-point aircraft-type bolts are used to assemble the big ends. The big end bores are flawlessly round, and the only attention that these rods require is fitting them to the piston pins if new pins and bushings are being installed.

The Cosworth pistons have an outstanding reputation with speed tuners, even those who use them in engines other than Cosworths. When BDA engines are obtained in kit form, the piston crowns must be machined to match them to the cylinder deck at TDC and to obtain the desired compression ratio. These kits include not only the engine components but the carbure-

Fig. 17-11. Cosworth connecting rod.

tors, pumps, and all other engine accessories. As Ken Duclos puts it, "Everything comes in the box but the oil."

A set of Cosworth pistons, with their full-floating pins installed, is shown in Fig. 17-12. These pistons have been machined to provide valve clearance, but the piston crowns are otherwise flat, and there is appreciable metal above the top ring groove. Later, these pistons will be machined down for correct deck height and compression ratio. Usually this can be done with a lathe because little or no dome is required, even at fairly high compression ratios. This is because the narrow valve angle of the pent-roof combustion chamber keeps the chamber volume small. In turn, the narrow valve angle is made possible by using four small valves instead of two big ones. Domes, of course, interefere with flame front travel in the combustion chamber, and Keith Duckworth places a great deal of importance on designing combustion chambers that burn well. It is interesting to compare the Cosworth four-valve combustion chambers and pistons with the irregular chambers and domed pistons that are needed to convert most stock engines for racing—and the BDA is a stock engine if we consider its use in the Escort RS.

Fig. 17-12. Cosworth racing pistons before crowns are machined.

As in the assembling of any engine, the piston pin fit must be checked carefully and corrected if it is not exactly to specifications. Cosworth pistons are not always used in these engines, and any replacement calls for the replacement and refitting of the connecting rod bushing as well. In addition to balancing, the pistons must be carefully measured with a micrometer to determine their exact diameters. Each cylinder is then honed to provide the Cosworth-specified clearance for each piston. During cylinder honing, Ken Duclos installs a dummy head on the block and torques all the head bolts so that the honed bores will be cylindrical when the actual cylinder head is installed.

Tuning

A tuner such as Duclos, who both builds engines and drives in competition, has a decided advantage over a tuner who limits his or her activity to the machine shop. Research and development modifications can be tested in one's own car for a season before they are used in a customer's engine. It is always much better to feel firsthand how a newly modified powerplant performs on the track than to rely on the secondhand impressions of a driver—who may be gifted with excellent physical reflexes yet know virtually nothing about tuning.

As in most other engines that have small, efficient combustion chambers, the BDA is sensitive to ignition timing adjustments, and the correct advance curve for an engine must be obtained on the dynamometer. This is done by operating the engine at full throttle at various points in the range of rpm that will be used in racing and then determining the degree of spark advance that produces the most power at that rpm. Then the distributor is removed and installed on a distributor testing machine so that the advance weights and springs can be calibrated to reproduce the correct advance at each rpm.

The total advance needed by the BDA falls in a very narrow range of from 30° to 34° before TDC. Again, this indicates how well the chamber burns. Some American V8 drag racing engines need 40° to 50° of total advance for best power, not only because the combustion chambers are large but because piston domes and pockets in the combustion chambers retard the burning process to a degree that must be compensated for by advancing the spark.

Carburetor jettings are also critical, and the right jets will vary with the altitude above sea level, the climate, and other factors that govern the density of the intake air. Jetting therefore must be done on a dynamometer, and, on engines assembled from a kit, both the main jets and the air correction jets will usually need to be changed. Only in emergencies should jet changes be made at the race track. It is very easy to lose 4 or 5 horsepower in a change of only one jet size.

In addition to shaping the tops of the pistons, there are only a few other areas where professional tuners can exercise their skills as machinists in preparing a BDA for the track. Compression ratios are one area, of course, and this goes hand-in-hand with what is done to the pistons. Different camshafts are a very fruitful area, and the last word in valve timing for the BDA has not yet been written. Still, by comparison to other popular competition engines, there is very little that can be done to improve a Cosworth.

18 / Datsun

The Datsun Heritage

Datsuns have a way of winning against larger and more expensive cars that has almost become a tradition. In the early 1970s, the 510 sedans defeated respected European marques such as Alfa Romeo and BMW with stunning regularity. Now, in the late 1970s, the little Datsun B210 with its tiny A13 pushrod engine is still beating BMW (not to mention American Motors, Chrysler, Ford, and General Motors) at the races, and the six-cylinder "Z" cars are doing the same thing to Porsche.

Many Americans are inclined to make an unwarranted assumption concerning Datsun; they think that it is American speed tuners, hired at great cost by Nissan Motor Corporation, who have made the cars successful on the track. These people would be surprised to learn that "DAT" (Den, Aoyama, Takeuchi) engines were winning races thirty years ago. Until recently, however, Datsuns raced only in Japan, just as American Motors cars race only in America.

Few racing successes have been recorded by Datsun engines outside Datsun cars. And while the engines are among the strongest and most reliable powerplants manufactured, one might think that the economy sedans in which most of them are installed would impose a handicap that not even the greatest engine could overcome. But though a stock B210 sedan shows no advantage in braking, handling, acceleration, or top speed over its market rivals, the racing heritage in every Datsun model makes them exceptionally amenable to racing preparation.

Expert Preparation

The foremost American experts in the preparation of racing Datsuns are at Bob Sharp Racing in Georgetown, Connecticut. None of the Bob Sharp cars is the work of a single person. But one person has to be in charge, and he is Gene Crowe, the team manager and crew chief, who knows every inch of the cars inside and out—although much of the actual work is entrusted to other highly skilled machinists and mechanics.

A surprising number of the parts that Gene puts into the Bob Sharp Datsuns are manufactured by Datsun. This is in vivid contrast to the average Chevrolet competition engine, which may use GM parts for only the major castings. Unlike General Motors, Nissan Motor Corporation has a vital competition department, and their American competition manager, Dick Roberts, is seen wherever Datsun cars are entered in important races. Thus, as with Porsche, it is seldom necessary to look outside the factory's part bins to obtain the components that are necessary for a winning engine.

L-Series Engines

Datsun's (and Bob Sharp Racing's) major competition efforts have always centered around the L-series engines (Fig. 18-1)—SOHC units of either four or six cylinders inline. These powerplants have been manufactured in a variety of piston displacements, each four-cylinder type generally paired with a six-cylinder sibling that uses the same pistons, connecting rods, and so forth. In addition, there are many parts that are common to all of the L-series engines, or are very similar and can in some cases be interchanged for purposes of competition preparation.

Each SOHC engine is designated by an "L" followed by a number that is based on the engine's displacement in liters. Taking the "Z" cars as an example, the 240Z has the L24 engine.

Fig. 18-1. Datsun L24 engine, showing design features typical of L-series units.

That is, it has an L-series engine of 2.4-liter displacement (actually, 2393 cm^3). The 260Z has an L26 engine, and from this it can be seen that "260Z" and "L26" are both based on the 2.6-liter (2565-cm^3) capacity. The 2753-cm^3 280Z has an L28 engine.

The L16 and L18 engines are, respectively, 1.6-liter and 1.8-liter powerplants. However, these are four-cylinder engines in the L-series. Whether "four" or "six", certain features of L-series engines immediately stand out. First, they have single-port-face cylinder heads with the valves inclined toward the ports. The intake ports, being higher in the head than the exhaust ports (Fig. 18-2), are as straight and direct as those of many DOHC designs. The combustion chambers are the classic "wedge" type, and the valves are operated by balljoint-pivoted rocker arms that are interposed between the cam lobes and the valve stems.

The camshaft is driven by a double-row roller chain, and the valve clearances are adjustable by changing the depth to which the rocker arm pivot ball studs are screwed into the cylinder head. (This layout is identical to that found on the finest Mercedes Benz.)

Fig. 18-2 also shows that the crankcase extends well below the crankshaft's centerline and has cast webs at the main bearings that extend fully down to the sump mounting flange. Each crankthrow has a main bearing on each side, so that the four-cylinder units have five main bearing crankshafts and the six-cylinder units have seven. The crankshafts are of sturdy but lightweight design. Thus the "bottom end" of the engine easily matches the quality that is lavished on the "top".

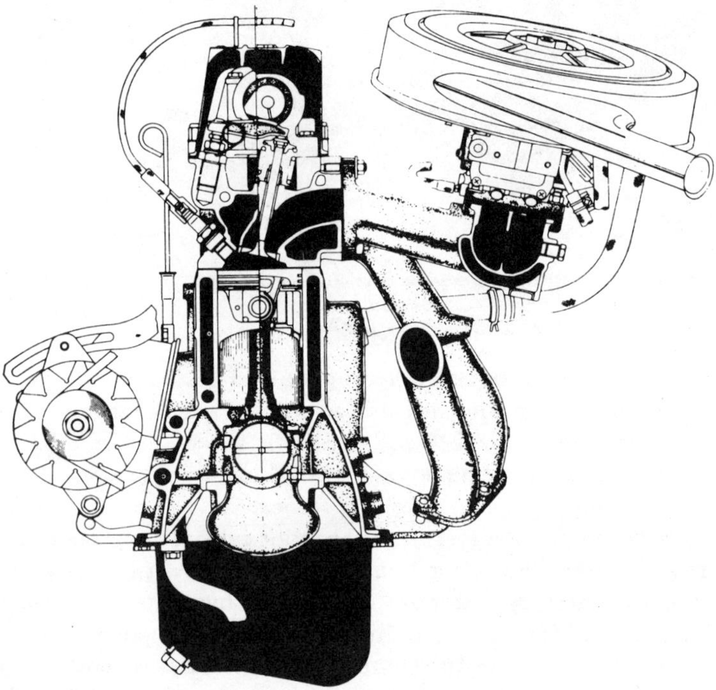

Fig. 18-2. End section of four-cylinder L-series engine.

The cylinder blocks are of cast-iron. However, as on the majority of modern engines—particularly those from Japanese designers—the cylinder head is aluminum alloy for improved cooling efficiency. Special aluminum bronze valve seats are installed for the intake valves, and cast-iron valve seats are used for the exhaust valves. These are not cast in but are a hot press fit. The camshaft bearing arrangement is commendably rigid. The camshaft rides directly in the cast-aluminum alloy brackets, but the brackets are full-ring components that bolt to the cylinder head rather than splitting in the middle. The result is that a cylinder head is not rendered unserviceable by a worn camshaft bearing.

There is little doubt that the extraordinary strength of the comparatively lightweight Datsun engines is responsible for their success. When tuned to deliver the highest attainable output, the basic structure is not overtaxed. This is especially praiseworthy at a time when some manufacturers are building weaker, hence cheaper, engines because they know that in stock form the reduced power resulting from lowered compressions and emission controls will not demand the strength that was required ten years ago.

Cylinder Head

First, there are the so-called FIA cylinder heads. Though no longer available from Nissan's competition department, a six-cylinder FIA head was used on the Bob Sharp 240Z that Sam Posey drove with brilliance in his quest for the 1977 Camel GT Series Under-2.5-liter Championship (Fig. 18-3). (Devastated remains of another six-cylinder FIA head lie moldering in the dyno room, a piston having apparently collided with a dropped valve.) Thus, because new six-cylinder FIA heads are unobtainable, all present development is centered around the production head.

The FIA heads which are thicker, are easily recognized by an external water manifold that carries coolant directly to each exhaust port. Stock six-cylinder heads have one water connection only, located at the front. The FIA valves, in addition to having longer stems, are larger in diameter. The ports are also larger, and the exhaust ports are round rather than rectangular as in

Fig. 18-3. Competition engine in Bob Sharp Datsun raced by Sam Posey in 1977 Camel GT series. Notice external water manifold.

production heads. Bob Sharp Racing has found that in some ways the ports of the stock heads are better—given the kinds of preparation that are commonly permitted in American racing classes.

FIA heads for the four-cylinder L-series engine are still available from Datsun, and these are generally used in all of the best OHC "fours". But in all other ways the competition preparation of four-cylinder and six-cylinder L-series engines is the same—with one other possible exception. Gene Crowe cautions anyone tuning a Datsun engine not to overdo the diameters of the exhaust header primary pipes. These should be no more than 1⅝ in. in diameter in order to maintain good pressure wave action. But on the two-liter "four", the primary pipes can be 1¾ in. in diameter. However 1⅝ in. is still a better upward limit even on this engine unless the remainder of the exhaust system is opened up accordingly.

In reworking L-series heads, the emphasis is on reshaping

and polishing the ports rather than enlarging them (Fig. 18-4). If racing class rules limit the valve diameters to stock dimensions, the standard Datsun valves are capable of competition service. It is very important to entrust head modifications to a good cylinder head service, and Datsun's competition department can supply a modified head that has proven very successful in racing. Tilton Engineering of El Segundo, California, manufactures many good Datsun competition parts, and its modified cylinder heads for Datsun cars are among the best available.

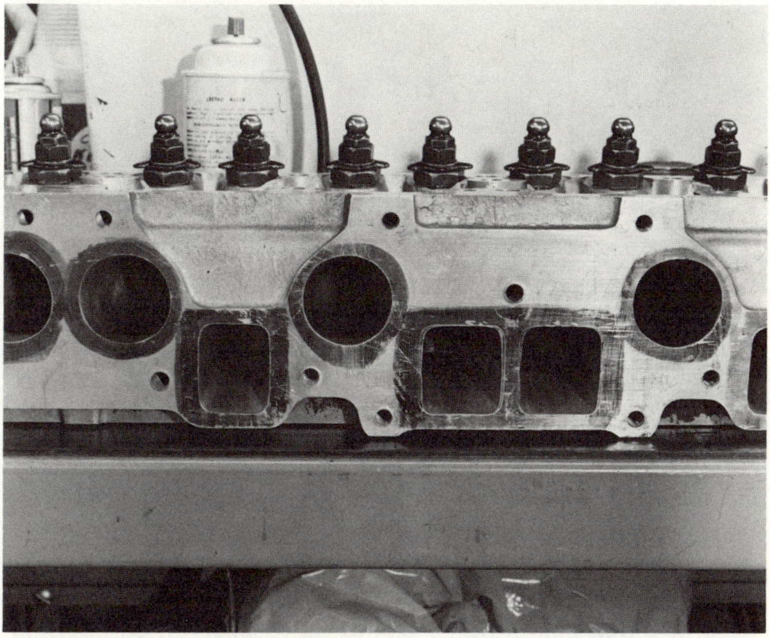

Fig. 18-4. Intake and exhaust ports of modified six-cylinder production head.

In addition to a competition cylinder head gasket, O-rings are usually used around the cylinder bores. When the rules permit, the combustion chambers are reshaped to reduce shrouding of the valves; this is especially important if the valve diameters are also being increased. The spark plug boss must not be cut away unduly (Fig. 18-5), and to prevent the threaded portion of the spark plug shell from projecting into the combustion chamber, it may be necessary to install a solid copper spacer gasket between the regular gasket of each spark plug and the cylinder head. The Champion gold palladium fine-wire electrode spark plugs have, incidentally, been found to work exceptionally well in Datsun competition engines (N-59G and N-2G are two representative types).

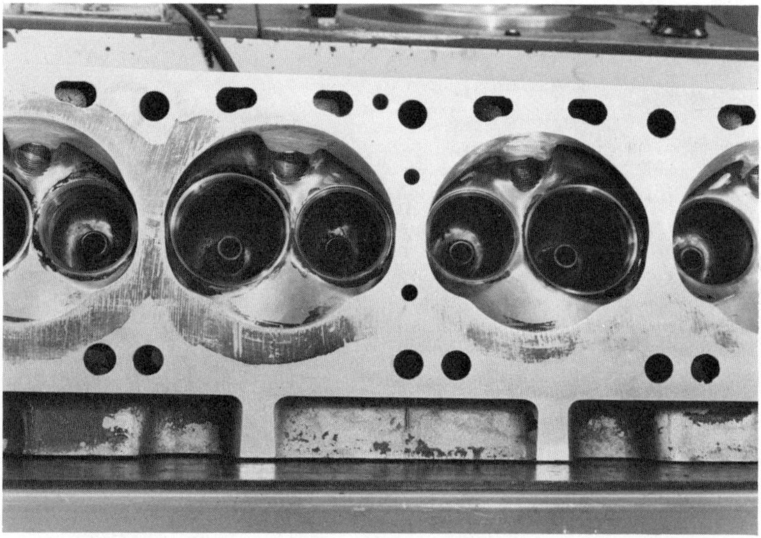

Fig. 18-5. Combustion chambers of modified six-cylinder production head.

When the combustion chamber is opened up, other steps must be taken to restore the compression ratio or to increase it to the practical limit. Some dome on the piston crowns is virtually unavoidable because the Datsun combustion chambers, once smoothed out, are not very compact. In addition, the cylinder head itself must be milled. This, of course, causes additional slack in the camshaft drive chain, and the correct way of accommodating this will be discussed later in conjunction with the camshaft and valve gear.

Induction System

The Solex sidedraft carburetors and intake manifolds that are standard Datsun competition items are unsurpassed. However, Yankee ingenuity frequently has its way on the four-cylinder units (Fig. 18-6). The electronic fuel-injection system used on the 280Z is the Bosch L Jetronic system, insofar as principles are concerned. As in the case of certain other production cars that use the L Jetronic design, carburetors are often substituted for competition work. The main difficulty with the injecion setup is that competition electronic control units (or "brains") are not available, so all tuning must be done by tinkering with the various sensors that transmit data to the control unit, thus "fooling the brain" into providing a nonstandard mixture.

Fig. 18-6. Four-cylinder Datsun engine equipped with single Holley four-barrel carburetor. This modification is popular with tuners who have had most of their experience with American engines.

Camshaft and Valve Gear

Gene Crowe's favorite camshaft for the OHC Datsuns is an Art Early design available from Iskenderian. When a camshaft is installed, it is important first to find true tdc for the engine and then to time the cam so that the overlap period is exactly split at tdc—or advanced up to 3° (the actual advance being determined largely by the exhaust system characteristics). If a very "wild" cam profile is used in the four-cylinder L-series engine, it is best to add a supplemental oiler for the cam lobes to prevent undue wear.

Stock Datsun rocker arms are always used, though these may be lightly polished to reduce stresses. Stock keepers can be used, but stock spring retainers do not allow the use of adequately long valve springs—necessary to accommodate increased valve lift. Therefore, special steel or titanium retainers from Tilton Engineering are used. Longer dual springs (Fig. 18-7) can be obtained from Datsun competition, Tilton, or Iskenderian (the choice generally being in that order).

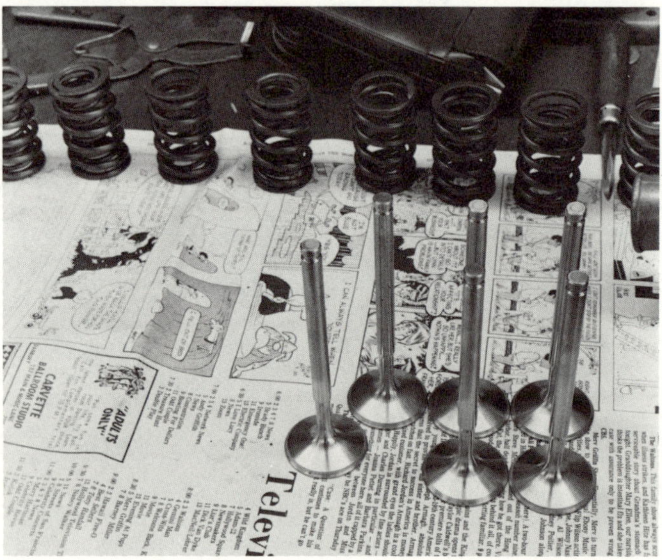

Fig. 18-7. Competition valves and valve springs for Datsun "six".

On the OHC engines it is extremely important that the timing chain does not flop because of the greater slack induced by milling the cylinder head and block. Under no circumstances should the automatic tensioner (Fig. 18-8) be depended on to take up the slack. When the engine is being assembled, the automatic tensioner's retracting clearance should always be adjusted to zero. If it is not, performance-robbing timing errors will occur, not to mention the possibility of increased wear or engine damage.

The curved timing chain guide should be moved toward the chain to take up the excess slack. If little metal has been milled from the head, it is sometimes possible to elongate the mounting holes of the stock curved guide. But for strength and added thickness, two guides can be installed, one atop the other, or the heavier 280Z curved guide can be used, with the mounting holes modified as necessary.

After advancing the curved guide toward the chain so that the excessive slack is removed, it is usually necessary to reshape the ends of the curved guide slightly in order to prevent a sharp change of direction in the chain as it enters or leaves the guide. If the chain leaves the guide at an angle, wear will take place and the chain slack will increase rapidly. If the automatic tensioner cannot be adjusted to zero protrusion, shim the space to zero by installing one or more solid copper Champion spark plug gaskets between the slipper pad and the tensioner body. If these things are done correctly, the chain will not cause trouble.

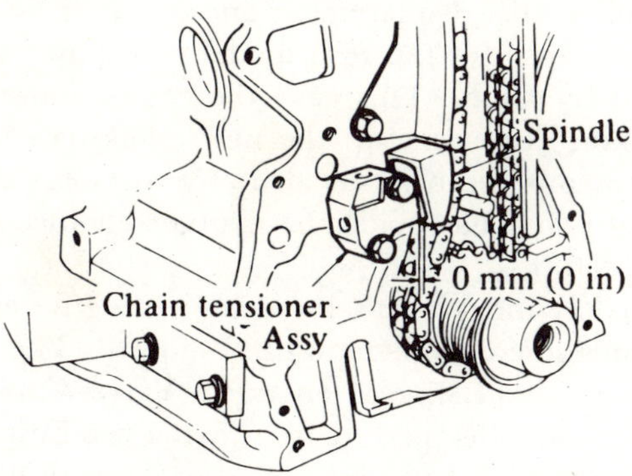

Fig. 18-8. Correct installation of automatic chain tensioner assembly. There should be no clearance between slipper pad and body of tensioner.

Pushrod Engine

The Datsun A13 pushrod engine used in the B210 has basically the same construction as the OHC engines with regard to the design of its cylinder block and moving parts. Many people have mistakenly concluded that this powerplant is one of the older Datsun pushrod units that was last used in the 1600 sports car. This is far from true.

From its founding in the early 1930s to the beginning of the 1960s, the Nissan Motor Corporation manufactured Austin cars under license from the English company. In fact, some of the first Datsun cars to reach America could be repaired with Austin parts. But as Austin began to falter, Nissan saw clearly that its future depended on designing its own cars—cars that would be better suited to the United States and other world markets. Datsun has been so successful that it now enjoys a worldwide popularity far in excess of Austin's in its heyday.

Datsun's home-grown engine designs continued for a few years to reflect the Austin parentage; for example, the rocker arm cover of a 1967 Datsun 1600 sports car and the rocker arm cover of today's MG are both held in place by two nuts atop the cover. But the present Datsun pushrod engine is altogether different and is totally Datsun in concept: it is strong, well made, and capable of very effective competition tuning.

The present 1397-cm^3 B210 engine has its roots in the 1200 engine and its smaller-displacement ancestor that was introduced about ten years ago. For the 1974 model year, however, the stroke was increased by 7 mm. This gave the engine more than an increase in piston displacement; the new crankshaft and connecting rods were considerably beefed up for awesome durability. It is this strength that has made the engine reliable even when it is highly supertuned.

Reworked cylinder heads for the A13 engine are not much different from those on other pushrod "fours". The ports are reshaped, polished, and slightly enlarged. Valve diameters are increased where the rules permit it, and the combustion chamber contours are smoothed out and polished. After milling the head either to restore or to increase the compression ratio, the rocker arm valve gear must be modified to correct its geometry. This is usually done with shortened pushrods. Competition dual valve springs are used, with different retainers, but the stock Datsun valves are wholly acceptable if the rules proscribe departures from the stock diameters.

Cylinder Block

The Datsun cylinder block (Fig. 18-9) is outstanding for its strength. The quality control is well above average at Datsun, and align boring, deck milling, cylinder boring, and similar blueprinting operations can easily be bypassed by a person who is building an engine on a tight budget. This is in direct contrast to the almost indispensible machine work that is needed to prepare American engines for racing. In fact, the Datsun blocks are seldom align bored even in the most perfectionist shops. At Bob Sharp Racing, the deck is given a clean-up cut; more metal is taken off only if it is necessary to correct the deck height above the main bearings. The cylinders are then bored to the class limit at a precise 90° to the deck.

Fig. 18-9. Six-cylinder Datsun racing short block assembly. Notice dry sump pan and adapter that replaces oil filter.

The lubrication system is also left amazingly stock. But because a wet sump will not work during high-G cornering and acceleration, a dry sump system is fitted. This consists mainly of an oil pan manufactured by Aviaid Metal Products of Van Nuys, California, and a new oil pickup that is bolted onto the regular Datsun competition oil pump. An adaptor for the hoses to the oil tank and, when fitted, an oil cooler and filter, is installed in place of the stock oil filter cartridge.

The oil pressure should be adjusted to 60 psi with the engine hot. This is near the stock setting, but keep in mind that in a competition engine the clearances will have been increased to what would almost be considered worn out in a production car. The pressure increase is obtained by installing a stronger regulating spring in the oil pump. These springs are a common speed shop item, and the tension can be varied with spacers for fine tuning of the pressure setting.

Two kinds of cylinder O-rings have been used in Bob Sharp Racing's engines (Fig. 18-10), though neither is the copper wire O-ring that is commonly added to competition-prepared engines

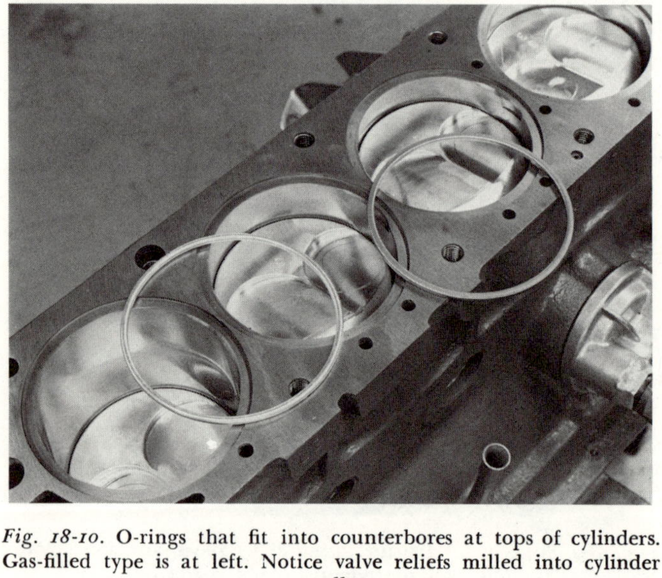

Fig. 18-10. O-rings that fit into counterbores at tops of cylinders. Gas-filled type is at left. Notice valve reliefs milled into cylinder walls.

by cutting grooves around the cylinder bores (see chapter 17). The O-rings are hollow and are installed in counterbores at the tops of the cylinders. The Datsun competition O-rings are of rolled steel construction. Recently, however, sealed gas-filled O-rings have been used. These are similar in design to those used in the short-lived Cosworth-Vega and are available from United Aircraft Products.

Each cylinder is finish-honed to obtain a .005- to .008-in. clearance with its individual piston. When larger valves are employed, it is necessary to cut small clearance reliefs in the tops of the cylinders. As is normally the practice on all competition engines, the head bolt holes are chamfered in order to prevent thread pull-up that could prevent accurate torque wrench readings from being obtained, as well as to reduce stresses near the sharp-edged junction of the bolt hole with the deck.

Crankshaft, Connecting Rods, and Pistons

The Datsun crankshaft is forged. But before exposing it to competition service, it should be magnafluxed and then Tuftrided or nitrided. Because the heat treatment will cause distortion, it is necessary to straighten the shaft afterward. No shot peening or similar treatment is required, although the oil holes should be chamfered and polished to a smooth radius, and the journals themselves should be polished. Balancing, as in any competition engine, is an absolute necessity, and all of this work can be carried out by a good crankshaft service.

The redline for these engines is very much determined by the harmonic vibration range of the crankshaft. On any engine, crankshaft breakage is always a possibility when the harmonic point is reached, and especially so on an inline "six". The stroke length determines the rpm at which the critical vibration occurs. So whereas a 240Z engine can safely rev to 8000 rpm, the 260Z and 280Z engines are redlined at 7500 rpm. Gene Crowe emphasizes that the Datsun competition harmonic balancer, or vibration damper, is indispensable—especially on the "six". Thus, it would be a serious mistake to substitute an ordinary pulley.

The running clearance for both the connecting rod big end

bearings and the main bearings should be .003 in. (compared to .0008- to .0028-in. main and .0010- to .0022-in. rod bearing clearances in a stock 280Z). The .003-in. clearance seems to be the optimum for these engines. Datsun competition bearings are available in many different thicknesses and diameters, so that the clearance can be adjusted precisely despite any metal removal that has occurred in align boring the block or in polishing the crankshaft journals. Standard bearings are not suitable because they are not available in this wide range of sizes.

Datsun connecting rods are normally used in Datsun competition engines. However, longer rods are desirable in the two-liter because of its relatively long stroke. The longer rods, supplied by Carillo, help to reduce angularity. Different pistons with a higher pin location make this modification possible.

Stock Datsun engines have intereference-fit piston pins that must be pressed into the bushingless connecting rod small ends. For racing, the rods are converted to accept full-floating piston pins. The rod-to-pin clearance should be .0010 to .0012 in., and no bushing is needed if a chrome pin is used. Either plain or chrome pins can be used if a suitable bushing is installed.

When connecting rods are prepared for competition, they should first be magnafluxed and then straightened. One or two oil holes must be drilled in the little ends because of the change to full-floating piston pins. The big end bolts should also be magnafluxed—or they can be replaced by bolts of greater diameter. Larger "SPS" bolts, available from Mr. Gasket, require that the bolt holes of the rod and cap be reamed to fit them. Afterward the connecting rod must be rebored, which brings the cost of preparing six rods to nearly $150. Normally the Datsun rods stand up well, occasionally needing to be reconditioned to size if the .0005-in. out-of-round limit is exceeded.

TRW, Venolia, Cosworth, and Arias pistons are all commonly used in Datsun competition engines. Plain cast-iron compression rings and three-piece chrome oil scraper rings are used, and the piston pin should fit the piston with an .0008-in. oil clearance. Some piston manufacturers tend to leave excessive end play between the piston pin and the circlips. If the end clearance is greater than .005 in., one circlip groove can be recut and two circlips installed to take up the excessive play. If the end play is greater than .005 in., there is danger that the circlips will be pounded out.

Cosworth circlips are probably the best. These are surface ground and have good end-play tolerance in the pistons. Complete sets of Cosworth pistons for the A13 (pushrod), L18, L20, and L28 (SOHC) engines, with pins, circlips, and piston rings, are available from Malvern Racing in Ivy, Pennsylvania. (Malvern is a good source for all kinds of Datsun competition parts.)

Generally tuners obtain their pistons with partially machined crowns. These can be milled to obtain clearance for any valve diameter and deck height, as well as domed to obtain any compression ratio (Fig. 18-11). As in the case of other engine components, it is important that the pistons be crack tested and discarded if any flaw is discovered after machining the crowns or after a period of use in competition.

Fig. 18-11. Two Datsun racing pistons, each with crown milled to a different shape.

Conclusion

On the basis of a cost/success ratio, it would be difficult to find better cars than the Datsuns—especially if one wants to be both competitive and not burdened by undue preparation costs or engine failures. It is almost newsworthy when a Datsun drops out of competition with mechanical difficulties, and when trouble does occur it is no problem to find replacement parts.

The main limitation of the Datsun engines is their size, which has kept them from being a factor in formula cars or in sports racing cars. The "six" is too long by today's design standards, and the "four" is rather tall and heavy by comparison to a Cosworth BDA, for example. But when installed in Datsun sedans and sports cars, the Datsun engines have been winners in every racing class where they are eligible. Undoubtedly they are the cars to beat.

19 / *Volkswagen Air-cooled*

Competition Beginnings

The first book on extracting extra power from the Volkswagen air-cooled engine appeared about fifteen years ago. Its author admitted from the outset, "There is not much that can be done to hop-up a VW." From this inauspicious beginning, we have progressed to a point where there is not much that *cannot* be done to hop-up a VW.

Until the mid-1960s, VW engines in competition were almost laughable, and if some bold prophet had stepped forward to predict that air-cooled VWs would someday make Offenhausers obsolete in USAC Midget racing, he would have been laughed into derision. Given this attitude, it is surprising that any progress was made toward making VW powerplants successful on the race track.

Formula Vee road racing marked the turning point in the VW's image. In this class, it became a competition engine without having to produce a great deal more power than it did in stock form, and tuners finally began to take seriously the matter of finding additional horsepower and torque. Nevertheless, the main boost that the VW had was the dune buggy craze of the mid-1960s. Despite occasional buggies with Simca or BMC powerplants, only the VW engine and chassis thoroughly met the buggy builders' needs. While a great many buggies retained stock powerplants in somewhat dressed-up form, the attention of America's hotrod community had finally been attracted. Now the real quest for power began, which culminated in today's VW drag racers that can do a standing quarter mile in less time than it took the V8 Chevies of only a few years ago. The supply and variety of available supertuning components for the air-cooled VW is monumental. But not all of the speed equipment and the know-how are applicable to every racing class where VW engines compete (Fig. 19-1). We will begin, therefore, with VW competition engines in their purest form: Formula Vee.

Fig. 19-1. Engine compartment of professional road racing VW Super Beetle.

State of the Art

One of the most successful Formula Vee engine builders in the United States is the AID organization, headed by John Harkness and located in Marblehead, Massachusetts. During the early part of the 1977 racing season, AID engines won every race they entered.

Though John Harkness is an expert machinist who has built many fine competition powerplants, impaired eyesight has forced him to limit his activities to administrative and theoretical concerns. Lee Goodwin, the chief machinist at the well equipped facility, is now in charge of producing the precision components that have made AID's VW engines the terrors of the track.

Fine Tuning

Formula Vee has always been an engine tuners' class; engine builders often have more to do with the outcome of the race than the drivers do. So critical are the fine points of Formula Vee preparation that John Harkness and Lee Goodwin talk enthusiastically about picking up half a horsepower by some new modification. This would be absurd in drag racing or in USAC Championship cars. But the combined pulling power of two or three "half-horses" can add up to decisive domination in this class.

Take, for example, the matter of intake valves. Brand new valves permit the greatest gasflow. One of the East's best Formula Vee drivers has found that, with a fresh AID engine, he can win his first race without trying very hard. Winning the second race means a fight for the lead, and in the engine's third race, the car is likely to finish second. The engine then goes back to AID for a valve grind, during which the intake valves are not ground at all but discarded and replaced by new ones. With this minor change only, the engine is once more ready to win two firsts and a second.

All components used in a Formula Vee engine (Fig. 19-2 and Fig. 19-3) must be VW parts. Some parts are better than others, but all must be original parts or standard VW replacement parts for the 1192-cm^3 1200 engine that was used in 1961 through 1965 Beetles. In addition to "rare" old parts that offer particular performance advantages, many of the currently manufactured replacement parts—some originating in Mexico—have helped to make these engines stronger now than when they were new.

Fig. 19-2. Typical Formula Vee engine installation with cooling fan retained.

Fig. 19-3. Typical Formula Vee engine with fan elimininated and ducted air used for engine cooling.

Cylinder Heads

The infamous "long stud" VW heads used from 1961 through 1964 are never installed on a Formula Vee engine—not even in modified, short-stud form. Nor are certain other heads that are scorned by tuners. The most competitive engines use one of three cylinder heads that show the most gasflow on the flow bench. The best of the three is the 101-D head, which is a new replacement part introduced in 1976. The second best is the 101 head (no letter suffix), and the third best is the 101-F.

The first step in Formula Vee cylinder head preparation, as well as in other VW racing applications, is to flycut the cylinder head (Fig. 19-4) to produce the correct combustion chamber volume, which is 43 cm^3 for Formula Vee engines. A milling machine is more accurate for this job than the less expensive tools that are hand operated or used with a drill press.

Fig. 19-4. VW 1600 cylinder head being flycut on milling machine.

Formula Vee Ports

The 1200 engine has a single siamesed intake port in each head, and the separation between the two valve seats, which is inside the head, must be machined to a knife edge. This streamlining helps to increase the gas velocity, thus imparting greater inertia to the mixture that tends to carry it into the correct cylinder instead of reversing direction around the separation and into the other intake valve. This keeps the cylinders from robbing one another, which can easily occur because of the consecutive firing of the two cylinders in each bank.

A pocket or relief is cut just inside the intake port, on the side toward the rocker arm shaft. This pocket is left somewhat rough to create turbulence. Thus, in addition to redirecting the mixture from the manifold toward the valves, fuel that tends to precipitate at this point is reatomized by the swirling air. The very long intake manifold branches and the abrupt change of direction near the ports combine to encourage "puddling" of the fuel globules, which has always been a problem in Formula Vee. The early solution was to use extremely rich carburetor jets so that adequate fuel would reach the cylinders despite the puddling. With the AID method of porting the head for turbulence, the main jets can be reduced from a Solex size 240 to a size 210 with an increase in power.

The exhaust port volume is increased at the point where the gases must pass around the valve guide, and both the intake and the exhaust ports are carefully matched to their respective manifolds. Incidentally, none of the port modifications described here shows up significantly on a flow bench, but the results show up conclusively on the dynamometer.

Porting for Other Purposes

Outside the Formula Vee class, porting (Fig. 19-5) tends to be carried to greater lengths, especially when larger valves are being installed. When very large valve seats are installed in a VW head, part of the spark plug boss is cut away, and a threaded steel insert must be installed in the spark plug hole. During radical porting, an old set of valve guides should be installed in the head. These guides will be ground away and destroyed during the porting operation, but their presence will prevent pockets from being inadvertently cut into the soft aluminum around the holes for the guides.

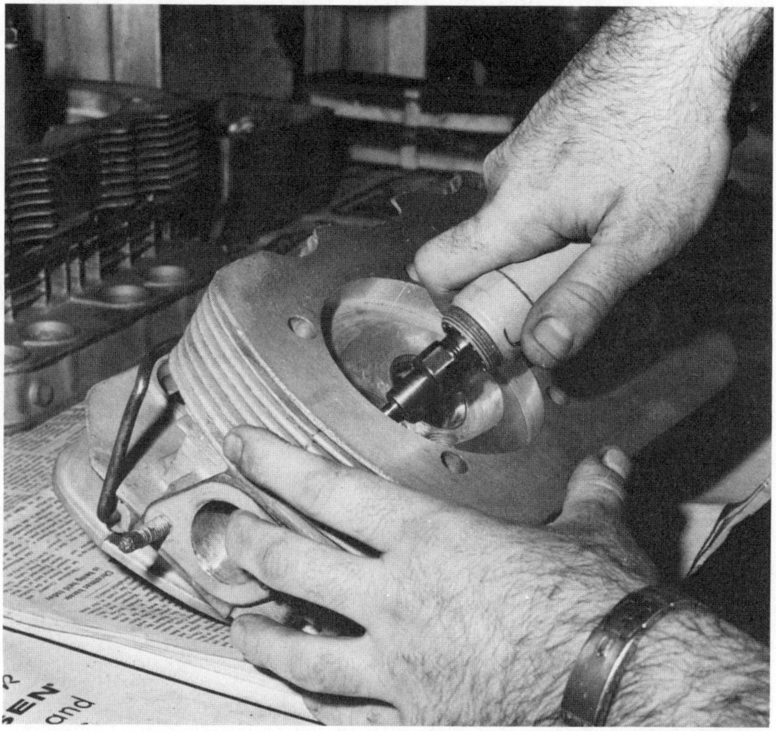

Fig. 19-5. VW 1600 cylinder head being ported with high-speed tool.

The flow through the ports can be improved by beveling the inner ends of the valve guides before installation. This can be done either by grinding or by turning them in a lathe. In addition, when a high-lift cam is to be used, the tops of the guides may have to be cut down, removing about .010 in. more than the lift increase provided by the cam. The part of the guide that is inside the port is cut off entirely in some drag racing engines for maximum gasflow, and guide wear is quite rapid.

Manifolds and Headers

Because of the vast array of intake manifolds and exhaust systems available for the VW air-cooled engines, it is impossible to cover them all here. Each kind of competition application requires a different setup, and in many cases the selection is limited by the rules of the racing class.

Formula Vee rules specify that a stock VW intake manifold be used. The best of these is the manifold designated by a part number ending with the digits 701-D. This component was used on 1963 and later VW 1200 engines. The inside diameter of the intake manifold can be increased by acid etching, so long as the weight is not reduced below the minimum specified by Formula Vee class rules.

Valves and Valve Gear

As in the case of exhaust and induction systems, a great many different valve gear setups are used. If dual valve springs are to be fitted, the cylinder head valve guide bosses must be machined so that the inner springs will fit over them; otherwise the springs will break. Proprietary components are available to replace every piece in the VW valve train—from timing gear to valve seat. The greater the power extracted, the more of these heavy-duty components are needed.

Formula Vee rules require that the valve gear use stock VW components. However, washers can be installed between the valve springs and the cylinder head in order to restore any spring tension that is lost through reconditioning of the valves and seats. Under the rules, the rocker arms can be balanced but not lightened. So the tuner must search not only for those production rocker arms that, owing to manufacturing variations, give greatest valve lift, but also for rocker arms that are unusually light. The other rocker arms in the engine can then be machined to match the weight of the lightest rocker. The engine is likely to be disqualified if there is not one unmachined rocker in the engine—indicating that the others were machined only to match its lighter weight.

The cam followers can be lightened legally by removing

metal from behind the foot. A two-piece cam follower used in 1961 is the lightest and most treasured kind. Pushrods, in addition to being selected for maximum lightness, are selected for maximum length. Longer pushrods mean that the valve clearance adjusting screws can be backed off, which changes the angles of the rockers, altering the geometry of the valve train and thereby increasing the valve lift.

The Formula Vee engine must use a stock camshaft. But for better high-rpm power, it is desirable to retard the valve timing. This cannot be done by relocating the timing gear on the camshaft because the gear is riveted on. The rules say that metal can be removed from but not added on to components. Thus, the timing gear rivets can be drilled out, but new ones cannot be installed.

There are two ways that the cams can be retarded; the ideal figure is 3°—though some tuners claim that more retard is better. First, a camshaft with a –3 (minus 3) gear can be used. This number indicates that the camshaft's gear is of the smallest-diameter production class. The undersize gear induces play in the timing drive, thus retarding the valve timing and also reducing internal friction. Further valve retard can be obtained by selecting a center camshaft bearing that has an unusually thin thrust face. This permits greater camshaft end play, allowing the helical gears to "unscrew" and thereby retard the valve timing.

Crankcase

At one time, tuners rarely found it necessary to align bore a VW crankcase. But current replacement crankcases, especially those from Mexico, should be align bored as part of competition preparation—Formula Vee is no exception. For most competition purposes, 1971 and later VW crankcases are best. They not only have a relocated oil cooler that does not block the flow of cooling air to the No. 3 cylinder but are made from a magnesium alloy that is more resistant to high-temperature strength loss than were earlier VW crankcase materials.

Replacement crankcases have camshaft bearings, which were not used in U.S. 1200 engines until the 1965 model year. But if an earlier case is used in Formula Vee, it should definitely be machined so that camshaft bearings can be used. The principal benefit of the use of bearings is that they reduce friction sufficiently to give the engine a solid one-half horsepower increase.

A windage tray must be installed in the crankcase for most kinds of competition to prevent the oil from being whipped to a foam by the crankshaft and reciprocating parts. This aeration results at high rpm from air currents inside the engine. An extension can be welded to the oil pickup if a larger-capacity dropped oil sump is used—and this is desirable because the stock oil capacity is limited. In Formula Vee, the dropped sump capacity cannot exceed 250 cm^3. Formula Vee engines operate at an oil temperature up to 260°F and, as on most competition VWs, the crankcase should be drilled to accept an oil temperature gauge sending unit.

Crankshaft, Connecting Rods, Pistons, and Cylinders

The most exciting recent development in Formula Vee is the 1977 rules change that permits the lightening of engine internal parts. Though it might seem practical to do some of this work with a bench grinder and other tools that may be found in one's own garage, it is doubtful that backyard modifications will deliver the speed and reliability that are offered by parts modified by expert machinists using precision machine tools.

The crankshaft can be lightened from its stock 18- to 18½-lb. weight to a trim 16 lbs., with a consequent improvement in acceleration and throttle response (Fig. 19-6). The crankthrow webs of the stock crank are roughly oval, viewed axially, and these can be machined to a more rectangular form to accomplish the major part of the weight reduction. However, where precision machine work really pays off is in the contouring of these crankthrow webs so that the crankshaft is more "aerodynamic". The discovery that wind resistance caused by the moving parts inside the crankcase can absorb a great deal of power has prompted the designers and tuners of all kinds of competition engines to rethink their approach to crankshaft and crankcase design.

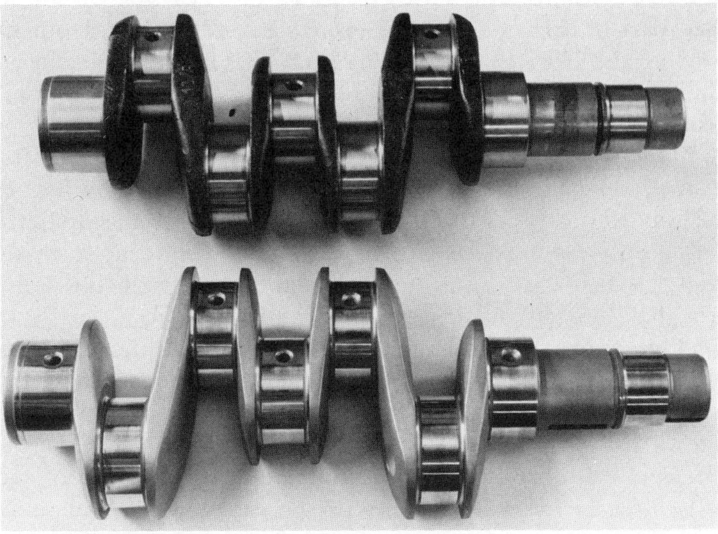

Fig. 19-6. Stock VW 1200 crankshaft (top) compared with AID's lightened, polished, and streamlined Formula Vee crankshaft (bottom).

Additional work on the Formula Vee crankshaft includes indexing the crank to its maximum allowable stroke, which is 2.525 in. + .005 in. This work decreases the crankpin diameter, thus reducing friction and eliminating mass from a point where it does most to diminish crankshaft inertia. In all cases, the oil holes are chamfered and radiused, and the crankshaft balanced dynamically.

Reducing the crankshaft and connecting rod weight can give a Formula Vee engine an additional 1 to 3 bhp—the better the workmanship, the better the output. AID does the complete job for $87.50 on a customer's magnafluxed crankshaft, finishing it off with a bead blasting that not only provides the best possible finish but helps to relieve internal stresses in the crankshaft material itself.

The best VW 1200 crankshafts are those with "rolled" journals, such as the modified unit shown in Fig. 19-6. On these crankshafts, there is a radiused groove on each side of the polished part of each journal, the groove extending to a depth that is below the level of the bearing surface. The bead blasting obtains throughout this groove, polishing being reserved to the actual bearing surfaces themselves.

The flywheel of Formula Vee engines can be lightened from its stock weight of about 18 lbs. to a racing weight of 12 lbs. The best way to do this is to remove metal at the extreme periphery, cutting away as much as half of the starter ring gear to remove weight as far from the center as possible, thus reducing the inertia that inhibits throttle response and acceleration.

Lightening the connecting rods is permitted under the 1977 rules. Normally this kind of work is not recommended for competition engines; however, because the rods are strong and the engines are not radically supertuned, it can be done in Formula Vee. Connecting rod lightening might seem to be a likely backyard chore; it has been done with no other equipment than a bench grinder, but the weight is usually removed in the wrong places (Fig. 19-7).

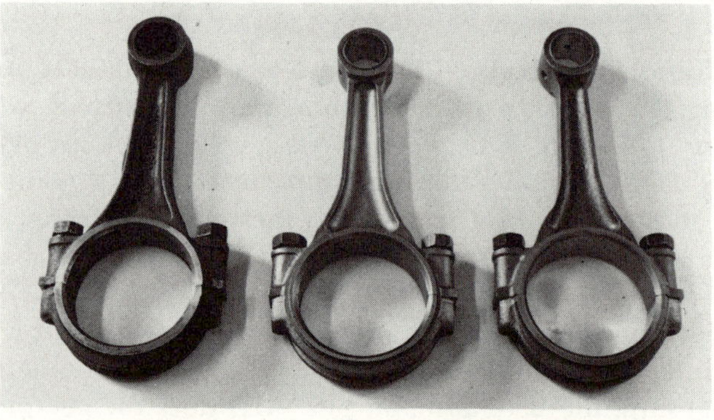

Fig. 19-7. AID lightened rod (center) compared with stock rod (left) and rod lightened by backyard method of grinding metal off I-beam section (right).

Considerable weight can safely be removed from the connecting rod small ends, leaving the piston pin bushing to protrude from each side as it does on stock Porsche connecting rods.

Though some tuners have machined most of the excess weight off the I-beam part of the rod, this is bound to induce weakness; when a VW rod bends or breaks, it is in this area. At AID, the connecting rods are lightened by machining metal from the bead that surrounds the big end eye using special machine tools (Fig. 19-8). The weight is removed where it will do the most good—from the end of the rod that goes around with the crankshaft—and at the point where it will do the least harm structurally.

Because VW rods are not noted for their center-to-center uniformity, the rods must be rebored so that the center-to-center lengths are absolutely uniform and to blueprint specs. Bead blasting is the final step after the rods have been straightened, bored, and balanced. This is very beneficial for improving resistance to metal fatigue.

There is currently some confusion as to whether the rules allow lightening of the pistons. If this work is permitted, Formula Vee pistons, instead of having their stock full skirts, will be converted into slipper-type pistons, as shown in Fig. 19-9. In Formula Vee, moly piston rings are always used for minimum friction and maximum wear resistance.

Fig. 19-8. **Big end of lightened AID rod. Bead surrounding big end bore has been narrowed and rod bead blasted. Compare with wide bead remaining on incorrectly lightened rod in Fig. 19-7.**

Fig. 19-9. Lightened VW 1200 piston for Formula Vee.

Formula Vee pistons tend to wear considerably, both on their thrust surfaces and in the piston ring grooves. Therefore, the pistons and cylinders should be replaced each season. New cylinders are honed to obtain a .0045-in. clearance with their individual pistons, which is only .0005 in. over the production minimum. At one time, tuners experimented with honing the cylinders to a taper so that when irregular heat expansion had taken place, which is common in air-cooled engines, the bores would be cylindrical. This fine point has not proven to be of any particular merit and is certainly not worth the trouble. By the rules, the pistons should have a 1-mm deck height, and in some cases it is necessary to install more than one paper gasket between a cylinder and the crankcase to meet this specification.

Formula Vee Ignition and Carburetion

As do many modified VW powerplants, Formula Vee engines use the Bosch 010 distributor. This is purely centrifugal advance unit that was originally installed in the 1200 VW Bus. The Formula Vee carburetor is the Solex 28 PCI, which can have its jets changed without removing or disassembling the unit (Fig. 19-10).

Fig. 19-10. Air correction jet being changed at race track without disassembling carburetor.

To improve airflow through the carburetor, the venturi is removed and machined to a wall thickness of only .015 in. Many competitors install velocity stacks (Fig. 19-11). However, considering the tendency for the pistons to wear rapidly, it is probably a better idea to install a small, low-restriction air cleaner.

Supertuning and Supercharging

Supercharging was once a common method of getting a bit more performance out of VW sedans and Karmann Ghia coupes. The premise was that you could install a supercharger without

Fig. 19-11. Formula Vee 28 PCI carburetor equipped with velocity stack.

removing and rebuilding the engine. As it usually turned out, those who installed superchargers were the people who spent the most time removing and rebuilding engines because the stock components were not truly up to the task of taking the extra pressures.

Supercharging is used on all of the fastest drag racing VWs, but these engines, in addition to being blown, are highly supertuned and rarely use any stock moving parts. For maximum outputs from any VW air-cooled engine, supertuning is the start, and there is a large array of speed equipment to choose from.

Among these items are fully counterweighted crankshafts that are ready to install and race. These are readily available in a variety of stroke lengths. Aluminum flywheels and connecting rods are widely applied, as are special lightweight forged-steel connecting rods and even connecting rod/crankshaft assemblies wtih roller bearings. Piston and cylinder sets can be purchased in any number of oversize bore dimensions. In combination with the long stroke crankshafts, big bore cylinders make it possible to increase the VW 1600's displacement to 1800 cm^3 or even 2000 cm^3. Crankshafts with bolted-on counterweights generally cost less than those with integral counterweights, which are forged.

High-performance cylinder heads, while available as already modified components from several suppliers, are often better obtained locally on a semicustom basis so that they can be tailored to your particular competition needs. High-strength and high-lift rocker arm setups—some with roller tips and frictionless bearings and machined from aluminum—are on the market in vast profusion, as are camshafts and other valve gear components. The number of intake and exhaust systems being sold defies description. Because it is often difficult to tell which equipment is best for your particular kind of racing, always seek professional advice from a company such as AID or one of the other advertisers in *Dune Buggies and Hot VWs* magazine.

20 / *Volkswagen Water-cooled*

New Force in Racing

The Volkswagen water-cooled engine (Fig. 20-1) has been around for about five years and has already had considerable impact on American racing. First, VW Dashers, Rabbits, and Sciroccos began to win regularly in SCCA Showroom Stock racing. Then VW launched a new racing class for Sciroccos only—the Bilstein Scirocco Cup Championship. Capping it all, a Volkswagen Scirocco with an engine supertuned by Bertil Solenskog won the 1976 Trans-Am Manufacturers Championship for Touring Cars under two liters. Front-drive cars show a considerable advantage on short, twisty circuits, and on tracks such as these, the Trans-Am Scirocco frequently finished not only well ahead of its class but in front of cars having several times its piston displacement.

By the end of 1977, the water-cooled VW powerplant became the newest racing engine for the masses. Formula Super Vee—long a highly competitive racing class—was changed from a class for supertuned VW 1600 air-cooled engines to a class for supertuned VW 1600 water-cooled engines. This caused some consternation among tuners who previously had specialized in modifying the air-cooled VW powerplants.

By virtue of the experience gained in building the winning

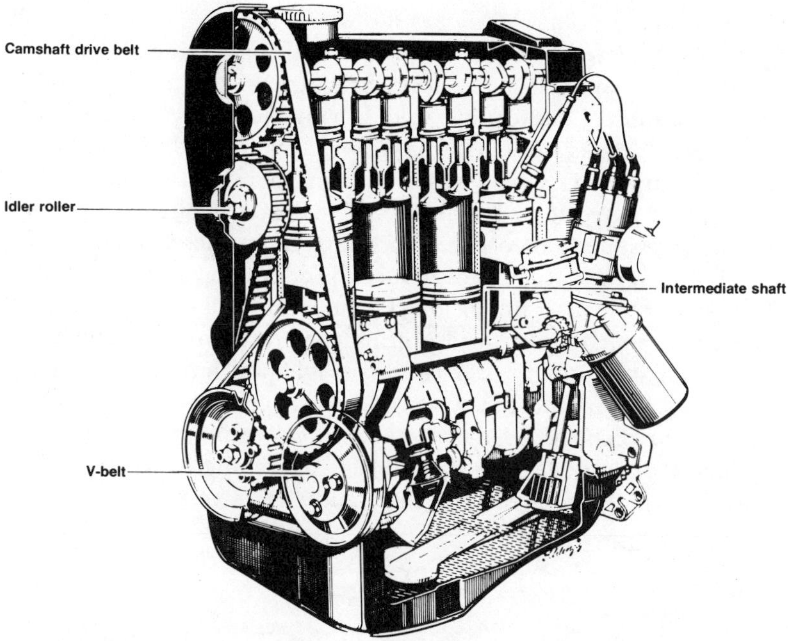

Fig. 20-1. Cutaway view of VW water-cooled engine. This powerplant is used in VW Rabbits, Sciroccos, and Dashers.

Trans-Am engines, and a good deal of development work since, Bertil Solenskog, who operates Bertil's Racing Engines in Libertyville, Illinois, found himself about a year and a half ahead of every other tuner in the country when it came to supertuning water-cooled VW engines. No newcomer to VW machinery, Solenskog built the U.S. Champion Super Vee winner's engines in both 1973 and 1975.

But while other tuners are still saying, "If Bertil can make the change, so can I", they say it nervously. Their composure has not been helped by SCCA and VWoA, who have been very slow in issuing definitive rules for the water-cooled Super Vee cars. So nobody—except Bertil—had started on an engine in 1977. And to obtain racing parts and to find out about supertuning the water-cooled VW, there is only one person to go to—the Swedish master builder himself, Bertil Solenskog.

Superior Strength

The VW water-cooled engine is a strong engine. There is a production 150-bhp fuel-injected Scirocco sold in Europe and a diesel version of the engine sold worldwide. Both use the same cylinder block, crankshaft, and connecting rods that are found in the ordinary U.S. water-cooled cars. Obviously a considerable amount of extreme supertuning can be carried out without imposing stresses equal to those encountered in diesel operation.

The engine is typical of modern production car powerplants. It has a cast-iron cylinder block, an aluminium alloy cylinder head, and a single overhead camshaft operated by a Gilmer belt. Somewhat surprisingly, VW chose a single-port-face cylinder head in an age when most new engines have crossflow heads. This, perhaps, is dictated by the use of vertical inline valves that are operated directly by the cam lobes rather than having rocker arms interposed between the cams and the valves.

Cylinder Head

The higher output achieved in the Trans-Am engines derived mainly from cylinder head modifications. However, some of the techniques used for Trans-Am racing probably will not apply to the Super Vee class. In Super Vee, the cylinder head will apparently have to be the stock thickness, which precludes the possibility of milling but eliminates the problem of camshaft drive belt slack. Whereas the air-cooled Super Vee powerplants were restricted to 41-mm intake valves, the valve size is rumored to be free in the water-cooled engines. However, the valves must remain inline so that no tricky and expensive valve gear modifications will become necessary in order to win.

In the Trans-Am units, 40-mm intake valves were used, which are a modest 6 mm larger than stock. The exhaust valve diameters were increased from 31 mm to 35 mm. The narrowest point in an induction system should be the valve seat, so valve diameters are limited by the maximum attainable diameters for the ports themselves. As on other modern single-port-face heads, the VW's intake ports are higher for a more direct route into the valve. As Fig. 20-2 shows, the ports are quite close together after they have been enlarged to their limits, and any significant increase in valve diameters would mean that the ports would become so close together that installing the manifolds would become a problem.

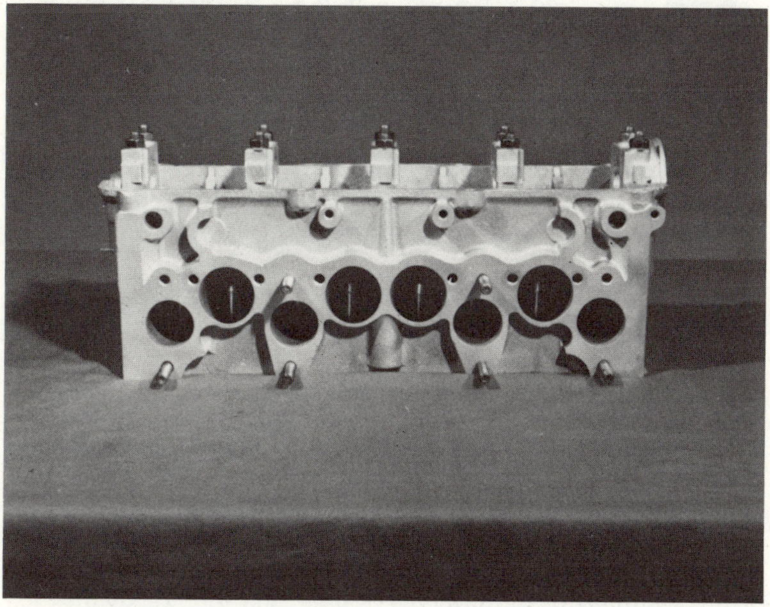

Fig. 20-2. Ports of modified cylinder head sold by Bertil's Racing Engines. Stock exhaust ports are elliptical. When all ports are enlarged and rounded, the port face becomes very crowded with holes.

The water-cooled VW engine has bathtub combustion chambers. The larger valves mean that the combustion chambers must be reshaped to prevent the valves from being shrouded (Fig. 20-3). In the Trans-Am engine, the compression ratio was brought up to 11.5:1 by milling the head. Bertil Solenskog has found that domed pistons interfere noticeably with flame travel in this engine, and he used flat-top pistons for the Trans-Am unit. If, as has been indicated, the new Super Vee rules prescribe that the head thickness remain stock, then domed pistons are perhaps the only alternative.

Fig. 20-3. Combustion chambers of one of Bertil's heads. Stock bathtub chambers are considerably opened up, unshrouding valves.

Camshaft and Valve Gear

As in most other competition engines, the intake valves should be swirl finished and as light as possible. One limitation of this engine is its seriously overweight cam followers with their large-diameter valve adjusting discs. Though the system makes valve adjustments extremely quick and simple for an engine with an overhead camshaft that operates the valves directly, each cam follower weighs approximately 71 grams in combination with its adjusting disc.

To avoid the need for stiff valve springs, which sap considerable power from a small-displacement engine, Bertil Solenskog has designed a special camshaft and modified valve gear (Fig. 20-4). The new cam followers weigh only 43 grams and have cap-type adjusting shims that fit atop the valve stems between the valves and the cam followers instead of between the cam followers and the cams. With the lightened valve gear, relatively soft dual valve springs can be used, and, despite their softness, the engine will rev to 9000 rpm without valve float.

Fig. 20-4. Bertil's complete big valve and valve gear package includes larger seat inserts, swirl-polished valves, special dual competition springs, custom cam, and lightweight cam followers. Notice the tiny adjusting shims at right.

Intake and Exhaust Systems

The Bosch K Jetronic fuel injection system is used in Showroom Stock Sedan and in Scirocco Cup racing. However, supertuned engines are equipped with dual side draft Weber 48 carburetors. These must be fitted to a manifold that matches both the carburetor bores and the reshaped and enlarged cylinder head ports. Because of the limited space on the cylinder head's manifold face, studs must be installed for mounting the exhaust headers. On the Trans-Am car, the headers were painted black to cut down the heat radiation. But in a Super Vee car the headers may have to be nearer the carburetors, and a heat shield between the exhaust primary pipes and the carburetor air intakes should be given serious consideration under these circumstances.

Crankshaft, Connecting Rods, and Pistons

The crankshaft should be magnafluxed. But because this is a very short, strong component, nitriding or other heat treatment is not absolutely necessary for all racing applications. The crankshaft should, however, be checked for straightness even if it has not undergone heat treatment and, if necessary, should be straightened. The index should also be corrected and all surfaces deburred and polished lightly to relieve stresses. This work can be done in conjunction with a very careful dynamic balancing job. The oil holes should be chamfered and radiused and all bearing journals polished. Bertil uses the standard main bearing clearance (.0012 to .0030 in.) and the standard connecting rod bearing clearance (.0011 to .0035 in.), though the larger end of the tolerance range is probably preferred.

The stock connecting rods are very strong and are used in diesel versions of this engine. Therefore, considerable lightening and polishing can be carried out without fear that the rods will be seriously weakened. After magnafluxing, the rods must be rebored to obtain the correct center-to-center length and perfect concentricity. Bertil removes 68 grams from each connecting rod, thoroughly polishing it in the process. Most of the metal is removed around the little end, though a considerable amount is also removed along the length of the rod and around the big end eye. A final polishing takes place during the balancing part of the operation, and the big end bore is honed for perfect contact between the rod and the bearing shell.

Lighter pistons and pins contribute a great deal toward reducing the loadings on the connecting rods—which makes the lightening of the rods a less hazardous undertaking. As previously mentioned, flat-top pistons should be given preference because they help the combustion chamber to burn well. For the Trans-Am car, Bertil used forged-aluminum pistons manufactured by Mahle. These are much shorter than the stock pistons (Fig. 20-5), to reduce both weight and friction losses. The steel piston pins have deep tapers at each end of their hollow centers for lighter weight with undiminished strength; an entire piston assembly, including rings, weighs about 425 grams, compared to 550 grams for the stock assembly.

Fig. 20-5. Lightweight pistons have pin bore near bottom of short skirt. Internal taper makes pins lighter.

The special pistons are available from Bertil and are fully described in his catalog. Most of the parts that he has developed are manufactured for him in Europe, where finer materials—particularly steels—are available than are in America. There are also big-bore pistons on hand that permit the capacity of the VW water-cooled engine to be raised to 1800 cm^3, which seems to indicate some thought of a possible challenge to the Cosworth BDA.

Cylinder Block

The VW water-cooled cylinder block is commendably short. The sides of the crankcase extend well below the main bearing centerline, and there are ribs reinforcing various stress points on both the inside and the outside of the casting. The engine is tall by comparison to a Formula Ford engine, especially in wet sump form.

The Trans-Am engine used a deeper sump, which required a lengthened oil pickup and baffles to keep the oil from sloshing away from the pickup during hard cornering. The deeper pan was not used solely to obtain greater oil capacity. It also kept the oil away from the fast-turning crankshaft and flailing connecting rods, which set up air currents that can cause oil aeration or foaming. The deeper sump—or even the stock sump—is undoubtedly out of the question in a formula car, so a dry sump system must be developed for use in Formula Super Vee. The stock oil pump is internally mounted and will probably be retained as a pressure pump. A separate scavenge pump will, however, be needed, and this could be external with a Gilmer belt drive.

Ignition

A Bosch electronic ignition system, which Bertil has found quite suitable for competition, is used on the latest VW cars. The distributor has no breaker points and is comparable to the Lucas Opus system in concept, though not in design. The stock breaker-point distributors found on 1973 through 1977 water-cooled VWs can also be modified for competition use.

21 / *Other Popular Competition Engines*

Wankel Rotary

When we talk about the Wankel engine in America, we are really talking about the Mazda rotary combustion (RC) engine. Apart from a brief appearance in the mid-1960s by the little single-rotor NSU Spider, the Mazda cars are the only rotary engine machines sold in the United States. The cars themselves are are a joy to drive, principally because of the turbine-like smoothness of the rotary engines. But despite the deceptive smoothness, few conventional engines of similar size can approach the excellent performance delivered by the Mazda.

The Mazda RC engine has suffered in competition because it was too good the first time out. The situation recalls the brief reign of the gas turbine at Indy; the turbine was so superior that the rules were changed to an extent that made using a turbine totally hopeless. The first appearances of the Mazdas in sedan racing were also disconcerting for the competition. Consequently the rules were tightened up too much. Thus, the very promising RC engine has not profited from the development that might have made it a good competition engine.

Nevertheless, one or two Mazdas have been competitive in recent IMSA events, particularly on longer courses where the Mazda's excellent acceleration and top speed show to a greater advantage. The cars have two chronic problems on corners. First, there is the limitation of the Mazda sedan itself, which does not have very wide wheels and has a suspension that, though excellent on the highway, permits a certain amount of roll, roll steer, and general imprecision. The second problem is its high-mounted four-barrel carburetor, which, by the rules, it must keep (Fig. 21-1).

Fig. 21-1. Mazda engine modified for competition in INSA road racing. Notice high carburetor position.

On the corners, the fuel tends to slosh to one side of the float chamber or the other, in some cases causing the mixture to go lean and in others causing fuel to be dumped into the intakes—which cuts the engine out altogether and produces extravagant fireballs from the exhaust pipe when the car straightens out and the engine begins firing again. The operation of the carburetor and induction system is unique (Fig. 21-2). The primary venturis of the carburetor serve the inboard side ports of each rotor chamber, and the secondary venturis serve the outboard side ports. There is little doubt that a pair of side draft Webers or a fuel-injection system would make a big improvement.

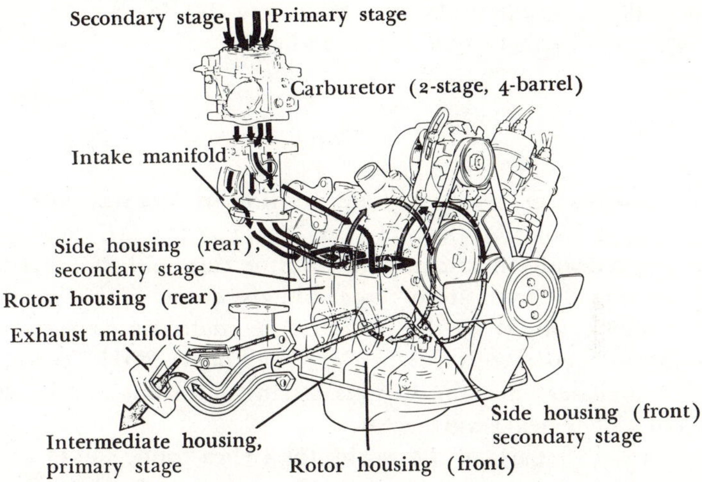

Fig. 21-2. Gasflow in the Mazda RC engine.

The Mazda RC engine has side intake ports, peripheral exhaust ports, and dual ignition—the latter mainly for emission control purposes (Fig. 21-3). It is obvious that, were the rewards worth it, a wise supertuning artist could do some interesting experiments in "valve" timing with no other equipment than a porting tool. It is also conceivable that inserts could be developed that would fit in holes milled in the castings, each interchangeable insert containing a different port profile for testing or for tuning the engine output for a particular kind of racing or for a particular race course.

Fig. 21-3. Exploded view of Mazda RC engine housing.

The Wankel's future in terms of racing development is uncertain. Given a class where uninhibited development is possible, the future might be bright indeed. Mazda is showing an interest in making high-performance versions of the RC engine, and perhaps the Wankel's best days lie ahead.

Toyota

Toyota cars are in an odd position in America with regard to competition. While these cars and engines are raced widely and successfully in other countries, the cars sold in America are somewhat different from those sold elsewhere. The Toyota Celica, which would be the obvious candidate for competition, is sold in America with a rather large and heavy SOHC powerplant that it shares with the Corona sedans. This engine also suffers from rather heavy valve gear.

In other parts of the world, the Celica coupe comes with the 1600 Hemi engine found in the American-market Corolla 1600. In 2T-C form, this engine develops 102 bhp in Australian tune, compared to 75 bhp for current U.S. Corollas. The 2T-B version develops 113 bhp and, if this is not adequate to propel an enthusiast in a Celica, there is the 2T-G DOHC version that cranks out 124 bhp at 6400 rpm (Fig. 21-4). So, because these "hot" engines have too little power once they are desmogged for the United States, the bigger sedan engine goes into the U.S. Celica and the reduced-power 2T-C Hemi goes into the super-economy U.S. Corolla. As a result, the least sporting cars in Toyota's U.S. line have been the models with the sportiest engines.

Not surprisingly, the Toyotas seen in competition are usually Corollas. These little cars have not done great things in sedan racing, but they have not done badly either. In 1976, Toyota introduced a pair of lighter, better-handling Corolla 1600s; one has a sport coupe body and the other a liftback. A sport coupe prepared by Buzz Marcus (Fig. 21-5) managed in 1977 to lead races and to finish well up, so there is a possibility that other competitive Corollas will soon be seen in racing.

Even in 2T-C pushrod form, the 1600 combustion chamber is excellent, and, with nothing more than high-compression pis-

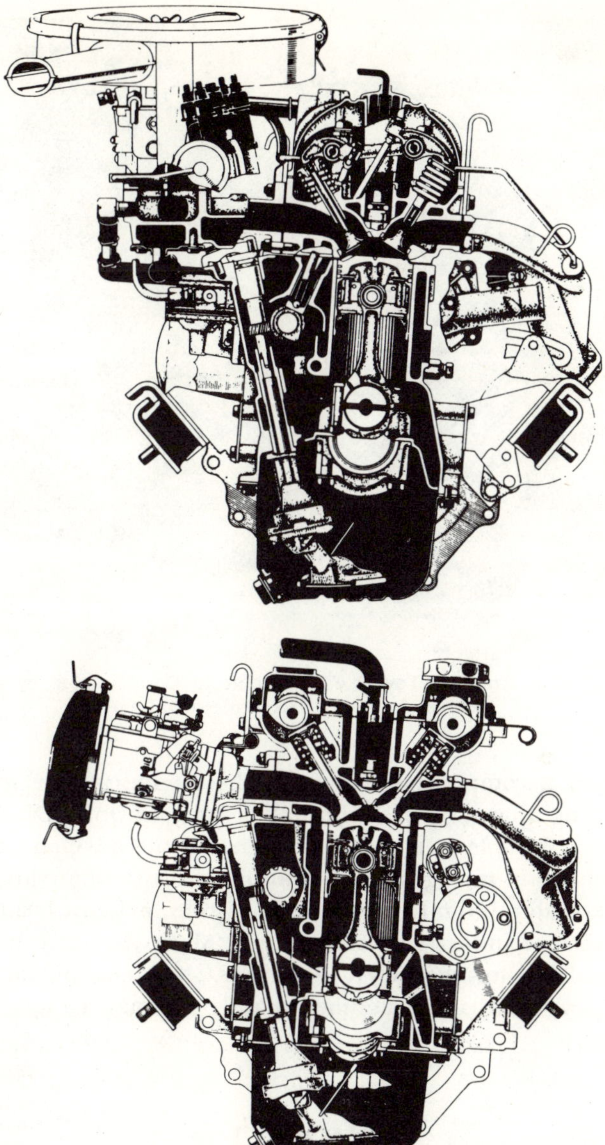

Fig. 21-4. Toyota 2T-C Hemi pushrod engine compared to 2T-G DOHC version that is not imported into the United States. Racing development of 2T-C is just getting started in America.

Fig. 21-5. Modified Toyota 2T-C engine in Corolla raced by Buzz Marcus in 1977 IMSA professional events.

tons and a competition camshaft, the engine is able to inhale everything that can pass through a pair of 40DCOE Webers. Readily available as speed equipment for the 2T-C are competition pushrods, racing cams, aluminum and titanium valve spring retainers, aluminum flywheels, 10:1 pistons, exhaust headers, and a number of induction system setups. Many similar components are available for the SOHC R and RC engines, but at present these powerplants are modified mainly for match racing on the drag strip.

Ford SOHC "Four"

A great number of Ford Pinto sedans and Capri coupes are seen in sedan racing. Some of these have the 1992-cm³ 2000 engine that was built in Germany, but beginning with the 1976 models all have had the American or Brazilian 2300 SOHC engines. Although the 2300 has automatic hydraulic valve adjusters, the cylinder head offers two advantages over that of the 2000. First, the 2300 port shape is better and more easily improved upon. Second, the camshaft can be taken out from the front so that it is not necessary to remove the head to change the camshaft.

Both engines have an extremely strong bottom end, which will absorb all the punishment that a supertuned induction system and cylinder head can deliver. Generally it is necessary to O-ring the cylinders for competition applications, and both dual side draft Webers and single Holley four-barrels have proved to be effective as competition carburetion setups.

There is a great deal of speed equipment available, and perhaps the only thing that keeps the Ford SOHC engines from being consistent winners is the comparatively large, heavy, and mediocre cars that they are installed in. There is a Formula Super Ford class in England that uses another version of the Ford SOHC engine. If this class comes to America, the SOHC will become one of the staples of American racing. A secondhand Pinto is a very inexpensive way to get started in racing, parts are available everywhere, and the engine is durable. But at present, they are more plentiful than victorious.

Porsche 911

There has never been a racing class that Porsche did not dominate, with the exception of Formula 1 where the marque enjoyed only limited success in the late 1950s and early 1960s. But even that may change. It is rumored that Porsche is preparing a Formula 1 car for the 1978 season. In the United States, the most widely seen Porsche in competition is the 911 model and its high-performance derivatives, the Carrera and the Turbo.

The Porsche "six" engine needs very little done to it to prepare it for competition. Factory high-performance components—many used in regular production—are readily available for the most exhaustive kinds of competition work. The engine compartment of a racing Porsche (Fig. 21-6) is not very different from that of a Porsche that you would see in a dealer's showroom.

Fig. 21-6. Two-liter Porsche 911 engine in competition tune.

Its remarkable strength and durability can be judged by looking at a cross-section of the flat-six powerplant (Fig. 21-7). The light-alloy crankcase is split vertically on the crankshaft centerline. Each of the six separate cylinders has a cast-iron liner on which is superimposed a light-alloy jacket, extensively provided with cooling fins (Fig. 21-8). The cylinders are deeply spigoted into the crankcase, each one being held by four long studs that pass outward through the detachable aluminum cylinder heads. Unlike VWs and Porsche "fours", individual heads are fitted to each cylinder.

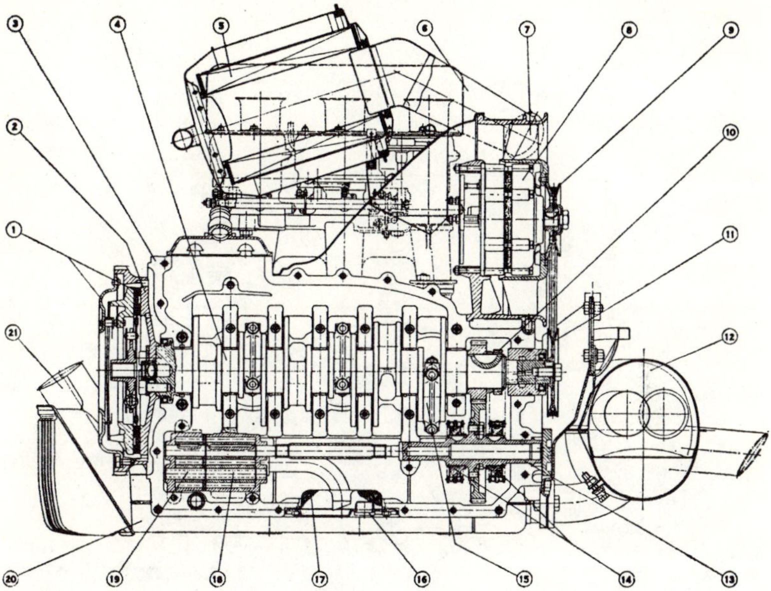

1. Clutch and pressure unit
2. Flywheel
3. Crankcase
4. Crankshaft
5. Air filter
6. Air filter casing
7. Fan and casing
8. Alternator
9. Pulley and belt
10. Distributor drive gear
11. Pulley
12. Silencer
13. Jackshaft
14. Chain sprockets
15. Big-end
16. Drain plug
17. Oil intake gauge
18. Scavenge pump
19. Oil pressure pump
20. Heat exchanger
21. Warm air outlet

Fig. 21-7. Side cross-section of Porsche flat-six. Notice the eighth main bearing at the rear and the jackshaft in the sump.

The fully-machined six-throw crankshaft is carried in seven thin-wall bearings supported in substantial bearing webs. There is also an eighth crankshaft bearing at the extreme rear of the crankcase, on the outside of the timing drive. The crankshaft's timing drive gear turns a jackshaft that is in the crankcase and that drives the oil pump and the chains for the single overhead

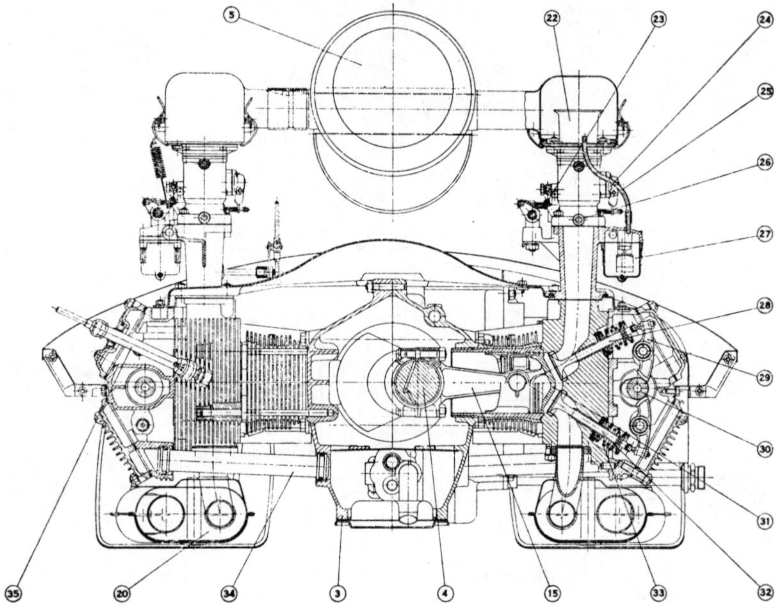

3. Crankcase
4. Crankpin
5. Air filter
15. Connecting rod
20. Heat exchanger
22. Carburetor air intake
23. Idling mixture adjustment
24. Accelerator pump
25. Vent pipe
26. Throttle shaft
27. Float chamber
28. Inlet valve
29. Rocker
30. Camshaft
31. Piston
32. Exhaust valve
33. Valve springs
34. Oil return pipe
35. Spark plug suppressor

Fig. 21-8. End section of Porsche flat-six. Combustion chamber details can be clearly seen.

camshafts. Actually, there are two oil pumps side by side. The engine employs a dry sump, and the smaller oil pump draws oil from the tank and supplies pressure to the bearings, while the larger pump scavenges the shallow sump and returns the oil to the tank.

Even in stock form the connecting rods are polished. The pistons are highly domed and are equipped with full-floating pins and three piston rings. The induction and exhaust systems are the main areas where a tuner can practice his skill; all other preparation consists of minor blueprinting and very careful assembly. The Carrera version of the engine is equipped with Bosch timed mechanical injection (Fig. 21-9). The Turbo engine compartment (Fig. 21-10) shows the greatest variation over the original 911. The turbocharger handles air only; the fuel is supplied by port injection.

Fig. 21-9. Engine compartment of a racing Carrera, showing fuel-injection pump and air intakes.

Fig. 21-10. Engine compartment of racing Turbo. Large ducts and boxes handle air only; fuel is injected at ports.

Amateur Racing

In this book we have concentrated on the engines that are used most often in American professional racing. Some have not been specifically mentioned, such as the American Motors "six", current BMWs, Dodge Colts, and Plymouth Arrows. These engines do appear in professional racing, as does the occasional SAAB, Volvo, or Renault. For the most part, however, the cars in this group obtain success only with fairly heavy factory backing or are entered on a shoestring by an enthusiast for the particular marque who doesn't mind finishing well back. Consequently, these engines have not had a great deal of development or success in the hands of any considerable number of racers.

There are, of course, a number of other engines that have long been used in amateur racing—particularly the MGs and Triumphs. These have not appeared in professional racing. Being in many ways antiquated designs, they long ago reached the peak of development; all of the various blueprinting and super-tuning techniques are well recorded. Persons interested in preparing these and other engines for amateur racing should obtain the readily available special tuning literature that is available from the competition departments of the car manufacturers.

Appendix / Definitions, Constants, and Formulas

Occasionally even the well informed find it difficult to convey in simple language the meaning of commonplace technical terms. The following definitions may prove useful in these situations.

Horsepower (hp) The traditional unit of work; 1 hp is equal to 33,000 ft. lb. per minute. This is the work done in lifting a weight of 33,000 pounds through a distance of one foot in one minute or any other quantities of pounds and feet that give 33,000 when multiplied.

Brake Horsepower (bhp) The actual power developed at the engine shaft, as measured on test by coupling the engine to a dynamometer or "brake" for absorbing the power.

Torque The effort applied to a shaft or wheel that tends to turn it. When the torque is of sufficient value to rotate the shaft through a definite distance in a given time, work is done, which is then stated in hp or kilowatts.

Kilowatt (kw) The SI (modernized metric system) unit of power. Kilowatt ratings have already replaced horsepower ratings in Europe.

Mean Effective Pressure (mep) Sometimes called *mean indicated pressure (mip)*. The average pressure produced in the cylinders on the expansion stroke and which results in the shaft power. It is measured at the cylinder itself by a scientific apparatus or indicator.

Brake Mean Effective Pressure (bmep) Sometimes called *brake mean pressure (bmp)*. A figure analogus to the above but obtained by calculation from the actual bhp. The figure, which allows for the mechanical losses in the engine, is imaginary. It does, however, enable useful comparisons to be made between designs in almost all aspects.

Mechanical Efficiency A factor expressed as a percentage, which shows how much of the explosion pressure is obtained as power at the engine shaft; it is the percentage difference between the mep and bmep. The loss is owed to friction, inertia, and so on in the engine's moving parts.

Thermal Efficiency A factor expressed as a percentage, which shows how much power is obtained from the heat energy in the fuel used; it is thus an indication of the effectiveness of the engine as an apparatus for converting heat into work.

Volumetric Efficiency A factor showing the degree of completeness with which a cylinder is charged, exhausted, and recharged during the operating cycle; it is a measure of the adequacy of the valves, ports, induction and exhaust systems, and so on.

Thermodynamics

The following definitions cover terms more commonly used in thermodynamics.

Specific Heat The ratio of the thermal capacity of a substance to the thermal capacity of water.

Unit of Heat (Btu or Chu) The heat given out by a unit volume of water in cooling through 1°.

Mechanical Equivalent of Heat (Joule's Equivalent) 1 btu = 778 ft./lb.

Calorific Value The capacity of a substance for giving up heat when a mass of it is burned (usually expressed as btus per pound).

Combustion The combination of different chemical elements resulting in heat emission.

Conduction of Heat The transfer of heat through a body from the hotter to the cooler part.

Conductivity The rate at which heat flows (in heat units per second) is directly proportional to the temperature difference between the two parts and to the conductivity of the substance.

Unit of Conductivity The quantity of heat, in heat units, that flows per second across a surface 1 square foot in area for each degree of temperature drop per foot of distance that the heat travels.

Convection Transfer of heat by virtue of the motion of the parts of heated bodies. The chief means of temperature equalization in liquids and gases. Cannot occur in solids.

Radiation Transfer of heat from a hot body to a cold one by waves of heat passing through the atmosphere.

Absolute Temperature When temperature is changed by 1°C (constant pressure), the amount by which a gas expands in volume is 1/273 part of its volume at 0°C. Therefore, at 273° below 0°C, the volume will be zero. Thus, absolute zero is –273°C. Temperatures from absolute zero may be stated in degrees K (Kelvin) or degrees R (Rankine). Thus the melting point of ice is 0°C, 32°F, 273°K, and 492°R.

Abbreviations and Symbols

Term	*Abbreviation or symbol*
Absolute	Abs.
British thermal unit	btu
Centigrade heat unit	Chu
Compression ratio	r
Efficiency	*n*
Enthalpy	H
Entropy	S
Efficiency (mechanical)	*nm*
Mechanical equivalent of heat (Joule)	J
Specific heat at constant pressure	^{c}p
Specific heat at constant volume	^{c}v
Specific heat, ratio of	Y
Temperature (abs.)	T

Useful Conversions

To convert from	*Multiply by*
Horsepower (550 ft. lbf/sec.) to kilowatts	.746
Metric (DIN) bhp to kilowatts	.736
Metric (DIN) bhp to SAE net bhp	.953
SAE net bhp to kilowatts	.769
Torque, kgf/m (mkg) to lbf/ft (ft. lb.)	7.233
Torque, lbf/ft (ft. lb.) to kgf/m (mkg)	.13825
Torque, kgf/m (mkg) to Nm (Newton-meter)	9.806
Torque, lbf/ft (ft. lb.) to Nm (Newton-meter)	1.356
Torque, Nm (Newton-meter) to kgf/m (mkg)	.102
Torque, Nm (Newton-meter) to lbf/ft (ft. lb.)	.7379
Cylinder capacity, cu. in. to cm^3 (cc)	16.39
Cylinder capacity, cm^3 (cc) to cu. in.	.06102

Note that these conversions are for use in comparing the specifications of different engines and are not intended for purposes where great scientific or mathematical accuracy is involved.

Useful Formulas

Swept volume of one cylinder $= 3.14 \times R^2 \times S$

$$\text{or} = \frac{3.14}{4} \times D^2 \times S$$

$$\text{or} = \frac{3.14 \times D^2 \times S}{4}$$

where D = cylinder bore; R = cylinder bore radius (one-half its diameter); S = piston stroke; 3.14 = pi; $\frac{3.14}{4} = .785$ (one-fourth pi).

Compression ratio $= \frac{V + v}{v}$ where V = swep volume
v = volume of space above piston at top of stroke.

Piston speed in feet per min. = 2 × rev/min × stroke in feet

$$\text{or} = \frac{\text{rev/min} \times \text{stroke in inches}}{6}$$

$$\text{or} = \frac{\text{rev/min} \times \text{stroke in mm.}}{152.4}$$

$\text{Bhp} = \frac{\text{plan}}{33{,}000}$ where p = brake mean effective pressure in lb/sq. in.
l = length of stroke in feet.
a = area of one piston in square inches.
n = number of power strokes per minute.

$\text{Bmep lb/sq in.} = \frac{\text{bhp} \times 33{,}000}{\text{lan}}$ for values of lan, see above.

$$\left.\begin{aligned}\text{Bhp} &= \frac{\text{rev/min} \times \text{torque}}{5{,}250} \\ \text{Torque} &= \frac{\text{bhp} \times 5{,}250}{\text{rev/min}}\end{aligned}\right\} \text{When torque is in lb/ft.}$$

$$\text{Mph} = \frac{\text{rev/min} \times \text{wheel dia. in inches}}{\text{gear ratio} \times 336} \text{ (approximate only)}$$

Mean gas velocity through port in ft. per sec.

$$= \frac{\text{Piston speed}}{60} \times \frac{D^2}{d^2}$$

Mean gas velocity through valve in ft. per sec.

$$= \frac{\text{Piston speed}}{60} \times \frac{D^2}{V \times L \times \frac{22}{7}}$$

Where D = diameter of piston.
d = diameter of port.
V = diameter at throat of valve.
L = lift of valve.
and flow coefficient is assumed as unity.

Torque calculations:
Where P = Bmep in lb/sq. in.
V = Swept Volume in cu. in.
T = Torque

$$T = \frac{PV}{4\pi} \text{ lb/in.}$$

$$\text{or } T = \frac{PV}{4\pi \times 12} \text{ lb/ft}$$

If volume is in cc (16.387 cc = 1 cu. in.)

$$T = \frac{PV}{4\pi \times 12 \times 16.387} \text{ lb/ft.}$$

$$\text{or } T = \frac{\text{bmep} \times \text{swept volume in cc}}{2473} \text{ lb/ft}$$

$$\text{and Bmep} = \frac{2473 \times T}{\text{swept volume in cc}} \text{ lb/in}^2$$

$$\text{or Bmep} = \frac{\text{bhp} \times 12{,}983{,}250}{\text{rev/min} \times \text{swept volume in cc}} \text{ lb/in}^2$$

The above equations are sufficiently accurate for general purposes.

The symbols used in formulas are not necessarily adhered to in all texts, etc.

Cylinder Head Milling

This formula, explained in chapter 12, is used to calculate the amount to be milled from a cylinder head to obtain a particular compression ratio.

Milled Amount in mm =

$$\frac{\text{New Disp. Ratio} - \text{Old Disp. Ratio}}{\text{New Disp. Ratio} \times \text{Old Disp. Ratio}} \times \text{Stroke in mm}$$

where the displacement ratio is 1.0 less than the compression ratio (the volume of one cylinder divided by the combustion chamber volume).

Index

C

D

E

F

G

H

I

J

K

L

M

W

X

Z